Logical Modalities from Aristotle to Carnap

Interest in the metaphysics and logic of possible worlds goes back at least as far as Aristotle, but few books address the history of these important concepts. This volume offers new essays on the theories about the logical modalities (necessity and possibility) held by leading philosophers from Aristotle in ancient Greece to Rudolf Carnap in the twentieth century. The story begins with an illuminating discussion of Aristotle's views on the connection between logic and metaphysics, continues through the Stoic and mediaeval (including Arabic) traditions, and then moves to the early modern period with particular attention to Locke and Leibniz. The views of Kant, Peirce, C. I. Lewis and Carnap complete the volume. Many of the essays illuminate the connection between the historical figures studied and recent or current work in the philosophy of modality. The result is a rich and wide-ranging picture of the history of the logical modalities.

MAX CRESSWELL has taught philosophy at Victoria University of Wellington since 1963. His publications include three widely used texts on modal logic with G. E. Hughes, and most recently, with A. A. Rini, *The World–Time Parallel* (Cambridge, 2012).

EDWIN MARES is Professor of Philosophy at Victoria University of Wellington. His publications include *Relevant Logic: A Philosophical Interpretation* (Cambridge, 2004), with Stuart Brock, *Realism and Anti-Realism* (2007), and *A Priori* (2011).

ADRIANE RINI is Associate Professor of Philosophy at Massey University in New Zealand. She is the author of *Aristotle's Modal Proofs* (2011) and, with M. J. Cresswell, *The World–Time Parawllel* (Cambridge, 2012).

Logical Modalities from Aristotle to Carnap

The Story of Necessity

Edited By

MAX CRESSWELL
Victoria University of Wellington

EDWIN MARES
Victoria University of Wellington

ADRIANE RINI
Massey University, Palmerston North

CAMBRIDGE
UNIVERSITY PRESS

CAMBRIDGE
UNIVERSITY PRESS

University Printing House, Cambridge CB2 8BS, United Kingdom

Cambridge University Press is part of the University of Cambridge.

It furthers the University's mission by disseminating knowledge in the pursuit of education, learning, and research at the highest international levels of excellence.

www.cambridge.org
Information on this title: www.cambridge.org/9781107077881

© Cambridge University Press 2016

This publication is in copyright. Subject to statutory exception and to the provisions of relevant collective licensing agreements, no reproduction of any part may take place without the written permission of Cambridge University Press.

First published 2016

A catalogue record for this publication is available from the British Library.

Library of Congress Cataloguing-in-Publication Data
Names: Cresswell, M. J., editor.
Title: Logical modalities from Aristotle to Carnap : the story of necessity / edited by Max Cresswell, Victoria University of Wellington, Edwin Mares, Victoria University of Wellington, Adriane Rini, Massey University, Palmerston North.
Description: New York: Cambridge University Press, 2016. |
Includes bibliographical references and index.
Identifiers: LCCN 2016024207 | ISBN 9781107077881 (hardback)
Subjects: LCSH: Necessity (Philosophy) – History. |
Modality (Theory of knowledge) | Modality (Logic)
Classification: LCC BD417.L64 2016 | DDC 123/.7–dc23
LC record available at https://lccn.loc.gov/2016024207

ISBN 978-1-107-07788-1 Hardback

Cambridge University Press has no responsibility for the persistence or accuracy of URLs for external or third-party internet websites referred to in this publication and does not guarantee that any content on such websites is, or will remain, accurate or appropriate.

Contents

Figures and Tables

Figures

Tables

Contributors

Peter R. Anstey is ARC Future Fellow and Professor of Philosophy at the University of Sydney. He specialises in early modern philosophy and is the author of *John Locke and Natural Philosophy* (2011).

Max Cresswell currently holds a part-time appointment at the Victoria University of Wellington, where he taught philosophy from 1963 until 1999. He has published eleven books (including three texts on modal logic with G. E. Hughes) and articles, mostly in modal logic, semantics and the history of philosophy.

Vanessa de Harven is Assistant Professor of Philosophy at the University of Massachusetts, Amherst. She is the author of *Nothing Is Something: The Stoic Theory of Void* (2015) and has published articles on Stoic metaphysics and on Plato.

Catherine Legg is Senior Lecturer in Philosophy at the University of Waikato. She is the author of a number of articles on aspects of Peirce's realism, including his accounts of truth, real universals and diagrammatic logic. She also publishes in computer science, on formal ontology.

Brandon C. Look is University Research Professor at the University of Kentucky. He has published articles on topics in the history of modern philosophy, and he is the author of *Leibniz and the 'vinculum Substantiale'* (1999), coeditor and translator of *The Leibniz–Des Bosses Correspondence* (2007), and editor of *The Continuum Companion to Leibniz* (2011).

Jack MacIntosh is a Professor of Philosophy at the University of Calgary. His publications include *The Excellencies of Robert Boyle* (2008) and *Boyle on Atheism* (2006), together with articles in the history of philosophy, particularly mediaeval and early modern, the philosophy of religion, and other topics.

Marko Malink is Associate Professor of Philosophy and Classics at New York University. He is the author of *Aristotle's Modal Syllogistic* (2013) and other publications on Aristotle and in linguistics.

Edwin Mares is a Professor of Philosophy at the Victoria University of Wellington. His publications include *Relevant Logic: A Philosophical Interpretation* (Cambridge, 2004), *A Priori* (2011) and, with Stuart Brock, *Realism and Anti-Realism* (2007).

Christopher J. Martin is Associate Professor of Philosophy at the University of Auckland. He is the author of the chapter on "Logical Consequence", in *The Oxford Handbook of Medieval Philosophy* (2012). He has published numerous papers on mediaeval logic and philosophy, in particular on the revolution which took place in logic in the twelfth century.

Cheryl Misak is Professor of Philosophy at the University of Toronto. She is the author of *The American Pragmatists* (2013); *Truth and the End of Inquiry: A Peircean Account of Truth* (1991); *Truth, Politics, Morality: Pragmatism and Deliberation* (2000); *Verificationism: Its History and Prospects* (1995), as well as articles on pragmatism, ethics and philosophy of medicine.

Calvin G. Normore is Professor of Philosophy at UCLA, Macdonald Professor of Moral Philosophy (Emeritus) at McGill, and Honorary Professor of Philosophy at the University of Queensland. He writes on medieval modal theory and is the author of chapters on logic in *The Cambridge Companion to Ockham* (1999) and in *The Cambridge Companion to Duns Scotus* (2002).

Adriane Rini is Associate Professor of Philosophy at Massey University. She is the author of *Aristotle's Modal Proofs* (2011) and, with M. J. Cresswell, *The World-Time Parallel* (Cambridge, 2012).

Robin Smith is an Emeritus Professor at Texas A&M University. He has translated *Aristotle's Prior Analytics* (translation with Introduction, notes, and commentary, 1989) and *Aristotle, Topics I, VIII, and Selections* (1997). He has published many articles and book chapters, including the chapter on Aristotle's logic for *The Cambridge Companion to Aristotle* (1995).

Nicholas F. Stang is Canada Research Chair in Metaphysics and Its History at the University of Toronto. His first book, *Kant's Modal Metaphysics*, was published by Oxford University Press in 2016.

Paul Thom is an Honorary Professor of Philosophy at the University of Sydney. His books include *The Logic of Essentialism: An Interpretation of Aristotle's Modal Syllogistic* (1996), *Medieval Modal Systems* (2003) and *Logic and Ontology in the Syllogistic of Robert Kilwardby* (2007).

Jonathan Westphal is Visiting Professor of Philosophy at Hampshire College. His publications include *Colour: A Philosophical Introduction* (1991) and articles on Leibniz and other topics.

Abbreviations

Several of the chapters of this volume refer to the works of classical authors (i.e. authors up to the nineteenth century). We include here a list of some of the abbreviations and titles which contributors have used in referring to these works. A year number after the reference indicates that the edition or translation will be found by year under that author's name in the Bibliography.

Aristotle

(*De Int.*) *On Interpretation*
(*An. Pr.*) *Prior Analytics*
(*An. Post.*) *Posterior Analytics*
(*Top.*) *Topics*
(*SE.*) *Sophistical Refutations*
(*Phys.*) *Physics*
(*De Caelo*) *On the Heavens*
(*De Gen. et Cor.*) *On Coming to Be and Passing Away*
(*Met.*) *Metaphysics*
(*Nic. Eth.*) *Nicomachean Ethics*

Translations may be found in *The Complete Works of Aristotle: The Revised Oxford Translation*, ed. Jonathan Barnes, 2 vols. (Princeton: Princeton University Press, 1985).

Philodemus

(*Sign.*) *On Signs*

Cicero

(*Div.*) *On Divination*, 1975
(*Fat.*) *On Fate*, 1975
(*De Nat. Deor.*) *On the Nature of the Gods*, 1933

Plutarch

(*Comm. not.*) *On Common Conceptions*, 1976a
(*St. Rep.*) *On Stoic Self-Contradictions*, 1976b

Epictetus

(*Diss.*) *Discourses*, 1916

Gellius

(*Noct.*) *Attic Nights*, 1968

Alexander of Aphrodisias

(*In Ar. An. Pr.*) *On Aristotle's Prior Analytics*, 1883
(*In Ar. Top.*) *On Aristotle's Topics*, 1891
(*Fat.*) *On Fate*, 1892

Sextus Empiricus

(*PH*) *Outlines of Pyrrhonis*, 1939–1949, Vol. 1
(*M*) *Against the Mathematicians*, 1939–1949, Vol. 2

Ammonius

(*In Ar. De Int.*) *On Aristotle's On Interpretation*, 1897

Diogenes Laërtius

(*DL*) *Lives of Philosophers*, 1964

Boethius

(*In Ar. De Int.*) *On Aristotle's On Interpretation*, 1877
De Hypotheticis Syllogismis, 1969
De Syllogismo Categorico, 2008

Anselm of Canterbury

Philosophical Fragments, 1936
(*De casu diaboli*) *The Fall of the Devil*, 1968, Vol. I

Garland

Dialectica, 1959

Abaelard

Dialectica, 1970
Glossae Super Peri Hermenias, 2010

Aquinas (bibliographical entries under 'Thomas Aquinas')

(*SCG*) *Summa contra gentiles*, 1946
(*QD De Anima*) *The Soul*, 1949
(*De Veritate*) *On Truth*, 1952a, 1953, 1954
(*De Pot.*) *On the Power of God*, 1952b
(*In Peri.*) *On Aristotle's* On Interpretation, 1962
(*ST*) *Summa Theologiae*, 1964–1981

Robert Grosseteste

De Libero Arbitrio, 1912

Gregori of Rimini

Lectures on the First and Second Sentences, 1979–1984

William of Ockham

(*SL*) *Summa Logicae*, Ockham 1974b, trans. Michael J. Loux, Part I 1974a, Part II, 1980a
(*Quodlibet*), *Quodlibetal Questions*, 1980b

Descartes

(*AT*) Adam and Tannery, 1964–1976 (ed.)
(*CSMK*) Cottingham, Stoothoff and Murdoch, 1985, 1991 (trans.)

Leibniz

(*Gerhardt*) 1961
(*Loemker*) 1969
(*New Essays*) 1981

Locke

(*Essay*) 1975

Boyle

(*Works*) 1999–2000
(*BOA*) 2006

Introduction

MAX CRESSWELL, EDWIN MARES, AND ADRIANE RINI

Philosophy investigates the basic structure of how things are. In addition, many philosophers have claimed that philosophy must also look at how things must be – that is, at what is necessary. But philosophers do not all agree about the nature of necessity and the related notion of possibility. In fact, these 'modal' notions are the focus of a debate which goes back to the earliest Greek philosophers, and many of the questions which the ancients asked about the nature of necessity can still be seen to occupy a central position in present-day philosophical discussion.

Some years ago when we three editors were reflecting together on current views about modality, possible worlds, and related issues, we started asking where philosophers through the ages have thought that necessity originates. What have they said necessity is? What do they say are its roots? We began to explore notions of necessity as they emerged in philosophers' and logicians' works over the centuries, and in doing so we were struck by the enormous range of different approaches to these topics, both from the point of view of contemporary metaphysics and from the point of view of the connection between logical inference and argument. We were also struck by the connection between the intuitive notions of necessity and the validity of 'logical' arguments in ordinary life. Even while these themes sit at the heart of much in philosophy, it seemed to us that the contemporary debates around them do not often suggest much awareness of the older philosophical discussions. And so we quickly came to feel the need for a more thoroughgoing study, one which is fundamentally historical and which brings together world experts on various philosophers and eras. We approached these experts seeking their guidance in looking back at the notion of necessity throughout its long history, and through close attention to the primary texts.

There are, we think, special advantages in taking the historical approach to a subject such as necessity. Not least among these advantages is the way in which the historical approach affords philosophers room for reflection, which allows us to sit outside the philosophical debates. However partisan one's views about modal reasoning might be, philosophers as a rule adopt a position of agnosticism when studying historical greats such as Aristotle or Abaelard or Locke or Leibniz. Philosophers do not study them to agree or disagree with them but to discover the structure, the strengths and weaknesses, and the influence of their views. This shared interest characterizes each of the chapters included here. And with this in mind we hope that the following chapters will provide a 'back to basics' foundation for philosophical discussion and help to make even contemporary discussions more readily accessible to philosophers and students.

The decision about which philosophers to include in such a study must of course be limited by a number of factors, not least by the demand to keep the book to a manageable size. There is good reason to think that the story of necessity extends back earlier than Aristotle; it is a theme to which even pre-socratic thinkers attended. Our choice to begin this volume with Aristotle is due to the fact that in his writings one can easily identify the ancient roots of our modern questions. Aristotle explicitly puts modal notions of necessity and possibility at the centre of both his metaphysical theory and his study of human reasoning. The centrality of necessity to Aristotle's views about human reasoning is neatly illustrated by his famous definition of what counts as a syllogism – that is, what counts as a valid deduction:

> A deduction is a discourse in which, certain things having been supposed, something different from the things supposed results of necessity because these things are so. (*Prior Analytics* A1, 24b18–22; Smith 1989)

In making this connection so explicitly Aristotle set the agenda for subsequent generations of philosophers. Because of this we have chosen Aristotle's definition as the focus of Chapter 1 in this volume. Here, Adriane Rini looks closely at Aristotle's *Prior Analytics* and asks what the textual evidence there can tell us about how Aristotle himself understood the necessity involved in deduction. In doing so, Rini illustrates the extent to which Aristotle's account can be understood in terms of matters of fact, as opposed to where it requires modal, counterfactual reasoning. She explains why Aristotle's notion of validity is in fact more sophisticated than what is captured by a simple 'substitutional' notion of validity.

It is known that Aristotle has two senses of possibility. One is possibility in the sense in which we think about it in current modal logic. Possibility in this sense includes what is necessary. In discussions of Aristotle this is sometimes called *one-sided* possibility. The other sense is what we usually call *contingency* (or *two-sided* possibility). A proposition p is contingent if it is neither necessarily true nor necessarily false. Aristotle speaks as though this latter sense is his 'official' definition of possibility. Yet there are many respects in which contingency is less tractable than one-sided possibility, and it is this question which is the focus of Marko Malink's analysis in Chapter 2. Malink explains that however much Aristotle's philosophy might suggest that two-sided possibility is basic, Aristotle's logic indicates that one-sided has to be his basic notion of possibility. Malink reveals Aristotle as a philosopher who begins with a view of possibility as contingency, but who is then compelled by the considerations of his logic to treat necessity and possibility as more basic notions. In Chapter 3, Robin Smith turns to the infamous question of why Aristotle developed the modal syllogistic at all. Smith makes the case that even though Aristotle's theory of science is demonstrative – i.e., it is the derivation of necessities from necessities – this requires only the simple assertoric syllogistic, and does not in fact require the modal syllogistic. The modal syllogistic arises, argues Smith, when Aristotle is concerned with the claim that there are things which are possible yet which may never happen – and Aristotle's interest in this is demonstrated in the *Metaphysics* in his objections to the views of 'the Megaric school', who, apparently, held that the possibility of something's being so is equivalent to its actually being so. From Aristotle's argument in the *Metaphysics* Smith teases out what a modal syllogistic needs to look like. Malink's and Smith's chapters both examine the question of the relation between the demands of a scientific theory and the demands of the formal properties of a logic which supports it. Whatever we may say about that question in respect of Aristotle, these two chapters illustrate one of the important themes of this volume, which is that a full understanding of the nature of necessity does require a consideration of its role in logic.

Although Aristotle's metaphysics of substance and accident presupposes a view on what necessity is, there is no real discussion in his works of this foundational question. That is to say, when Aristotle uses modal notions he never defines them except in terms of other modal notions. He does, as we have seen, make clear that the basic sense of possibility is contingency, but even when he is talking about this he is using modal notions to explain what he means. When we move to the Stoics and to the mediaeval period what we find is more explicit discussion of various sense of necessity and possibility. In her account of necessity in Stoic thought in Chapter 4 Vanessa de Harven finds three kinds. She calls them logical necessity, metaphysical necessity and providential necessity. For her, metaphysical necessity is the kind of necessity exhibited by the laws which constrain the way the world works – perhaps something like what some current philosophers call physical necessity. Providential necessity is the kind of necessity which arises as the result of particular situations which agents may find themselves in. This is the kind of necessity assumed by those who say that the past is necessary while the future is open. Such discussions go back to a passage in Aristotle's *On Interpretation* 9, where he makes remarks which have been interpreted by some to entail that many statements about the future have no truth value. They think that assigning them a truth value entails that the future is determined. While this issue seems not to have been important to Aristotle it was important to the Stoics, and became even more important in mediaeval thought, where the worry was that if what will happen is already determined, humans have no free will. If so there is a real question about whether they are responsible for their actions.

An appreciation of what logicians have meant by necessity and possibility can be furthered by looking at the Arabic tradition. In Chapter 5 Paul Thom concentrates on Avicenna and Averroes, and their thirteenth century successors. A distinction that we meet here and which becomes important in mediaeval thought is the distinction between the necessity of a 'class of beings', as Thom puts it, and a proposition. The reason for this importance is that in Islamic as in Christian thought the nature of a necessarily existing God assumes a central role. One claim, which it appears that Avicenna may have endorsed, is that to say that a being is necessary is to say that the proposition that that being exists is necessarily true. While, as Thom explains, the necessity of God's existence is for Avicenna not *just* the necessity of the proposition that God exists, yet within the history of necessity, the idea of necessity itself as a property of propositions appears in the Arabic tradition. Treating the objects of necessity as propositions allows you to make sense of 'scope' distinctions. Avicenna is a thinker who appears to have seen the importance of the distinction. He produces six different kinds of 'relative necessity'. Thom provides evidence to suggest that he thought that the necessity of a conditional was to be construed in terms of the necessity of the judgement that the consequent follows from the antecedent. Thom is a bit more sceptical, though, about just how Avicenna would handle the question concerning Aristotle on the necessity of the connection between premises and conclusion in a valid syllogism. A later Arab logician, Averroes, seems to have held the view that necessary statements assert a connection between universals, while purely assertoric statements assert a connection between individuals.

The question of scope is at the heart of a distinction made in mediaeval logic between what can be called *adverbial* sentences about necessity and *nominal* sentences. Consider the difference between

(1) It is necessary that every running horse is running

and

(2) Every running horse is necessarily running.

Take (1) first. (1) seems to be true in the sense that what it says is that a sentence – later called a *dictum* – is necessarily true. And indeed the sentence

(3) Every running horse is running

does indeed seem a necessarily true sentence. Now (2) *could* also attribute necessity to (3), as Aristotle seems to have realised about his examples, but (2) is more naturally interpreted in such a way that it is false. For even if a horse is running it surely has the ability to be standing still, and therefore its running, however true it actually is, is not *necessary*. It is this distinction which Chris Martin in Chapter 6 sees as exercising the twelfth century logicians whose work influenced Abaelard. Abaelard's own theory seems more complex, and we leave its discussion to Martin's chapter.

When we turn to late mediaeval logic in Chapter 7 Calvin Normore explains that the same questions arise. Up to the mid- to late thirteenth century, possibility was intimately connected with power – some state of affairs was thought possible just in case there was some agent which could bring it about. This view remained central until the middle of the fourteenth century, when a number of writers broke the connection between possibility and power, and eventually replaced it with a conception of possibility as a semantic or logical matter.

With Jack MacIntosh's contribution in Chapter 8 we find a bridge between mediaeval and early modern views about necessity. Aquinas discusses a distinction between 'absolute' and 'relative' necessity. Absolute necessity is a logical notion for Aquinas, in the sense that it applies to a proposition if its negation is contradictory. This is contrasted with relative necessity. One very important kind of relative necessity is tensed necessity, whereby the past and present are held to be necessary while the future is still open. The problem for Aquinas, as for mediaeval thought more generally, arises in relation to God's omniscience. For if a statement about the future is definitely true or false, then God knows *now* whether it is, and if God knows this now, it seems as though it is now determined, and that there is therefore nothing we can do about it.

In the early modern period we find the beginnings of a tension between the new science and the Aristotelian claim that the world is composed of substances which have natures that are not only necessary to them but can also be known, and that it is the job of science to identify the necessary truths which emerge from these natures. That does not mean that the early modern philosophers had no place for necessity. One area in which there was strong debate concerned the connection between necessary truths and God's omnipotence. Descartes, for example, appeared to flirt with the idea that God might, if he wished, choose to make it so that $2 + 2 \neq 4$. Such views were not really popular, maybe because they undercut attempts to use logic in debating the very existence of God. Robert Boyle produces an argument which MacIntosh suggests requires a system of modal logic which contains the principle that if the truth of a proposition logically entails the necessity of that proposition, then the possibility of that proposition entails its truth.

The theme of how early modern philosophers thought of the connection between knowledge and necessity is investigated by Peter R. Anstey in Chapter 9, where he suggests that the way to classify early modern thinkers about necessity is to see which of the claims they rejected in the following inconsistent tetrad:

1. We have knowledge of necessary facts about the world.
2. We can have knowledge of necessary facts about the world only if we acquire it by observation or reason.
3. We cannot acquire knowledge of necessary facts about the world from observations.
4. We cannot acquire knowledge of necessary facts about the world by reason.

The latter part of Anstey's chapter uses John Locke as an example. Locke thought that the view called 'corpuscularianism' – according to which everything that happens is determined by the behaviour of the minute corpuscles of which all material bodies are composed – was the most plausible of the new philosophies, though he was careful not to commit himself to it. He also thought that we humans, at least in the present state of our knowledge, or perhaps even in principle, could have no knowledge of just how the corpuscles worked. So that, although he seems to have thought that there had to be necessity in the world, he was less confident that it was the kind of necessity that we could know. Locke's view is that our knowledge of the external world is indirect. We humans only have direct knowledge of the contents of our minds. As far as particular individuals in the external world go, it is not clear that he thought that what are called *de re* necessities are anything that we can know. Although that seems to be his belief in the case of things in the external world, it seems not to apply to ideas. For it seems that for Locke ideas have to be just as they are, so that, for instance, being white is essential to all ideas of white.

Of the philosophers we are discussing in this book there is a sense in which we might consider it to be Leibniz who first produces a theory designed to tell us what necessity *is* rather than a discussion of what kind of necessities there are. Brandon Look's contribution in Chapter 10 links Leibniz' accounts of necessity with the principle of sufficient reason. He reminds us of Leibniz's worry that the use of the principle of sufficient reason might all too easily lead to the view that everything is absolutely necessary. It is generally recognized that Leibniz developed two accounts of our modal notions of necessity, possibility and contingency. According to his first account, propositions can be considered necessary or contingent per se. Leibniz seems to have abandoned this account in favour of a second, according to which a proposition is necessary if it can be resolved into a statement of identity in a finite number of steps, while a proposition is contingent if its analysis can go on to infinity without arriving at a statement of identity. In his analysis, Look addresses such issues as this.

The other logical feature for which Leibniz is known is the treatment of an individual as the sum of all its properties. A consequence of such a view seems to be that any truth about an individual must be necessary. Leibniz seems to think that this is correct from God's perspective, but not from ours, because we cannot process the infinitely many properties a thing has. But, as Jonathan Westphal notes in Chapter 11, there seems to be a problem in the form of what has been called the 'lucky proof'. Here is a quotation from Robert Adams:

> Even if infinitely many properties and events are contained in the complete concept of Peter, at least one of them will be proved in the first step of any analysis. Why couldn't it be Peter's denial? Why couldn't we begin to analyze Peter's concept by saying 'Peter is a denier of Jesus and ....'?

Westphal's solution is that while the 'lucky proof' may establish that *someone* denied Jesus, it does not tell us that it was *Peter*, since Peter is an individual constituted by infinitely many properties.

There are no chapters on Hume in our book, but it is not difficult to see Hume as a successor of Locke, with even less sympathy for any notion of necessity which applies to entities outside the mind. Like Locke, Hume presumably believes that there are logical relations among ideas, but it is less clear whether Hume thinks that there is any kind of necessity at all in the external world. In the Enquiry, iv.20, (Hume 1902, p. 25) Hume introduces his distinction between, on the one hand, relations of ideas and, on the other hand, matters of fact. The propositions of mathematics are of the first kind, and Hume is explicit that the reason for their certainty is that they are

> discoverable by the mere operation of thought, without dependence on what is anywhere existent in the universe. Though there never were a circle or triangle in nature, the truths demonstrated by Euclid would for ever retain their certainty and evidence.

By contrast

> Matters of fact, which are the second objects of human reason, are not ascertained in the same manner; nor is our evidence of their truth, however

great, of a like nature with the foregoing. The contrary of every matter of fact is still possible; because it can never imply a contradiction, and is conceived by the mind with the same facility and distinctness, as if ever so conformable to reality.

In Chapter 12 Nicholas Stang's discussion of Kant contrasts two notions of necessity and the corresponding two notions of possibility. Kant thinks of necessity and possibility as applying to the existence of things. Kant's notions of possibility are more straightforward than his concepts of necessity and so we concentrate on those here. A thing's existence is logically possible if its non-existence would entail a contradiction. But a thing is really possible if there are grounds for its existence. For example, triangles in various sizes and acuity could really exist because the nature of space allows them to exist. We can know that they could really exist because we have an intuition of space and can construct them mentally using that intuition. But Stang thinks that, for Kant, the notion of real possibility reaches beyond experience. Kant distinguishes between schematised and unschematised categories. A schematised category is one that is adapted to fit with objects in time. So, for example, the schematised notion of possibility is of an object that exists at some time. The unschematised categories, however, are pure concepts that can be applied to noumenal objects as well as to empirical ones. Stang argues that Kant thought that there is a notion of noumenal object that is really possible. Its ground is whatever causes it to exist (here the notion of a cause is the unschematised notion too). This means that Kant's conceptual framework supports a metaphysics of modality that is far richer than is often supposed.

Necessity is a touchstone issue in the thought of C. S. Peirce, not least because his account of meaning relies upon modal terms. Catherine Legg and Cheryl Misak offer an overview of Peirce's take on the matter in Chapter 13. They outline the source of Peirce's theory of representation in his categories of Firstness, Secondness and Thirdness, (monadic, dyadic and triadic relations). The first category corresponds to possibility, the second to mechanical necessity and the third to a kind of semantic or intentional necessity. Legg and Misak then turn to Peirce's modal epistemology and finally to Peirce's formal logic, focusing on his diagrammatic system of Existential Graphs. They discuss Peirce's modal metaphysics and its implications for determinism and realism about universals.

With the advent of the twentieth century we see the emergence of what is now known as modal logic. Modern modal logic is usually dated from a 1912 paper in *Mind* by C. I. Lewis. In Chapter 14 Edwin Mares follows the progression of Lewis's philosophy of logic from the origin of his theory of strict implication in 1911 to the end of his career. Contrary to the received view, Lewis did articulate semantic interpretations of his logical system. Although these interpretations are incomplete in the sense that they do not fully characterise his chosen logics in the formal sense, they do show how the logics are to be integrated into Lewis's wider epistemological and metaphysical programmes.

In two works in 1946 and 1947 Rudolf Carnap developed a semantics for modal logic which depends on the claim that, where $\Box$ (Carnap's '*N*') means 'it is necessary that', $\Box\ \alpha$ is true iff α is L-true, i.e., valid. Max Cresswell in Chapter 15 looks at what motivated Carnap to develop a modal logic. An important concern is to consider how Carnap's logic fares in the light of the criticisms of W. V. Quine. Cresswell argues that the use of meaning postulates, together with Carnap's views about analyticity, enables the 1946 semantics to provide an alternative, and arguably superior, answer to Quine than the answer Carnap provided in the 1947 book in terms of individual concepts.

The chapters in this volume have all been specially written, and have not been previously published, though of course they link with other work by the contributors. Many of them began life as contributions to a workshop organised in Wellington, New Zealand, in December 2012 on the history of necessity, and we are grateful to the New Zealand Government's Marsden Fund, administered by the Royal Society of New Zealand, for financial support. Additional support was provided by the Victoria University of Wellington, Massey University and the University of Auckland, together with a number of the universities of the workshop participants. Extra contributions were subsequently solicited to fill gaps in the fascinating history of this subject.

1 Aristotle on the Necessity of the Consequence

ADRIANE RINI

1. Introduction

In *Prior Analytics* A1, Aristotle explains what a deduction (a syllogism) is, and he makes clear that a syllogism involves 'necessity'[1]:

(1) A deduction is a discourse in which, certain things having been supposed, something different from the things supposed results of necessity because these things are so. (*Prior Analytics* A1, 24b18–22)[2]

Aristotle's explicit description of the conclusion of a valid syllogism as 'resulting of necessity' makes it look as though he must have in mind a modal notion. But can we say with any certainly what Aristotle understands by the 'necessity' here? Scholars have of course asked this many times, but there is no consensus and a variety of interpretations have been suggested.[3] The aim of the present paper is to focus on a source which sometimes is overlooked but which can be used to help shed light on Aristotle's understanding of the necessity in (1). This source is the examples Aristotle uses to establish where we do *not* have a syllogism. These examples are sometimes taken to be more trouble than help, and yet because they are Aristotle's own they are among the most direct textual evidence we have of his reasoning about his logic.[4]

[1] This definition appears also in *Topics* I.1, 100a25–27.

[2] Except where otherwise indicated, I have used Robin Smith's translation of the *Prior Analytics* (Smith 1989).

[3] Rini 2013 outlines various answers which scholars have suggested were available to Aristotle.

[4] van Rijen 1989 is cautious about Aristotle's examples. Ross 1957 and Łukasiewicz 1957 almost give the impression that they are embarrassed by the very idea that Aristotle might have produced counterexamples. See also Patzig 1968, who lays out a range of early views.

This research is funded by a Marsden Grant from the Royal Society of New Zealand (2011–2013). Parts of the project also received support from the Royal Flemish Academy for Science and the Arts (2010).

While modern philosophers and logicians might have a fairly sure idea of what today we mean by logic, we have to remember that when Aristotle was setting out the material which we know as the *Prior Analytics*, he was inventing logic. And though there are passages in *An. Pr.* where he does reflect on his invention, in setting out the syllogistic he is more focused on *doing* logic, less on talking about it or reflecting on the nature of his discovery. In Chapter 3 of this volume, Robin Smith investigates passages where Aristotle is reflecting on his logic, but these are special passages. For the most part reading Aristotle's discussion of the syllogistic feels like reading an introductory logic exercise book. There is in fact so little explicit reflection on the methods that some commentators have supposed that Aristotle must have proceeded simply by trial and error. Some more explicit discussion of his overarching concept of logic would perhaps help to answer this charge, but Aristotle does not give one. Because he does not in fact tell us much about his definition of a syllogism, this makes it difficult to say precisely how he understands that feature of his system which today we call logical consequence. If we want to understand more than the most basic mechanics of the syllogistic, then we have to piece together the available evidence which we find in the *Prior Analytics*. And in doing so we have to be careful about how we use our modern tools, lest they begin to colour our picture. For example, we should not suppose that because Aristotle is after all doing *logic* then he must have a concept of logic which is relevantly like ours; rather, we need to consider whether there is textual evidence to show that his concept of logic is like ours. But this makes the 'trial and error' suggestion especially interesting. Can we find evidence to show that Aristotle had (or that he lacked) the kind of overarching appreciation of logic needed to carry his own syllogistic beyond mere trial and error? Can we find evidence that Aristotle reached any deeper understanding of the consequence relation? This chapter will show how textual evidence from *An. Pr.* provides answers to such questions and helps to show what precisely is at stake in the 'trial and error' criticism.

If we want to cash these questions out in modern terms, we might ask whether Aristotle's understanding of a syllogism extends beyond a simple substitutional view of logic. On a substitutional view of logic, an inferential schema is valid if and only if every possible substitution instance of it is truth-preserving. When we put terms into the valid schemas and generate premises involving actual truths, then the conclusion itself is always going to be an actual truth. In order to know whether such a substitutional view is what Aristotle has in mind, we need to look not just at the (valid) syllogisms but also at the examples he offers when there is no syllogism – when a schema must be rejected. In such cases Aristotle tells us we cannot syllogize. As a rule, it is only in his instructions for constructing such examples that Aristotle offers terms – 'man', 'animal', 'moving' are among his favourites. By contrast, in cases where we can syllogize – that is, where there is a valid syllogism – Aristotle's discussion relies on the more abstract term-variables A, B and C, and he does not suggest terms. (It is worth noting that Aristotle does not have our modern labels 'valid' and 'invalid'. He expresses the distinction between what today we call a valid syllogistic schema and an invalid one by describing when 'there is a syllogism' and when 'there is no syllogism', or equivalently, when we can syllogize and when we cannot.)

Aristotle's syllogistic unfolds in stages and it will help to follow the stages of development as laid out in *An. Pr.* My initial focus is, therefore, the non-modal (or assertoric) examples. As this paper will show, in these Aristotle can always use terms which generate instances of schemas which involve only *actual* truth. Section 2 of this chapter sets out the textual evidence for this claim. But as later sections show, in the modal syllogistic, when Aristotle gives instructions for constructing examples to establish when a conclusion does not follow, the results often go beyond actual truth and falsity, and instead require non-actual possibilities. *This suggests that at least in the modal syllogistic Aristotle has a modal view of validity which goes beyond the notion that any substitution of terms preserves actual truth.* But, first, what should we say about the non-modal? How well can the substitutional view capture what Aristotle describes there?

2. Non-modal Reasoning in *An. Pr.* (the Assertoric Syllogistic)

In order to begin to answer, we need to look at Aristotle's method for demonstrating when a premise-pair will not yield a syllogism. The first such non-modal (assertoric) example occurs in *An. Pr.* A4:

However, if the first extreme [A] follows all of the middle [B] and the middle [B] belongs to none of the last [C], there will not be a deduction of the extremes [A, C], for nothing necessary results in virtue of these things being so. For it is possible for the first extreme to belong to all as well as to none of the last. Consequently, neither a particular nor a universal conclusion becomes necessary; and, since nothing is necessary because of these, there will not be a deduction. Terms for belonging to every are animal, man, horse; for belonging to none, animal, man, stone. (*An. Pr.* A.4, 26a2–9)[5]

Aristotle's point is that this is a case in which from the given premise-pair – 'A belongs to all B' and 'B belongs to no C' – we have no guarantee of *any* conclusion. (Some scholars say that such a premise-pair is 'inconcludent'.) That *no* proposition can be obtained is important, and reflects the nature of syllogistic reasoning. In Aristotle's system there are just four basic forms which propositions can take. Medieval scholars used the vowels *i*, *o*, *a*, and *e* to label the different forms:

'A belongs to some C' is a particular affirmative – i.e., an *i*-proposition.
'A does not belong to some C' is a particular privative – i.e., an *o*-proposition.
'A belongs to every C' is a universal affirmative – i.e., an *a*-proposition.
'A belongs to no C' is a universal privative – i.e., an *e*-proposition.

In the syllogistic, Aristotle routinely relies on the lesson from his square of opposition, that *i* and *e* are contradictories, and that *o* and *a* are contradictories. The premises described in the preceding passage are an *a*-proposition 'A belongs to all B' and an *e*-proposition 'B belongs to no C.' By showing that there is no conclusion from these premises, Aristotle argues that there is no syllogism. He is not looking for just any conclusion but a conclusion of a specified form – in this passage the purported conclusion must itself be a proposition which links an A *predicate* to a C *subject*.

So, we need to consider whether either an affirmative conclusion 'A belongs to some C' or a privative conclusion 'A does not belong to some C' follows from the premises. This means we are looking at the following two schemas, where (*i*) and (*o*) indicate the type of the purported conclusion:

A belongs to every B		A belongs to every B	
B belongs to no C		B belongs to no C	
A belongs to some C	(*i*)	A does not belong to some C	(*o*)

[5] I have inserted the A, B and C to make clear the structure of Aristotle's text.

Although Aristotle does not explicitly state this, as proof that these premises do not yield a syllogism it clearly suffices for him to show that neither an 'affirmative particular'(*i*) nor a 'negative particular' (*o*) can be obtained. Indeed he seems to have in mind to show both that you cannot get 'A belongs to some C' as the conclusion, and that you cannot get 'A does not belong to some C.' He rules out the *i*-proposition 'A belongs to some C' as a conclusion by offering a set of terms which makes its contradictory true. These are the 'terms for belonging to none': animal, man, and stone. They give the true *e*-proposition 'Animal belongs to no stone.' Aristotle rules out a conclusion in the form of the *o*-proposition 'A does not belong to some C' by offering a set of terms which makes its contradictory 'A belongs to every C' true. These are the 'terms for belonging to every': animal, man, and horse. These give the true *a*-proposition 'Animal belongs to every horse.' When we put in the terms which Aristotle recommends, we get the following:

'terms for belonging to every'		'terms for belonging to none'	
Animal belongs to every man		Animal belongs to every man	
Man belongs to no horse		Man belongs to no stone	
Animal belongs to every horse	(*a*)	Animal belongs to no stone	(*e*)

From the given premises, no conclusion of the required form 'becomes necessary' because the different sets of terms give different (true) results. One set of terms gives a true *a*-proposition, which shows that you cannot obtain an *o*-conclusion (and so not an *e*-conclusion). The other set gives a true *e*-proposition, so you cannot obtain an *i*-conclusion (and so not an *a*-conclusion). So you cannot obtain *any* conclusion of the required form. That is to say, there is no conclusion relating an A predicate to a C subject. So Aristotle says 'there is no syllogism'. The rejected conclusions – both the *i* and the *o* – are particulars, and showing that no particular AC proposition logically follows *also* establishes that no universal AC proposition follows, but Aristotle does not comment on this.

Aristotle uses this method right through the assertoric syllogistic of *An. Pr.* A4–6, in order to establish which non-modal, *assertoric* premise-pairs do not yield a syllogism. Chapter A4 deals with schemas in what Aristotle calls the 'first figure'. (The preceding schema is in the first figure.) Chapter A5 deals with schemas in the 'second figure', and chapter A6 deals with the 'third figure'. Taking A, B and C as our terms, we can represent the figures schematically:

First Figure	Second Figure	Third Figure
Pred-Subj	Pred-Subj	Pred-Subj
A-B	A-B	A-C
B-C	A-C	B-C
A-C	B-C	A-B

Aristotle calls the term which occurs twice in the premises the 'middle term'. The other terms he calls the 'extremes'. So in the first figure, the B term is the middle and drops out of the conclusion. In the second figure, the A term is the middle, and in the third figure, C is the middle.

Triples of terms together with the definitions of the three figures give us Aristotle's 'recipe' for constructing examples to show which schemas do not syllogize. Some awkwardness arises right away. As a rule Aristotle does not himself actually put the terms into these examples for us; instead, he sketches what he reasons we need to have in order to be able to do so ourselves. He offers proofs of validity of the valid syllogisms, but leaves the final business of showing when we cannot syllogize as homework for us, his readers. So, the evidence of these examples is not straight, direct evidence. We have to do our homework, and that involves some pencil work and therefore some small amount of interpretation on our own part. So if these examples do provide evidence of how Aristotle was thinking about the nature of his logic, then in order to access that evidence that we have to unpackage it. The tables which follow give a complete list of the term-triples which Aristotle offers for showing which assertoric, non-modal schemas do not let us syllogize and should be interpreted in the way I have described. These schemas are listed by the *An. Pr.* chapter in which they appear – that is, A4, A5, or A6, in bold – and by the line numbers of the relevant explanations. The term-triples are listed in the top line of each entry. The term-triples are numbered using square brackets, for greater ease of reference. And where Aristotle indicates that we should construct an example using the term-triples, the example is included. The terms give two true premises, but in each case no conclusion follows from the premises, and Aristotle's point is that in none of these can we syllogize.

Rejected assertoric schemas

A4: (First Figure)

26a2–9		animal-man-horse;	animal-man-stone	[1]
	a	Animal belongs to all men	Animal belongs to all men	
	e	Man belongs to no horse	Man belongs to no stone	
		Animal belongs to all horses	Animal belongs to no stone	
26a11–12		science-line-medicine;	science-line-unit	[2]
	e	Science belongs to no line	Science belongs to no line	
	e	Line belongs to no medicine	Line belongs to no unit	
		Science belongs to all medicine	Science belongs to no unit	
26a34–36		good-condition-wisdom;[a]	good condition-ignorance	[3]
	i	Good belongs to some condition	Good belongs to some condition	
	a	Condition belongs to all wisdom	Condition belongs to all ignorance	
		Good belongs to all wisdom	Good belongs to no ignorance	
26a38		white-horse-swan;	white-horse-raven	[4]
	i	White belongs to some horse	White belongs to some horse	
	e	Horse belongs to no swan	Horse belongs to no raven	
		White belongs to all swans	White belongs to no raven	
26b6–10		animal-man-white; [swan and snow in place of white]		[5]
	a	Animal belongs to every man	Animal belongs to every man	
	o	Man does not belong to some swan	Man does not belong to some snow	
		Animal belongs to every swan	Animal belongs to no snow	
26b10–14		inanimate-man-white [swan and snow in place of white]		[6]
	e	Inanimate belongs to no man	Inanimate belongs to no man	
	o	Man does not belong to some snow	Man does not belong to some swan	
		Inanimate belongs to some snow	Inanimate does not belong to some swan	
26b24–25		animal-white-horse	animal-white-stone	[7]
	i	Animal belongs to some white	Animal belongs to some white	
	i	White belongs to some horse	White belongs to some stone	
		Animal belongs to every horse	Animal belongs to no stone	
	o	Animal does not belong to some white	Animal does not belong to some white	
	i	White does not belong to some horse	White does not belong to some stone	
		Animal belongs to every horse	Animal belongs to no stone	
	i	Animal belongs to some white	Animal belongs to some white	
	o	White does not belong to some horse	White does not belong to some stone	
		Animal belongs to every horse	Animal belongs to no stone	
	o	Animal does not belong to some white	Animal does not belong to some white	
	o	White does not belong to some horse	White does not belong to some stone	
		Animal belongs to every horse	Animal belongs to no stone	

[a] Aristotle's terms in [3] and [4] will show that *oa* and *oe* are also inconcludent.

In [5] and [6] Aristotle first offers white as one of the recommended terms but then in elaborating on the proof, he suggests that we "let swan and snow also be selected from among those white things of which man is not predicated." Where white names an accident, swan and snow name substances and so generate examples in which the truth value of the purported conclusions is easy to appreciate. As the terms in [5] and [6] make clear, we never need move beyond simple actual-world truths.

The terms offered in [7] are given to show that a family of schemas can all be quickly rejected. These all involve combinations of various particular premises. Aristotle uses the same terms to establish that in each such case there is no deduction. Here is the passage:

Nor will there be a deduction in any way if both the intervals are particular, whether positively or privatively, or if one is stated positively and the other privatively, or if one is indeterminate and the other determinate, or both are indeterminate. Common terms for all are animal, white, horse; animal, white, stone. (*An. Pr.* A4, 26b22–25)

A5: (Second Figure)

27a18–20		substance-animal-man;		substance-animal-number[a]	[8]
	a	Substance belongs to every animal		Substance belongs to every animal	
	a	Substance belongs to every man		Substance belongs to every number	
		Animal belongs to every man		Animal belongs to no number	
27a20–23		line-animal-man;		line-animal-stone	[9]
	e	Line belongs to no animal		Line belongs to no animal	
	e	Line belongs to no man		Line belongs to no stone	
		Animal belongs to every man		Animal belongs to no stone	
27b4–6		animal-substance-raven;		animal-white-raven (27b5-6)	[10]
	o	Animal does not belong to some substance		Animal does not belong to some white	
	a	Animal belongs to every raven		Animal belongs to every raven	
		Substance belongs to every raven		White belongs to no raven	
27b6–8		animal-substance-unit;		animal-substance-science	[11]
	i	Animal belongs to some substance		Animal belongs to some substance	
	e	Animal belongs to no unit		Animal belongs to no science	
		Substance belongs to all unit		Substance belongs to no science	
27b16–23		[no terms for belonging];[b]		black-snow-animal	[12]
			e	Black belongs to no snow	
			o	Black does not belong to some animal	
				Snow belongs to no animal	
27b26–28		[no terms for belonging];[c]		white-swan-stone	[13]
			a	White belongs to every swan	
			i	White belongs to some stone	
				Swan belongs to no stone	

27b28–32	white-animal-raven;	white-stone-raven	[14]
i	White belongs to some animal	White belongs to some stone	
e	White belongs to no raven	White belongs to no raven	
	Animal belongs to every raven	Stone belongs to no raven	
27b33	white-animal-swan;	white-animal-snow	[15]
i	White belongs to some animal	White belongs to some animal	
a	White belongs to all swans	White belongs to all snow	
	Animal belongs to all swans	Animal belongs to no snow	
27b36–38	white-animal-man;	white-animal-inanimate	[16]

[a] Aristotle calls substance the middle.

[b] As Ross explains, no terms are given here since Aristotle has already shown that second figure *ee* is inconcludent, so you would only need a proof that *eo* is inconcludent if the *o* was not an *e*. But if the *o* is not an *e* then there must also be an *i* (since if the *e* is false – that it is not so that none – then it is so that some). But if there is an *i* as well as an *o* we get Festino, and then Aristotle puts it this way: "But it is not possible to find terms of which the extremes are related positively and universally, if M belongs to some O, and does not belong to some O." [This is Jenkinson's translation, in Aristotle 1985.]

[c] Here we require the second *i* not to be an *a*, so we have to have an *o* as well as an *i*. But second figure *ao* is *not* inconcludent since we have Baroco.

The terms in [16] occur in a brief passage in which – as in [7] earlier – we find a sweeping description of several separate premise combinations, none of which produces a valid syllogism. In [16] Aristotle explains that all such second-figure combinations can be shown to be invalid by the same sets of terms – and the terms here in [16] are strikingly similar to the terms in [7]. The full passage in which he offers the term-triples for the second figure is as follows:

But neither does a deduction come about if the middle term belongs or does not belong to some of each extreme, or belongs to one and does not belong to the other, or not to all of either, or indeterminately. Common terms for all these are white, animal, man; white, animal, inanimate. (*An. Pr.* A5, 27b36–39)

With white as the middle term, each of these various premise combinations is straightforward and does not involve any modal suppositions.

i	White belongs to some animal	White belongs to some animal
i	White belongs to some man	White belongs to something inanimate
	Animal belongs to every man	Animal belongs to nothing inanimate
i	White belongs to some animal	White belongs to some animal
o	White does not belong to some man	White does not belong to something inanimate
	Animal belongs to every man	Animal belongs to nothing inanimate
o	White does not belong to some animal	White does not belong to some animal
i	White belongs to some man	White belongs to something inanimate
	Animal belongs to every man	Animal belongs to nothing inanimate

o	White does not belong to some animal	White does not belong to some animal
o	White does not belong to some man	White does not belong to something inanimate
	Animal belongs to every man	Animal belongs to nothing inanimate

A6: (Third Figure)

28a30–33		animal-horse-man;	animal-inanimate-man	[17]
	a	Animal belongs to every man	Animal belongs to every man	
	e	Horse belongs to no man	Inanimate belongs to no man	
		Animal belongs to every horse	Animal belongs to nothing inanimate	
28a33		animal-horse-inanimate;	man-horse-inanimate	[18]
	e	Animal belongs to nothing inanimate	Man belongs to nothing inanimate	
	e	Horse belongs to nothing inanimate	Horse belongs to nothing inanimate	
		Animal belongs to every horse	Man belongs to no horse	
28b22–24		animate-man-animal;	["We cannot get terms"] (because of Datisi)	[19]
	a	Animate belongs to every animal		
	o	Man does not belong to some animal		
		Animate belongs to every man		
28b36-38		animal-man-wild;	animal-science-wild	[20]
	i	Animal belongs to something wild	Animal belongs to something wild	
	e	Man belongs to nothing wild	Science belongs to nothing wild	
		Animal belongs to every man	Animal belongs to no science	
29a2		animal-science-wild;	animal-man-wild	[21]
	o	Animal does not belong to some wild	Animal does not belong to some wild	
	e	Science belongs to nothing wild	Man belongs to nothing wild	
		Animal belongs to no science	Animal belongs to all men	
29a3		["We cannot get terms"] (because of Ferison)	raven-snow-white;	[22]
			e White belongs to no raven	
			o Snow does not belong to some white	
			Raven belongs to no snow	
29a9–10		animal-man-white;	animal-inanimate-white	[23]

In [23], we again have a passage in which Aristotle gives a sweeping description of a number of separate premise combinations, none of which produces a valid syllogism and all of which Aristotle says can be shown to be invalid by the same terms. The full passage in which these term-triples appear is as follows:

Nor will there be a deduction in any way if each term belongs or does not belong to some of the middle, or if one belongs and the other does not belong, or if one belongs to some and the other not to every, or if they belong indeterminately. Common terms for all these are animal, man, white; animal, inanimate, white. (*An. Pr.* A6, 29a7–10)

Putting the terms in place shows we cannot syllogize in any of these cases.

i	Animal belongs to some white	Animal belongs to some white
i	Man belongs to some white	Inanimate belongs to some white
	Animal belongs to all men	Animal belongs to nothing inanimate
i	Animal belongs to some white	Animal belongs to some white
o	Man does not belong to some white	Inanimate does not belong to some white
	Animal belongs to all men	Animal belongs to nothing inanimate
o	Animal belongs to some white	Animal belongs to some white
i	Man does not belong to some white	Inanimate does not belong to some white
	Animal belongs to all men	Animal belongs to nothing inanimate
o	Animal does not belong to some white	Animal does not belong to some white
	animal belongs to all men	animal belongs to nothing inanimate

Ross (p. 302) notes: "where A. states that a particular combination of premises yields no conclusion he gives no reason for this, e.g. by pointing out that an undistributed middle or an illicit process is involved; but he often points to an empirical fact which shows that the conclusion follows." While Ross does not particularly *like* the introduction of empirical fact, he carefully notes throughout most of his discussion of these cases that the 'conclusion' expresses a relation between subject and predicate which is *in fact* true. This is an important point, a point which makes our tables needed evidence, but it seems to be a point whose significance Ross has not entirely grasped. If I have represented Aristotle's plan as he intended, then the terms he offers for constructing the assertoric examples leave open the possibility that he understands the necessity in (1) – that is, the necessity in his definition of a syllogism – as indicating only that *any substitution* of terms in a (valid) schema which gives true premises will also give a true conclusion. His examples never take us beyond what is *in fact* true.

Aristotle clearly thinks that he has shown that his fourteen assertoric syllogisms are the only assertoric syllogisms. But in addition to the assertoric, non-modal syllogisms, Aristotle considers two kinds of modal syllogisms. *Apodeictic* syllogisms have at least one premise explicitly involving necessity. *Problematic* syllogisms have at least one premise involving possibility. Aristotle's definition of a syllogism as it is stated in (1), above, is clearly intended to cover all syllogisms including the apodeictic and problematic, so we need to consider the rejected modal schemas as well. The rejected apodeictic and problematic modal schemas raise different considerations. Section 3 looks at counterexamples in the apodeictic syllogistic, in A9–11. Section 4 looks at counterexamples in the problematic syllogistic, in A14–22.

3. Deductions Involving Necessary Premises in *An. Pr.* (the Apodeictic Syllogistic)

The shift to syllogistic schemas with necessary premises also brings with it a shift in Aristotle's counterexamples. Where in the non-modal counterexamples actual facts are all that is needed, this is not the case in the apodeictic counterexamples, and Aristotle is clearly aware that apodeictic counterexamples require more. To see why, let us turn to the text and his discussion, and let us begin with the schema known as Barbara XLL. (These names – 'Barbara', 'Datisi', 'Ferison', etc. – encode instructions for the construction of the proof. See Rini 2011 for an account of the basic mnemonics. When we move to modal syllogistic, scholars often add to the old mnemonics, using 'X' to indicate a non-modal, assertoric proposition, 'L' to indicate a premise about necessity, and 'Q' and 'M' to indicate various kinds of possibility. The name Barbara XLL then tells us that we are considering a schema in the form of Barbara in which the first *a*-premise is assertoric (X), the second *a*-premise is about necessity (L), and that we are considering whether we can syllogize to a conclusion which is itself an *a*-proposition about necessity (L).)

Barbara XLL
A belongs to all B
B belongs of necessity to all C
A belongs of necessity to all C

Aristotle explains that this does not produce a syllogism – that is, Barbara XLL is invalid:

However, if AB is not necessary but BC is necessary, the conclusion will not be necessary. For if it is, it will result that A belongs to some B of necessity, both through the first and through the third figure. But this is incorrect: for it is possible for B to be the sort of thing to which it is possible for A to belong to none of. It is, moreover, also evident from terms that the conclusion can fail to be necessary, as, for instance, if A were motion, B animal, and C stood for man. For a man is of necessity an animal, but an animal does not move of necessity, nor does a man. (A9, 30a24–32)

When we put Aristotle's terms – A motion, B animal, and C man – into Barbara XLL, we get the following instance:

Motion belongs to all animals
Animal belongs of necessity to all men
Motion belongs of necessity to all men

This first premise, however, is not actually true. Aristotle is aware of this, and his careful explanation marks an important shift beyond the kind of simple factual reasoning which characterizes the examples in the non-modal counterexamples. In the discussion of Barbara XLL, Aristotle wants to make clear that the initial AB premise is not a modal one. Initially, he describes the AB premise as 'not necessary' in order to distinguish it from the second premise and the conclusion, both of which are about necessity. Today, we simply mark such a premise with an 'X' to indicate its assertoric nature, as distinct from an apodeictic premise, which we mark with an 'L' to indicate necessity. Aristotle's explanation, however, is more complicated. He says, "*it is possible* for B to be the sort of thing to which *it is possible* for A to belong to none of" – and so uses *two* occurrences of a modal expression to describe what is a non-modal premise. It is much easier to explain the premise's assertoric nature by putting Aristotle's terms in place: 'it is possible for animal to belong to nothing in motion.' That is, there is nothing about being an animal which prevents animals from being in motion, so we can suppose that the premise is true – that all animals are in motion. The terms allow a step away from the formal discussion. Aristotle's simple point is that animals can be but do not have to be moving. For the purpose of the proof, he wants us to take as an AB premise the *supposition* that all animals are moving. We are to assume that the premise is true, even when strictly – that is, even when *in the actual world* – it is not.

The need for such assumptions characterizes many of Aristotle's apodeictic counterexamples and indicates an increasing sophistication in the move from actual truths in the assertoric syllogistic to counterfactual assumptions in the apodeictic syllogistic.

As the modal syllogistic unfolds, Aristotle becomes more explicit about the counterfactual assumptions sometimes required by his apodeictic counterexamples. When he argues, in *An. Pr.* A10, that the schema Camestres LXL does not produce a syllogism, he introduces a new distinction between what is 'necessary without qualification' and what is 'only necessary when these things are so'. When a conclusion is only necessary when these things are so, we are instructed to assume that an assertoric premise which is actually false were true. Here is how he explains:

> nothing prevents the A having been chosen in such a way that it is possible for C to belong to all of it. And moreover, it would be possible to prove by setting out terms that the conclusion is not necessary without qualification, but only necessary when these things are so. For instance, let A be animal, B man, C white, and let the premises have been taken in the same way (for it is possible for animal to belong to nothing white). Then, man will not belong to anything white either, but not of necessity: for it is possible for a man to become white, although not so long as animal belongs to nothing white. Consequently, the conclusion will be necessary when these things are so, but not necessary without qualification. (*An. Pr.* A.10, 30b30–40)

Camestres LXL has the following form:

A belongs of necessity to all B
A belongs to no C
B belongs of necessity to no C

Aristotle's terms bring out the need for a counterfactual assumption.

Animal belongs of necessity to every man
Animal belongs to nothing white
Man belongs of necessity to nothing white

Again, the non-modal, assertoric premise is strictly false, because, of course, there are many things which are animals but which are nonetheless white. But for the counterexample to work, we must suppose that the premise were true. That is, we must reason counterfactually. When we make this counterfactual assumption, then we *can* validly conclude that 'man belongs to nothing white' (an assertoric conclusion) and so we do get a syllogism Camestres LXX. By contrast, Camestres LXL is not a syllogism, and so we cannot validly conclude that 'man belongs of necessity to nothing white' (an apodeictic conclusion). It is in order to capture this difference between Camestres LXX and LXL that Aristotle introduces the expressions 'necessary without qualification' and 'necessary when these things are so'. The assertoric conclusion 'man belongs to nothing white' is only 'necessary when these things are so' – that is, it follows of necessity only from the counterfactual assumption that animal belongs to nothing white. This allows us to syllogize to the assertoric conclusion, but not to the apodeictic proposition 'man belongs of necessity to nothing white.' So when we do syllogize according to our counterfactual assumption, the conclusion we reach is *not* necessary without qualification – that is, it is not itself a modal proposition.

4. Deductions Involving Premises about Possibility in *An. Pr.* (the Problematic Syllogistic)

In *An. Pr.* A14–22, Aristotle focuses on the syllogistic involving possibility. And he carefully distinguishes between two senses of possibility: 'possibility according to the definition' – that is, that which is contingent, 'neither necessary nor impossible' – and 'possibility not according to the definition' – that is, that which is not necessarily not the case. Scholars often call the first 'two-sided possibility' and the second 'one-way possibility'. Aristotle does not in fact spend much time discussing the effect of one-way possibility on the syllogistic – his account of these syllogisms is largely contained in *An. Pr.* A8. On the other hand, the effect of two-sided possibility on syllogistic reasoning clearly interests him, and drives his study through *An. Pr.* A14–22.[6]

One of the themes which Robin Smith explores in this volume is the importance of Aristotle's method of proof involving the realization of a possibility. This realization method makes its first appearance in *An. Pr.* A15, in a discussion of a schema called Barbara XQM, where the 'Q' indicates a premise about contingency or two-sided possibility, and the 'M' indicates one-way possibility. The schema is as follows:

Barbara XQM
A belongs to every B
B contingently belongs to every C
A possibly belongs to every C

[6] Marko Malink's chapter in this volume is a careful discussion of the relation between these two senses of possibility in Aristotle's logical (and other) works.

Here is Aristotle's explanation:

Now, with these determinations made, let A belong to every B and let it be possible for B to belong to every C. Then it is necessary for it to be possible for A to belong to every C. For let it not be possible, and put B as belonging to every C (this is false although not impossible). Therefore, if it is not possible for A to belong to every C and B belongs to every C, then it will not be possible for A to belong to every B (for a deduction comes about through the third figure). But it was assumed that it is possible for A to belong to every B. Therefore, it is necessary for it to be possible for A to belong to every C (for when something false but not impossible was supposed, the result is impossible). (An. Pr. A15, 34a34–b2)

The BC premise is about contingency, but in his instructions for construction of the proof Aristotle carefully signals the need for counterfactual assumption, telling us to "put B as belonging to every C (this is false although not impossible)." So from the given premise, 'B contingently belongs to every C', we are told to suppose that 'B belongs to every C' – that is, we are to suppose the possibility in the initial premise is realized. Aristotle coaches us through by suggesting how we might put terms (animal, moving and man) in place. Following his instructions, we begin with a premise about possibility: 'motion contingently belongs to all men.' When we suppose that what is possible were actual, then we obtain the non-modal proposition 'motion belongs to all men', which Aristotle says "is false although not impossible". At 34b11–12, he explains: "For perhaps nothing prevents man from belonging to everything in motion at some time (for example, if nothing else should be moving)."

The expression 'nothing prevents' reminds us that our reasoning here is counterfactual. Aristotle understands that if something is possible then it *might* be true, and that supposing that it is true might lead to a falsehood, but since it is possible then supposing that it is true cannot lead to an absurdity. I have discussed this schema more thoroughly in Rini (2011) and (2013), looking closely at the details of Aristotle's method. But what makes Aristotle's method of proof through realization important in the present discussion is that it demonstrates the extent to which Aristotle comes to embrace explicitly counterfactual reasoning in the syllogistic.

Consider a similar example in the second figure. In chapter A19, 37b36–38, Aristotle explains why there is not a syllogism from two affirmative premises, where one premise is non-modal and the other contingent. This is not surprising because there are no second-figure syllogisms from two affirmative premises even in the non-modal cases, and so we might expect that the non-modal results will carry over to the modal case. Aristotle, at any rate, suggests terms for the construction of the proof and leaves the actual construction to the reader. This is all that we find in the text:

> But if both premises are put as positive, there will not be a deduction. Terms for belonging are health, animal, man; terms for not belonging are health, horse, man. (37b36–38)

These instructions give the following results:

Every animal is contingently healthy
Every man is healthy
Some man is possibly not an animal

Every animal is healthy
Every man is contingently healthy
Some man is possibly not an animal

Every horse is contingently healthy
Every man is healthy
Some man is a possible horse

Every horse is healthy
Every man is contingently healthy
Some man is a possible horse

If these are to stand as proof that there is no second-figure syllogism from one assertoric premise and one premise about contingency, then we must suppose that the premises are in fact true. But of course it is not true that every man is healthy, nor that every animal is healthy, nor that every horse is healthy. The supposition that these premises are true then involves a counterfactual assumption. That is, for these terms to work, we must suppose that every man is healthy, that every animal is healthy, that every horse is healthy.

Aristotle's own reasoning is not restricted to actual-world truth, and because it is not, the necessity involved in his discussion moves beyond what a purely substitutional notion of validity can accommodate. And so this stands as further evidence that what Aristotle means by 'necessity' in (1) is a genuinely modal notion.

Aristotle's method exhibits a sureness and confidence about counterfactual reasoning and shows how counterfactual reasoning does in fact play a central role in his understanding of syllogistic. I would hazard to say that *probably* we should read this counterfactual reasoning all the way back into the assertoric syllogistic as well, and simply attribute to Aristotle a sophisticated counterfactual notion of logical consequence. But as noted in Section 2, above, we lack the concrete *textual* evidence to establish that in the earliest –that is, the assertoric – part of the syllogistic Aristotle had reached the sophisticated counterfactual notion of validity that he later exhibits in the modal syllogistic. It could be that in the assertoric syllogistic of *An. Pr.* A4–6, he has at most a substitutional notion. And it could be that the more sophisticated counterfactual understanding only arises through his efforts to explain the first invalid modal schemas in A9. But one thing we can clearly say is that when he reaches that discussion in the modal syllogistic, his understanding does rely on counterfactual notions, and establishing this gives us the answer to our initial question about what Aristotle understands by the necessity in his definition of a syllogism. At the very least, by the time he gets to the apodeictic syllogistic his definition of a syllogism involves a genuinely counterfactual understanding of validity. Aristotle's understanding of syllogistic reasoning is not just about what happens to be so as a matter of actual fact, but about what must be so.

2 Aristotle on One-Sided Possibility

MARKO MALINK

The term 'possible', in Aristotle's view, is ambiguous.[1] It has two senses, known as one-sided possibility and two-sided possibility (or contingency). Being two-sided possible means being neither impossible nor necessary, and being one-sided possible simply means being not impossible. Aristotle defines the two senses in *Prior Analytics* 1.13 (32a18–21). This definition is followed by a controversial passage which states several equivalences between modal expressions (32a21–28). It is often thought that this passage is spurious and should be excised even though it is found in all manuscripts.[2] I argue that the passage is not spurious, but contains a coherent argument justifying the one-sided sense of 'possible' (Section 1). This argument follows a pattern of proof explained by Aristotle elsewhere in the *Prior Analytics* and *De caelo* (Section 2). Moreover, it is closely related to a similar argument concerning one-sided possibility in *De interpretatione* 13 (Section 3). Examining this latter argument will help us better understand the controversial passage in *Prior Analytics* 1.13 (Section 4).

1. The Problematic Passage

At the beginning of *Prior Analytics* 1.13, Aristotle introduces the two senses of 'possible' as follows:

[i] I use the expressions 'to be possible' and 'what is possible' in application to something if it is not necessary but nothing impossible will result if it is

[1] *Prior Analytics* 1.3 25a37–38.
[2] Becker 1933: 11–14, Ross 1949: 327–328, Mignucci 1969: 112 and 295–296, Seel 1982: 163, Smith 1989: 18 and 125–126, Patterson 1995: 270, Huby 2002: 92, Striker 2009: 18 and 128–129.

I would like to thank Simona Aimar, David Charles, Max Cresswell, Brad Inwood, and Ben Morison for helpful comments on an earlier draft of this paper.

put as being the case; [ii] for it is only equivocally that we say that what is necessary is possible.[3] (*An. Pr.* 1.13 32a18–21)

In point [i] of this passage, Aristotle characterizes two-sided possibility. Something is two-sided possible just in case it is not necessary and nothing impossible results from putting it as being the case. This characterization involves the term 'impossible'. Unlike 'possible', 'impossible' is not ambiguous in the *Prior Analytics* but is consistently used in the sense of 'not one-sided possible' as opposed to 'not two-sided possible'.[4] Aristotle takes the condition that nothing impossible results from putting X as being the case to be equivalent to the condition that X is not impossible.[5] Thus, he holds that something is two-sided possible just in case it is neither necessary nor impossible.

In point [ii] of the passage, Aristotle turns to one-sided possibility. He does not offer an explicit definition of one-sided possibility, but characterizes it by pointing out the specific case of application that distinguishes it from two-sided possibility. Thus, he states in point [ii] that whatever is necessary is one-sided possible. Accordingly, something is one-sided possible just in case it is not impossible. Two-sided possibility is, so to speak, bounded on two sides by impossibility and necessity, whereas one-sided possibility is only bounded on one side by impossibility.

Aristotle often does not explicitly indicate whether a given occurrence of 'possible' is to be understood in the two-sided or one-sided sense, but in the *Prior Analytics* it is usually clear from the context

[3] [i] λέγω δ' ἐνδέχεσθαι καὶ τὸ ἐνδεχόμενον, οὗ μὴ ὄντος ἀναγκαίου, τεθέντος δ' ὑπάρχειν, οὐδὲν ἔσται διὰ τοῦτ' ἀδύνατον· [ii] τὸ γὰρ ἀναγκαῖον ὁμωνύμως ἐνδέχεσθαι λέγομεν.

[4] In the *De interpretatione*, by contrast, 'impossible' (ἀδύνατον) is occasionally used in the sense of 'not two-sided possible' (see n. 38).

[5] See Ross 1949: 327, Łukasiewicz 1957: 154–55, Ebert & Nortmann 2007: 470–471, and Beere 2009: 120–121. Aristotle holds that if something impossible results from putting X as being the case, then X is impossible (*De caelo* 1.12 281b14–15 and 23–25; *Physics* 7.1 243a1–2, *An. Pr.* 1.15 34a25–33; cf. Alexander *in An. Pr.* 157.7, Beere 2009: 121–122, Rosen & Malink 2012: 185–187, 196–200, and 210–211). Moreover, he is committed to the converse, that if X is impossible then something impossible results from putting X as being the case. Otherwise there would be an X which is impossible but satisfies Aristotle's two conditions for being two-sided possible stated in point [i] (since nothing impossible is necessary; *De int.* 13 22b13–14). Consequently, something would be both impossible and two-sided possible, which is absurd (see *De int.* 13 22a15–17). This leaves open the question of whether, if X is impossible, X itself can be regarded as something impossible that results from putting X as being the case. Ebert & Nortmann (2007: 470–471) think that the answer is yes, whereas Beere (2009: 120 n. 4) is sceptical.

which of them he means. In point [ii], the one-sided sense is qualified as being merely equivocal (ὁμωνύμως). This qualification indicates that the one-sided sense is not the preferred sense of 'possible' in the system of modal syllogisms expounded in *Prior Analytics* 1.13–22.[6] Modal syllogisms concerning one-sided possibility play only a minor role in these chapters.[7] Instead, Aristotle focuses on modal syllogisms concerning two-sided possibility.[8] Accordingly, he refers to the characterization of two-sided possibility in point [i] as his official 'definition' (διορισμός) of possibility in the modal syllogistic.[9]

Having introduced the two senses of 'possible' in points [i] and [ii], Aristotle proceeds as follows:

> [iii] That this [viz., one-sided possibility] is what is possible is evident from opposed pairs of denials and affirmations. [iv] For 'it is not possible to belong' and 'it is impossible to belong' and 'it is necessary not to belong' are either the same or follow one another, [v] so that their opposites, 'it is possible to belong' and 'it is not impossible to belong' and 'it is not necessary not to belong', will also either be the same or follow one another; [vi] for either the affirmation or the denial holds of everything. [vii] Therefore, what is possible will not be necessary and what is not necessary will be possible.[10] (*An. Pr.* 1.13 32a21–29)

[6] For this use of ὁμωνύμως, see Bonitz 1870: 514a49–61 and Striker 2009: 128.

[7] Aristotle generally does not consider any modal syllogisms that contain a one-sided possibility premiss (the only exception is his use of a syllogism called Barbara MXM at 1.15 34b2–6; see Malink & Rosen 2013: 971–973). Aristotle discusses a number of modal syllogisms that have a one-sided possibility conclusion, but he does so only when the modal syllogism in question is established by *reductio ad absurdum*.

[8] Thus, two-sided possibility is the preferred notion of possibility in the modal syllogistic in *Prior Analytics* 1.3 and 8–22. By contrast, one-sided possibility is preferred in the framework of propositional modal logic developed in *Metaphysics* Θ 4 and *Prior Analytics* 1.15. This framework is based on the principle that if B follows from A then the possibility of B follows from the possibility of A (Rosen & Malink 2012: 179–195). The principle is correct for one-sided possibility but not for two-sided possibility (Hintikka 1973: 59–60). Thus, the adjective 'possible' (δυνατόν) must be understood as expressing one-sided possibility in *Metaphysics* Θ 4 and *Prior Analytics* 1.15 (1047b14–30 and 34a5–33). On the other hand, the adjective is never used to indicate possibility in the premisses and conclusions of modal syllogisms in *Prior Analytics* 1.3 and 8–22 (instead, Aristotle uses the verbs ἐνδέχεσθαι and ἐγχωρεῖν; see Malink & Rosen 2013: 959–961).

[9] *An. Pr.* 1.14 33b23, 1.15 33b28, 33b30, 34b27; see Malink 2013: 257 n. 14.

[10] [iii] ὅτι δὲ τοῦτ᾽ ἔστι τὸ ἐνδεχόμενον, φανερὸν ἔκ τε τῶν ἀποφάσεων καὶ τῶν καταφάσεων τῶν ἀντικειμένων· [iv] τὸ γὰρ οὐκ ἐνδέχεται ὑπάρχειν καὶ ἀδύνατον ὑπάρχειν καὶ ἀνάγκη μὴ ὑπάρχειν ἤτοι ταὐτά ἐστιν ἢ ἀκολουθεῖ ἀλλήλοις, [v] ὥστε

This passage has been the subject of controversy. The main question is whether the pronoun 'this' in point [iii] refers to two-sided or one-sided possibility. Since Philoponus, most commentators have taken it to refer to two-sided possibility.[11] Thus, they take points [iii]–[vii] to justify the notion of two-sided possibility introduced in point [i]. However, this is problematic because Aristotle's claims in point [v] are true for one-sided but not for two-sided possibility (e.g., the claim that 'it is possible to belong' is equivalent to 'it is not necessary not to belong'). Because of this problem, it is often thought that points [iii]–[vii] are spurious and should be excised even though they are found in all manuscripts (see n. 2).

On the other hand, some have argued that the passage can be retained if the pronoun 'this' in point [iii] is taken to refer to one-sided possibility.[12] On this interpretation, points [iii]–[vi] are intended to explain the notion of one-sided possibility introduced in [ii]. Aristotle does so by establishing, in point [v], a number of theses that are characteristic of one-sided possibility but do not hold for two-sided possibility. By contrast, point [vii] is again about two-sided possibility, drawing a consequence from the initial characterization of two-sided possibility in [i].[13] Thus, points [i] and [vii] deal with Aristotle's preferred notion of possibility, whereas [ii]–[vi] constitute a parenthetical argument introducing and justifying the secondary notion of one-sided possibility.

On this interpretation, there is no reason to think that points [ii]–[vi] are spurious, even if they are parenthetical. In what follows, I elaborate

καὶ τὰ ἀντικείμενα τούτοις, τὸ ἐνδέχεται ὑπάρχειν καὶ οὐκ ἀδύνατον ὑπάρχειν καὶ οὐκ ἀνάγκη μὴ ὑπάρχειν, ἤτοι ταὐτὰ ἔσται ἢ ἀκολουθοῦντα ἀλλήλοις· [vi] κατὰ παντὸς γὰρ ἡ φάσις ἢ ἡ ἀπόφασις. [vii] ἔσται ἄρα τὸ ἐνδεχόμενον οὐκ ἀναγκαῖον καὶ τὸ μὴ ἀναγκαῖον ἐνδεχόμενον. In point [vi] of this passage, φάσις is used in place of κατάφασις to mean 'affirmation' (Waitz 1844: 403, Bonitz 1870: 813a16–23).

11 Philoponus *in An. Pr.* 148.6–11, Pacius 1597a: 146, Waitz 1844: 402–403, Becker 1933: 11, Mignucci 1969: 295–296, Hintikka 1973: 32, Smith 1989: 125, Nortmann 1996: 162–165, Ebert & Nortmann 2007: 474–475.

12 Gohlke 1936: 66, Hintikka 1977: 81, Buddensiek 1994: 21–23, Thom 1996: 13, King 2014: 487–488.

13 Maier 1900: 140 n. 1, Buddensiek 1994: 21–23. Thus, the connective 'therefore' (ἄρα) at the beginning of [vii] indicates a consequence, not of the preceding argument in [iii]–[vi], but of [i] (*pace* Becker 1933: 12). Point [vii] leads into the discussion of two-sided possibility propositions at 32a29–b3. The phrase 'what is not necessary' in [vii] should presumably be understood as shorthand for 'what is not necessary either way', excluding both what necessarily belongs and what necessarily does not belong (see Hintikka 1973: 32–34, 1977: 81, Patterson 1995: 270, Nortmann 1996: 167).

Table 2.1 *Opposed pairs of affirmations and denials at* Prior Analytics *1.13 32a21–28*

it is not possible to belong	it is possible to belong
it is impossible to belong	it is not impossible to belong
it is necessary not to belong	it is not necessary not to belong

and vindicate this interpretation by showing how Aristotle's argument in [iii]–[vi] succeeds in establishing the one-sided sense of 'possible', and how it is related to a similar argument concerning one-sided possibility in *De interpretatione* 13.

2. Establishing One-Sided Possibility in *Prior Analytics* 1.13 (32a21–28)

As Aristotle indicates in point [iii], his argument for one-sided possibility appeals to opposed pairs of denials and affirmations. The denials he has in mind are linguistic expressions such as 'it is not possible to belong'; the corresponding affirmations are expressions such as 'it is possible to belong'.[14] In Table 2.1, Aristotle considers the six expressions in points [iv] and [v]. Each row of Table 2.1 contains an opposed pair of affirmation and denial. The denials in the left-hand column are the contradictory opposites of the corresponding affirmations in the right-hand column, and vice versa. Aristotle does not simply assume that these expressions are contradictory opposites, but argues for this in detail in *De interpretatione* 12.[15]

In the earlier chapters of *De interpretatione*, affirmations and denials are defined as complete declarative sentences affirming or denying something of something.[16] In *De interpretatione* 12–13 and in points [iii]–[vi] above, however, Aristotle uses the terms 'affirmation' and 'denial' to refer not to complete sentences but to incomplete expressions that can be completed into sentences.[17] Thus, the six expressions

[14] τὸ (οὐκ) ἐνδέχεται ὑπάρχειν. The combination of the definite article τὸ with the finite verb ἐνδέχεται in points [iv] and [v] indicates that Aristotle is concerned with linguistic expressions rather than with non-linguistic items signified by expressions; cf. *Categories* 10 12b5–16.

[15] 21b23–26 and 21b34–22a13.

[16] *De int.* 4 17a2–3, 5 17a8–9, 6 17a25–26; see Crivelli 2004: 86–89.

[17] See Ackrill 1963: 150, Weidemann 2002: 402–403, and Striker 2009: 128.

listed in Table 2.1 are sentential fragments that can be completed into sentences of the form 'It is possible for A to belong to B' and 'It is not possible for A to belong to B' (or, in other words, 'It is possible for B to be A' and 'It is not possible for B to be A'). In what follows, I use 'affirmation' and 'denial' to refer to such incomplete expressions rather than to complete sentences.

The affirmations and denials listed in Table 2.1 can be regarded as sentential forms such as 'It is possible for … to belong to —' and 'It is not possible for … to belong to —'. Each of them has two free slots. Let X be a pair of expressions such that the result of filling the two slots with these expressions is a well-formed sentence. If the resulting sentence is true, we say that the affirmation or denial in question *holds of* X. For example, if X is the pair of expressions 'walking' and 'Socrates', then the affirmation 'it is possible to belong' holds of X, whereas the corresponding denial does not hold of X (since it is possible for Socrates to be walking).

In point [vi], Aristotle states a version of the law of excluded middle (LEM), according to which 'either the affirmation or the denial holds of everything'. The same version is stated in *De interpretatione* 12 (21b4).[18] This version of LEM seems to presuppose that the affirmations and denials in question are not complete sentences, but incomplete expressions that are capable of *holding* of something. In *De interpretatione* 12, Aristotle adds a corresponding version of the principle of non-contradiction (PNC), according to which 'it is impossible for opposite expressions (τὰς ἀντικειμένας φάσεις) to be true of the same thing' (21b17–18). Thus, each of the contradictory pairs listed in Table 2.1 satisfies the following two principles:

Where X is a pair of expressions as described above:

PNC: Not both the affirmation and the denial hold of X
LEM: Either the affirmation or the denial holds of X

As we will see shortly, both PNC and LEM play an important role in Aristotle's argument in points [iii]–[vi].

In point [iv], Aristotle states that the three expressions on the left-hand side of Table 2.1 are 'the same or follow one another (ἀκολουθεῖ ἀλλήλοις)'. By this he means that the expressions are mutually equivalent.

[18] Similarly, *An. Pr.* 1.46 51b32–33, 2.11 62a13–14, *Topics* 6.6 143b15–16.

In the *Prior Analytics*, Aristotle uses the verb 'follow' (ἀκολουθεῖν or ἕπεσθαι) to express that a term is predicated universally of another term in the sense that the former term holds of everything of which the latter holds.[19] For example, when Aristotle says that animal follows man, he means that 'animal' holds of everything of which 'man' holds (*An. Pr.* 1.27 43b31). Correspondingly, A and B follow one another just in case they are equivalent in the sense that, for any X, A holds of X just in case B holds of X.[20] In point [iv], Aristotle states that the three expressions on the left-hand side of Table 2.1 are mutually equivalent in this sense. He does not argue for these equivalences but seems to take them for granted (see Section 4).

Aristotle goes on, in point [v], to infer that the three expressions on the right-hand side of Table 2.1 are mutually equivalent.[21] He infers their equivalence from that of their contradictory opposites on the left-hand side by appealing to LEM in point [vi]. Aristotle does not spell out the details of this inference in *Prior Analytics* 1.13. However, he provides an account of the inference elsewhere, in chapters 1.46 and 2.22 of the *Prior Analytics* and in *De caelo* 1.12.[22] Let us consider these accounts in turn, beginning with the one in chapter 1.46:

Whenever A and B are so related that it is not possible for them to belong to the same thing at the same time, but necessarily one or the other of them

[19] See, e.g., *An. Pr.* 1.27 43b3–4, 43b22–32; cf. Striker 2009: 192.

[20] For this use of 'follow one another' (ἀκολουθεῖ ἀλλήλοις), see *Metaph.* Γ 2 1003b22–25 and *De caelo* 1.12 282a30–b2; similarly, *De int.* 12 21b35–36.

[21] These equivalences hold for one-sided possibility but not for two-sided possibility. Accordingly, some commentators who take points [i]–[vi] to be about two-sided possibility suggest that Aristotle does not state any equivalences between modal expressions in [iv] and [v] (Hintikka 1973: 33, Nortmann 1996: 162–166, Ebert & Nortmann 2007: 474–475; cf. n. 11above). However, once [ii]–[vi] are taken to be about one-sided possibility, there is no reason to deny that Aristotle states these equivalences in [iv] and [v].

[22] In points [iv] and [v], Aristotle states that the expressions in each column of Table 2.1 'are either the same or follow one another'. If they are taken to be the same (or to signify the same), Aristotle's inference is underwritten by the principle that 'if things are the same, their opposites also will be the same, in any of the recognized forms of opposition; for since they are the same, there is no difference between taking the opposite of the one and that of the other' (*Topics* 7.1 151b33–36). Aristotle's account in *Prior Analytics* 1.46, 2.22, and *De caelo* 1.12 justifies the inference on the weaker assumption that the expressions are not the same (or do not signify the same) but are equivalent in the sense that they follow one another.

belongs to everything, and again C and D are related in the same way, and A follows C, ... then D will follow B... For, since necessarily one or the other of C and D belongs to everything, but it is not possible for C to belong to anything to which B belongs, because C brings along with it A, and it is not possible for A and B to belong to the same thing, it is evident that D will follow B. (*An. Pr.* 1.46 52a39–b8)

In this passage, Aristotle considers four terms A, B, C, and D. He assumes that A and B do not both hold of the same thing (Premiss 1), but that one or the other of them holds of everything (Premiss 2). The same is true for C and D (Premisses 3 and 4). Moreover, Aristotle assumes that A follows C, that is, that A holds of everything of which C holds (Premiss 5). He concludes that D follows B:

Premiss 1:	For any X, not both A and B hold of X
Premiss 2:	For any X, either A or B holds of X
Premiss 3:	For any X, not both C and D hold of X
Premiss 4:	For any X, either C or D holds of X
Premiss 5:	For any X, if C holds of X then A holds of X
Conclusion:	For any X, if B holds of X then D holds of X

This argument is valid. Its conclusion follows logically from Premisses 1, 4, and 5. Premisses 2 and 3 are dispensable in the present argument, but they are required for the corresponding argument inferring the converse of the conclusion from the converse of Premiss 5.[23] Although Aristotle does not mention this converse argument in *Prior Analytics* 1.46, it is a straightforward version of the one he presents in the passage just quoted. Taken together, the two arguments establish that, given Premisses 1–4 and given that A and C follow one another, it may be inferred that B and D follow one another:

Premiss 1:	For any X, not both A and B hold of X
Premiss 2:	For any X, either A or B holds of X
Premiss 3:	For any X, not both C and D hold of X
Premiss 4:	For any X, either C or D holds of X
Premiss 5*:	For any X, A holds of X if and only if C holds of X
Conclusion*:	For any X, B holds of X if and only if D holds of X

[23] In this converse argument, the conclusion 'For any X, if D holds of X then B holds of X' is inferred from Premisses 2–3 and the converse of Premiss 5 (i.e., 'For any X, if A holds of X then C holds of X').

Aristotle justifies the inference from Premisses 1–5* to Conclusion* in *De caelo* 1.12 as follows:

It has been shown that A and C follow one another. When terms stand to one another as these do, that is, when A and C follow [one another], B and A never belong to the same thing,[24] but one or the other of them belongs to everything, and D and C likewise, then it is necessary that B and D follow one another. For suppose that B does not follow D. Then A will follow, since either B or A belongs to everything. But C belongs to [everything] to which A belongs. Consequently, C will follow D; but it was assumed that this is impossible. And the same argument shows that D follows B. (*De caelo* 1.12 282b25–283a1)

This argument proceeds by *reductio ad absurdum*. The assumption for *reductio* is that B does not follow D. In other words, the assumption is that it is not the case that, for any X, if D holds of X then B holds of X. This means that there is something – call it Y – such that D but not B holds of Y. Aristotle proceeds to draw a consequence from the assumption for *reductio*. He describes this consequence by the phrase 'A will follow' (τὸ ἄρα Α ἀκολουθήσει). Contrary to what might be thought, this does not mean that A will follow D; for this would not be warranted by the assumption for *reductio*. Instead, Aristotle's intended meaning is that A will hold of Y, of which D holds. Similarly, Aristotle concludes the *reductio* by claiming that 'C will follow D' (τῷ ἄρα Δ τὸ Γ ἀκολουθήσει). Again, this should not be taken to mean that C holds of *everything* of which D holds. Rather, Aristotle uses the phrase 'C will follow D' to express that C holds of something, namely, Y, of which D holds. By contrast, the phrase 'D follows B' in the last sentence of the passage just quoted expresses the claim that D holds of *everything* of which B holds. Thus, Aristotle's use of the verb 'follow' in the *reductio* at 282b30–32 is potentially misleading (as we will see, the same misleading use of 'follow' occurs in *De interpretatione* 13). Nevertheless, his argument is a lucid proof, in two halves, of the inference from Premisses 1–5* to Conclusion*.

[24] This translates τὸ δὲ Ε καὶ τὸ Ζ μηθενὶ τῷ αὐτῷ. Here and elsewhere in this passage I supply ὑπάρχει; see 282a14–21. For ease of exposition, I write 'A', 'B', 'C', 'D' for Aristotle's 'Z', 'E', 'Θ', 'H' in this passage.

Aristotle presents the same proof in *Prior Analytics* 2.22 without the misleading use of 'follow':

> If either A or B and either C or D belong to everything, but they cannot belong at the same time, then, if A and C convert, B and D also convert. For if B does not belong to something to which D belongs, it is clear that A belongs to it; and if A belongs, then also C does; for they convert. Consequently, C and D belong to it together; but this is impossible. (*An. Pr.* 2.22 68a11–16)

In this version of the argument, Aristotle uses the verb 'convert' instead of the phrase 'follow one another'. He avoids the misleading use of 'follow' in the *reductio* by writing 'A belongs to it' instead of 'A will follow', and writing 'C and D belong to it together' instead of 'C will follow D'.

In sum, then, Aristotle has demonstrated that Premisses 1–5* entail Conclusion*. In *De caelo* 1.12, he proceeds to use this pattern of argument to establish the equivalence of the terms 'imperishable' and 'ungenerable' based on the equivalence of their contradictory opposites. To this end, he interprets the schematic letters as follows:[25]

A generable
B ungenerable
C perishable
D imperishable

In *Prior Analytics* 1.13, he uses the same pattern of argument in point [v] to establish the equivalence of modal affirmations and denials. In this application of the pattern, the quantified variable 'X' ranges over suitable pairs of expressions that yield complete sentences when combined with incomplete affirmations and denials, as explained above. The schematic letters are interpreted as follows:

A it is not possible to belong
B it is possible to belong
C it is impossible to belong
D it is not impossible to belong

In this interpretation of the schematic letters, A is the contradictory opposite of B, and C is the contradictory opposite of D. Consequently,

[25] *De caelo* 1.12 282b23–25 and 283a1–3. Similarly, *An. Pr.* 2.22 68a8–11; cf. Ross 1949: 479.

Table 2.2 *Aristotle's square of modal expressions* (De interpretatione *13 22a14–31)*

possible to be not impossible to be not necessary to be	not possible to be impossible to be necessary not to be
possible not to be not impossible not to be not necessary not to be	not possible not to be impossible not to be necessary to be

Premisses 1–4 of the inference are guaranteed by LEM and PNC, respectively. Thus, Aristotle's appeal to LEM in point [vi] of his argument is a statement of Premisses 2 and 4 of the inference. Aristotle does not explicitly assert PNC in *Prior Analytics* 1.13, but takes it for granted since he regards it as the firmest of all principles.

3. Establishing One-Sided Possibility in *De interpretatione* 13 (22b11–14)

In *De interpretatione* 13, Aristotle discusses various implications (ἀκολουθήσεις, 22a14) between modal expressions. He presents these implications by means of the square in Table 2.2.[26] For the sake of brevity, I will write 'necessary' as shorthand for 'necessary to be', 'possible not' as shorthand for 'possible not to be', and so on.

Aristotle asserts that the first expression in each quadrant of the square implies the second and third expressions (22a14–23). For example, 'possible' implies 'not impossible' in the sense that, for any X, if the former holds of X then the latter holds of X.[27] Likewise,

[26] Aristotle's square contains separate entries for two distinct expressions of possibility: δυνατὸν εἶναι and ἐνδεχόμενον εἶναι (22a24–31). Aristotle takes these two expressions to be equivalent in *De interpretatione* 13 (22a15–16; see Ammonius *in De int.* 231.9–16, 237.14–15, Seel 1982: 149, Weidemann 2002: 424–425); thus it suffices for our purposes to consider only one expression of possibility.

[27] I use 'imply' as the converse of Aristotle's 'follow' (ἀκολουθεῖν) at 22a14–37. Thus, an expression implies another expression just in case the latter follows from the former. Hintikka's (1973: 45–47) suggestion that ἀκολουθεῖν at 22a14–37 signifies the mutual equivalence of expressions is not convincing; cf. Bluck 1963: 216–17, Frede 1976: 239–240, Brandon 1978: 175–177, Seel 1982: 148–149, Weidemann 2002: 425, 2012: 109.

'possible' implies 'not necessary', and so on. Aristotle does not assert that the expressions in each quadrant are mutually equivalent; nor does he assert that the second or third expression implies the first expression.[28] Rather, his focus is on determining what follows from the first expression in each quadrant.

Unfortunately, as Aristotle goes on to explain at 22b10–28, the square in Table 2.2 turns out to be incoherent. The incoherence is best described as being due to the ambiguity of 'possible' between the two-sided and the one-sided sense. Although Aristotle does not explicitly distinguish between these two senses in *De interpretatione* 13, the distinction underlies his discussion of the incoherence. In the left-hand column of the square, 'possible (not)' is taken to imply 'not necessary (not)'. This implication requires that 'possible' be understood not in the one-sided sense but in the two-sided sense. In the right-hand column, by contrast, 'not possible' is taken to imply 'necessary not', and 'not possible not' is taken to imply 'necessary'. These implications require that the occurrences of 'possible' in these phrases be understood not in the two-sided but in the one-sided sense. For if 'possible' is understood in the two-sided sense, then 'not possible' does not imply 'necessary not' but only the disjunctive phrase 'necessary or necessary not'.[29] Thus, 'possible' expresses two-sided possibility in the left-hand column of the square, but one-sided possibility in the right-hand column.[30]

This does not mean that Aristotle's use of 'possible' in the square is random or haphazard. There is a systematic difference between the two columns that helps to explain the distribution of one-sided and two-sided uses of 'possible': in the left-hand column 'possible' occurs unnegated in affirmations, but in the right-hand column it occurs negated in denials. When introducing the square at the beginning of chapter 13, Aristotle states the implications in question without argument (22a14–23). Thus he takes it for granted that 'possible' expresses two-sided possibility when it occurs unnegated in affirmations, whereas

28 Brandon 1978: 176; *pace* Hintikka (1973: 45–47). However, Aristotle asserts that the second expression implies the third expression in the two top quadrants; see 22b3–10 and 22b15–16; cf. Seel 1982: 147–150, Weidemann 2002: 425, 2012: 102.

29 Cf. *An. Pr.* 1.17 37a9–26.

30 See Ackrill 1963: 151, Bluck 1963: 215–216, Seel 1982: 160, Whitaker 1996: 162–164, Weidemann 2002: 442.

it expresses one-sided possibility when it occurs negated in denials. He does not, at this point in chapter 13, recognize any unnegated one-sided uses of 'possible', nor any negated two-sided uses of 'possible'.

The double use of 'possible' threatens not only the coherence of Aristotle's square, but also the main results of chapter 12. There, Aristotle argues at length that 'possible' and 'not possible' constitute an opposed pair of affirmation and denial.[31] He requires that such opposed pairs satisfy LEM, according to which 'the affirmation or the negation holds of everything'.[32] However, Aristotle's square in Table 2.2 violates LEM. For example, consider the affirmative sentence 'It is possible for every human to be an animal' and the corresponding negative sentence 'It is not possible for every human to be an animal'. LEM requires that at least one of them be true. But according to Aristotle's square, 'possible' is to be understood as expressing two-sided possibility in the affirmative sentence, and as expressing one-sided possibility in the negative sentence. As a result, LEM is violated because neither sentence is true (since it is necessary for every human to be an animal).

Because of the homonymy of 'possible', the occurrences of 'possible' and 'not possible' in Aristotle's square fail to constitute a genuine opposed pair of affirmation and denial.[33] For in a genuine opposed pair the same thing must be denied and affirmed of the same thing *not homonymously*:

> Let us call an affirmation and a denial which are opposite a contradiction (ἀντίφασις). I speak of sentences as opposite when they affirm and deny the same thing of the same thing – not homonymously, together with all other such conditions that we add to counter the troublesome objections of sophists. (*De int.* 6 17a33–37)

According to this definition, 'not possible' fails to be the contradictory opposite of 'possible' in Aristotle's square. On the other hand, Aristotle holds that for every affirmation there is a corresponding denial which is its contradictory opposite, and that for every denial there is a corresponding affirmation which is its contradictory opposite (17a30–33).

31 *De int.* 12 21b23–25, 21b37–22a1, 22a11–13.
32 *De int.* 12 21b4; see n. 18.
33 Weidemann 2002: 443–444.

Hence, if 'possible' has the two-sided sense in a given affirmation, there must be a corresponding denial in which it has the same sense. And if 'possible' has the one-sided sense in a given denial, there must be a corresponding affirmation in which it has the same sense. Given this, Aristotle is in a position to construct a revised square in which 'possible' has the same sense throughout, and which therefore does not suffer from the problems described previously.

In *De interpretatione* 13, Aristotle chooses to construct a revised square for one-sided possibility (he does not give a square for two-sided possibility). As a first step toward this revised square, Aristotle argues that 'necessary' implies 'possible', as follows:

> [i] What is necessary to be is possible to be. [ii] For otherwise the denial will follow, [iii] since it is necessary either to affirm or to deny; [iv] and then, if it is not possible to be, it is impossible to be; [v] so what is necessary to be is impossible to be – [vi] which is absurd.[34] (*De int.* 13 22b11–14)

In point [i] of this passage, Aristotle states the thesis to be proved, that everything necessary is possible: for any X, if 'necessary' holds of X then 'possible' holds of X. Aristotle's proof of this thesis proceeds by *reductio*. The assumption for *reductio* is that there is something – call it Y – such that 'necessary' holds of Y but 'possible' does not hold of Y.

In point [ii], Aristotle draws a consequence from the assumption for *reductio*. He indicates this consequence by the phrase 'the denial will follow' (ἡ ἀπόφασις ἀκολουθήσει). Contrary to what is sometimes thought, this is not intended to mean that 'not possible' follows 'necessary' in the sense that the former holds of *everything* of which the latter holds; for this is not warranted by the assumption for *reductio*.[35] Rather, point [ii] exhibits the same slightly misleading use of 'follow' that we saw in *De caelo* 1.12 (282b30–32). When Aristotle writes that 'the denial will follow', his intended meaning is that 'not possible' holds

[34] [i] τὸ μὲν γὰρ ἀναγκαῖον εἶναι δυνατὸν εἶναι· [ii] εἰ γὰρ μή, ἡ ἀπόφασις ἀκολουθήσει· [iii] ἀνάγκη γὰρ ἢ φάναι ἢ ἀποφάναι· [iv] ὥστ' εἰ μὴ δυνατὸν εἶναι, ἀδύνατον εἶναι· [v] ἀδύνατον ἄρα εἶναι τὸ ἀναγκαῖον εἶναι, [vi] ὅπερ ἄτοπον.

[35] *Pace* Ackrill (1963: 152), who holds that when Aristotle 'claims that 'necessary' must imply 'possible' since otherwise it would have to imply 'not possible' he is, of course, misusing the principle of excluded middle'. Similarly, Fitting & Mendelsohn (1998: 34) think that in point [ii] 'Aristotle confused the negation of the conditional with the negation of the consequent'.

of something, namely, Y, of which 'necessary' holds.[36] Aristotle justifies this claim by appealing to LEM in point [iii]: since the assumption for *reductio* implies that 'possible' does not hold of Y, it follows by LEM that 'not possible' holds of it.

In point [iv], Aristotle goes on to infer that, since 'not possible' holds of Y, so does 'impossible'. This is justified by one of the implications stated in Aristotle's original square (in the top right-hand quadrant, 22a19–20). Finally, Aristotle infers in point [v] that 'what is necessary to be is impossible to be'. Again, this does not mean that 'impossible' holds of everything of which 'necessary' holds. Rather, it means that 'impossible' holds of something, namely, Y, of which 'necessary' holds.[37] In point [vi], Aristotle states that this last consequence is absurd, thereby concluding the *reductio*.

Aristotle's argument in points [i]–[vi] can thus be reconstructed as follows:

1.	There is an X such that 'necessary' holds of X and 'possible' does not hold of X	[assumption for *reductio*]
2.	'Necessary' holds of Y and 'possible' does not hold of Y	[from 1: existential instantiation]
3.	'Necessary' holds of Y	[from 2]
4.	'Possible' does not hold of Y	[from 2]
5.	Either 'possible' holds of Y or 'not possible' holds of Y	[premiss: LEM]
6.	'Not possible' holds of Y	[from 4, 5]
7.	For any X, if 'not possible' holds of X then 'impossible' holds of X	[premiss: 22a19–20]
8.	'Impossible' holds of Y	[from 6, 7]
9.	There is an X such that 'necessary' holds of X and 'impossible' holds of X	[from 3, 8]
10.	Not: there is an X such that 'necessary' holds of X and 'impossible' holds of X	[premiss]
11.	Not: there is an X such that 'necessary' holds of X and 'possible' does not hold of X	[*reductio*: 1–9, 10]
12.	For any X, if 'necessary' holds of X then 'possible' holds of X	[from 11]

[36] Weidemann 2002: 437–441. The same use of 'follow' is found at 22b30, where Aristotle recapitulates point [ii] of the present argument.

[37] The use of the phrase 'what is necessary to be' (τὸ ἀναγκαῖον εἶναι) in [v] differs from its use in [i]. In [i], this phrase indicates a general term true of everything

This argument relies on three premisses. First, it relies on LEM (line 5, stated in [iii]). The second premiss is that 'not possible' implies 'impossible' (line 7, stated in [iv]). The third premiss is that 'impossible' and 'necessary' are disjoint in that they do not hold of the same thing (line 10, stated in [vi]). This last premiss makes it clear that 'impossible' is taken to mean 'not one-sided possible' rather than 'not two-sided possible'. This is a substantive assumption because 'impossible' is also occasionally used in the latter sense in *De interpretatione* 13.[38] It follows that 'not possible', since it implies 'impossible', is disjoint from 'necessary'. This means that the negated occurrence of 'possible' in 'not possible' does not have the two-sided sense but the one-sided sense; for if it had the two-sided sense, 'not possible' would mean 'not two-sided possible' and would fail to be disjoint from 'necessary' (in fact, it would follow from it). Thus, the second and third premisses taken together entail that a *negated* occurrence of 'possible' has the one-sided sense.

On this basis, Aristotle's argument establishes, by means of LEM, that an *unnegated* occurrence of 'possible' has the one-sided sense. For if the argument's conclusion in line 12 is to be true, the unnegated occurrence of 'possible' in it cannot have the two-sided sense but must have the one-sided sense.[39] This is the first unnegated occurrence of 'possible' that is used in the one-sided sense in *De interpretatione* 13. Prior to Aristotle's argument at 22b11–14, all unnegated occurrences of 'possible' in this chapter are understood in the two-sided sense.[40] Aristotle's argument shows that unnegated occurrences of 'possible' can also have the one-sided sense.

4. The Revised Square of Modal Expressions (*De interpretatione* 13 22b10–28)

Having shown that 'necessary' implies 'possible', Aristotle goes on to argue that the left-hand column of his original square is incorrect and in

of which 'necessary to be' holds. In [v], by contrast, the phrase picks out a particular item, Y, of which 'necessary to be' holds (Weidemann 2012: 441).

38 For example, 'impossible' presumably means 'not two-sided possible' in Aristotle's claim that 'not impossible' implies 'not necessary' (22b15–16); see Seel 1982: 161–163, Weidemann 2002: 443.

39 Provided that 'necessary' holds of something; otherwise the conclusion in line 12 would be vacuously true.

40 See *De int.* 13 22a15–31; cf. Seel 1982: 160.

Table 2.3 *Revised square of modal expressions* (De interpretatione *13 22b10–28)*

possible to be not impossible to be not necessary not to be	not possible to be impossible to be necessary not to be
possible not to be not impossible not to be not necessary to be	not possible not to be impossible not to be necessary to be

need of revision. In particular, he argues that, contrary to what he said at the beginning of chapter 13, 'possible' does not imply 'not necessary'. For otherwise – since 'necessary' implies 'possible' – it would follow that 'necessary' implies 'not necessary', and that is absurd (22b14–17). Nor does 'possible' imply 'necessary' or 'necessary not' (b17–22). From this Aristotle concludes that the only expression concerning necessity that can be implied by 'possible' is 'not necessary not' (b22–28).

Thus, Aristotle revises his original square (Table 2.3) by transposing 'not necessary' and 'not necessary not' in the left-hand column. In this revised square, 'possible' is not ambiguous but is consistently used in the one-sided sense throughout. Accordingly, 'impossible' means 'not one-sided possible'. As before, Aristotle holds that the first expression in each quadrant implies the second and third expressions.

The top half of the revised square corresponds exactly to the two groups of modal expressions discussed by Aristotle in his argument for one-sided possibility in *Prior Analytics* 1.13 (see Table 2.1). This argument can easily be adapted to capture the bottom half of the revised square. Thus, Aristotle's discussion in *Prior Analytics* 1.13 is closely connected to the revised square in *De interpretatione* 13. Nevertheless, there are differences between his arguments in the two chapters. In *De interpretatione* 13, Aristotle does not give a full direct proof showing that 'possible' implies 'not necessary not'. Instead, he argues that 'possible not' does not imply any of the other three expressions concerning necessity in the square. From this he infers that it implies 'not necessary not'. However, since he has not shown that it must imply at least one of the four expressions, his argument remains incomplete. In *Prior Analytics* 1.13, by contrast, Aristotle gives a full proof that 'possible' implies 'not necessary not'. In fact, he proves that the two expressions are equivalent.

In *De interpretatione* 13, Aristotle states that 'not possible' implies 'necessary not' (in the top right-hand quadrant of the square). By contraposition, it follows that 'not necessary not' implies 'possible', and hence that the two expressions are equivalent. Crucially, however, Aristotle does not state the latter implication; nor does he perform such inferences by contraposition in *De interpretatione* 13. Although he is committed to the equivalence of the three expressions in each quadrant of the revised square, Aristotle does not explicitly state these equivalences in *De interpretatione* 13.[41] Thus, the argument in *Prior Analytics* 1.13 is not only more concise and straightforward than the one in *De interpretatione* 13; it is also more powerful in that it establishes the equivalence of the three expressions in each quadrant of the square. In particular, it establishes the two equivalences that guarantee the interdefinability of one-sided possibility and necessity, namely, the equivalence of 'possible' and 'not necessary not', and – provided that the argument is extended to the bottom half of the square – the equivalence of 'necessary' and 'not possible not'.

Despite these differences, the arguments in *Prior Analytics* 1.13 and *De interpretatione* 13 share important similarities. First, they are both of an extensional nature in that they prove implications (or equivalences) between modal expressions by showing that the one expression holds of everything of which the other holds (and vice versa). Moreover, they both aim to establish the one-sided sense of *unnegated* occurrences of 'possible' by appealing to the one-sided sense of *negated* occurrences of 'possible'. They do so by invoking the law of excluded middle (LEM). Among other things, this law encodes Aristotle's commitment to the principle that for every denial in which something is denied of something there is a corresponding affirmation in which the same thing is affirmed of the same thing *not homonymously*. It follows from this principle that for every negated occurrence of 'possible' in a denial there is a corresponding unnegated occurrence in an affirmation such that both occurrences have the same sense. If the negated occurrence has the one-sided sense, the unnegated occurrence must have the one-sided sense too.

[41] See Seel 1982: 157, Weidemann 2002: 425 and 456, 2012: 102; cf. n. 28 above. When Fitting & Mendelsohn (1998: 7) claim that Aristotle identifies 'not necessary not' as the equivalent of 'possible' in *De interpretatione* 13, this claim must be qualified accordingly.

Aristotle's strategy in both arguments shows that, in his view, negated occurrences of 'possible' are most naturally interpreted in the one-sided sense, whereas unnegated occurrences are more naturally interpreted in the two-sided sense. For Aristotle, one-sided readings of unnegated occurrences are somewhat artificial and stand in need of justification. Aristotle does not explain why this is so. Modern theorists have provided an explanation on his behalf by appealing to the Gricean concept of conversational implicature. These theorists disagree with Aristotle that there are two distinct senses of 'possible'. Instead, they argue that there is only the one-sided sense, and that the two-sided interpretation is a pragmatic phenomenon resulting from a conversational implicature. Thus, Lauri Karttunen writes:

> Aristotle distinguished between one-sided possibility and two-sided possibility... However, I doubt that there is any need to postulate these two distinct senses. As far as I understand it, Aristotle's distinction is designed to account for the same facts that are also covered by Grice's conversational postulates. Assuming that the speaker is following the cooperative principle by saying that something is possible, he indicates that he is not in the position to make a stronger statement [to the effect that that thing is necessary]... Therefore, for all he knows, the contrary is also possible. The two-sided interpretation of *possible* arises from these considerations; it is not part of the meaning of *possible*.[42] (Karttunen 1972: 6 n. 2)

One of Grice's postulates governing cooperative communication is the requirement that the speaker is to provide the addressee with the strongest relevant information available. Thus, if the speaker asserts that something is one-sided possible she must have reasons to withhold from the stronger claim that it is necessary. On the assumption that the speaker knows whether or not the stronger claim is true, her assertion of the weaker claim gives rise to a conversational implicature to the effect that the stronger claim is false.[43] This helps explain why

[42] Similarly, Horn 1973: 207, 1989: 211–212, Levinson 1983: 163–164, van der Auwera 1996: 184–185.

[43] For the same reason, an assertion of a particular affirmative sentence such as 'Some As are B' gives rise to a conversational implicature to the effect that the corresponding universal sentence 'All As are B' is false. Accordingly, particular affirmative sentences can be used in a two-sided sense ('Some but not all As are B') and in a one-sided sense ('Some, perhaps all, As are B'). In *Prior Analytics* 1.1–22, particular affirmative sentences are used only in the

the two-sided interpretation of unnegated occurrences of 'possible' is usually preferred to the one-sided interpretation in ordinary language. It may also help explain why Aristotle felt the need to include an argument justifying the one-sided sense of unnegated 'possible' in *Prior Analytics* 1.13.

On the other hand, negation in ordinary language typically operates exclusively on the meaning of expressions but not on conversational implicatures that might be associated with them on a given occasion of use.[44] Thus, if a negation is applied to 'possible', it only negates its one-sided meaning but not the implicature that gives rise to its two-sided interpretation. Consequently, 'not possible' is usually understood as 'not one-sided possible' rather than 'not two-sided possible' in ordinary language. This is to say, negated occurrences of 'possible' are usually interpreted as expressing one-sided possibility in ordinary language.

In the *Prior Analytics*, Aristotle sometimes uses the phrase 'not possible' in the sense of 'not two-sided possible'. However, these uses are restricted to special contexts in which Aristotle explicitly indicates that the kind of possibility that is being negated is two-sided possibility. For example, he often says that 'what is necessary was not possible' in chapters 1.14–19 of the *Prior Analytics*.[45] By using the past tense form

one-sided sense, whereas 'possible' is frequently used in the two-sided sense (see Horn 1989: 209–210). In the second book of the *Prior Analytics*, however, particular affirmative sentences are sometimes used in the two-sided sense, when Aristotle takes 'Some As are B' to imply the particular negative sentence 'Some As are not B' (*An. Pr.* 2.2 55a15, 2.3 56a15; see Waitz 1844: 490–491, Ross 1949: 431 and 433, Smith 1989: 188). Even in chapters 1.1–22, Aristotle is trying to avoid examples in which a particular sentence is true together with the corresponding universal sentence (*An. Pr.* 1.4 26a39–b10; see Patzig 1968: 177–178, Brunschwig 1969: 14–15, Ebert & Nortmann 2007: 304–305). When he must use such examples, he finds it necessary to explain this fact by pointing out that particular sentences are to be understood in the one-sided sense rather than in the two-sided sense (1.4 26b14–20, 1.5 27b20–23, 1.6 28b28–30), just as he finds it necessary to explain the one-sided sense of 'possible' in chapter 1.13. It is sometimes thought that Aristotle connects the one-sided use of 'possible' to the one-sided use of particular sentences at *De interpretatione* 13 23a16–18 (Seel 1982: 167–169, Whitaker 1996: 169; *pace* Pacius 1597b: 104, Weidemann 2002: 455). Whatever the correct interpretation of 23a16–18, the parallel between the one-sided uses of 'possible' and particular sentences is an apt one.

44 See, e.g., Horn 1989: 362–389.

45 1.14 33b17, 1.17 37a8–9, 37b9–10, 1.19 38a35–36. See also 37a14–24, where it is clear from 37a8–9 that two-sided possibility is under consideration.

'was', Aristotle indicates that he is referring to his official definition of two-sided possibility as the preferred notion of possibility introduced in chapter 1.13 (32a18–20).[46] Absent such explicit markers, the usual interpretation of negated occurrences of 'possible' is the one-sided one. Aristotle takes advantage of this fact in *De interpretatione* 13 when he takes it for granted that 'not possible' implies both 'impossible' and 'necessary not', and in *Prior Analytics* 1.13 when he takes it for granted that these three expressions are equivalent.

Finally, we are in a position to respond to an objection raised by Gisela Striker, who rejects this interpretation of the argument in *Prior Analytics* 1.13 on the following grounds:

> One might then think that these lines [1.13 32a21–8] are intended to explain the one-sided sense of 'possible'. If so, however, one would expect something like the consideration mentioned at *De Int.* 13, 22b11–14 or 29–33: it would seem absurd to say that what is necessary is not possible (and hence impossible). But this does not seem to have worried Aristotle in the *Analytics* (cf. 33b17; 37a8–9; 38a35). (Striker 2009: 129)

First, it is important to note that the passages mentioned by Striker at the end of this quotation are the ones in which Aristotle uses the past tense 'was' to make it clear that he is negating two-sided possibility but not one-sided possibility (see n. 45). These passages do not show that Aristotle is comfortable calling what is necessary 'not possible' when he has not made it clear that what is being negated is two-sided possibility. Moreover, there is no need for Aristotle in *Prior Analytics* 1.13 to invoke the consideration from *De interpretatione* 13, because the structure of his argument differs in the two chapters. The two arguments share a common goal: to establish the one-sided sense of unnegated occurrences of 'possible'. In the *De interpretatione*, Aristotle achieves this goal by invoking Striker's consideration to argue that 'possible' does not imply 'not necessary' but 'not necessary not'. In the *Prior Analytics*, he achieves the goal in a more powerful and elegant way by showing that 'possible' is equivalent to 'not necessary not'.

[46] See n. 9.

3 *Why Does Aristotle Need a* Modal *Syllogistic?*

ROBIN SMITH

To supply a connection between the modal syllogistic developed in Aristotle's *Prior Analytics* and the rest of his philosophy interpreters frequently assume that the modal syllogistic must somehow have been intended for use in Aristotle's theory of demonstration as found in the *Posterior Analytics*, and, on this basis, try to find explanations for some of its peculiarities. According to the most common such proposal, Aristotle's claims about necessity syllogisms are often explained by making a distinction between essential and accidental predication and supposing that the necessary premises of demonstrations are necessary because they rest on definitional or essential predications. The evidence adduced for such claims is essentially that if we suppose Aristotle to have some notions in mind that appear in the *Posterior Analytics* when he develops the modal syllogistic, then we can see why he makes the claims that he does about what is or is not a necessary proposition, or a valid inference, etc. However, these are at best speculative answers to the question why Aristotle developed the modal syllogistic as he did, *given that he had already decided to develop it.* They do not explain why he might have wanted to develop such a theory in the first place. They do not tell us what use the modal syllogistic might have been to him in his theory of demonstration – or, more precisely, what use he might have had in mind for it. In order even to speculate productively about that, we should really like to identify some puzzles or difficulties concerning demonstrative science that results of the modal syllogistic might help resolve. At the very least, we would hope to find some evidence in the *Posterior Analytics* of use of even some general results found in the modal syllogistic. However, all we see is the claim that if what is demonstrated is itself necessary, then the premises from which it is demonstrated must also be necessary.

In order to link the modal syllogistic with the rest of Aristotle's philosophy we must look elsewhere. It turns out that there is a passage in the *Prior Analytics* in which Aristotle articulates a principle about possibility, which is almost word for word identical with a passage which appears in book Θ of the *Metaphysics*. In *Met* Θ Aristotle is concerned to argue against contemporaries who appear to have believed that only what is actual is possible. The principle in question states that what follows from what is possible is itself possible, and this passage provides a textually supported link between Aristotle's modal logic and his metaphysics.

1. Syllogistic in the *Posterior Analytics*

Aristotle's *Analytics* is about scientific knowledge[1] (*episteme*), which for Aristotle means demonstrative knowledge: that form of knowledge which consists in the possession of a demonstration or proof so that the propositions involved in such knowledge not only are true but are necessary. Aristotle declares this in the opening sentence of the treatise:

> We must first state what our inquiry is about and what its object is, saying that it is about demonstration and that its object is demonstrative science. (*An. Pr.,* 24a10–11)

A demonstration or proof (*apodeixis*) is a type of argument or deduction (*sullogismos*). Therefore, says Aristotle, he must give an account of arguments before turning to demonstrative arguments in particular. The former is contained in the first part of the treatise, known to us as the *Prior Analytics*, and the latter in the second part, which we call the *Posterior Analytics*.

This description is more than superficial. The *Posterior Analytics* relies on results established in the *Prior Analytics* to reject a challenge, presented in *An. Post.* I.3, to Aristotle's conception of demonstration. Since a demonstration or proof must be a deductive argument from premises known to be true, we may ask, how are the premises themselves known to be true? One tentative answer is that they are known by demonstration from still further premises. According to *An. Post.* I.3, some of Aristotle's contemporaries objected that this conception

[1] I translate ἐπιστήμη as 'scientific knowledge' rather than simply 'knowledge' because it is really a term of art for Aristotle. The alternative translation 'understanding', proposed by Burnyeat, seems to me misleading and ambiguous (unless it is given a technical sense – but then we might as well use 'science' with the same sense).

relies on an incoherent view of scientific knowledge. If all scientific knowledge must be demonstrated from other scientific knowledge, then a vicious regress results. If this regress continues indefinitely, then ultimately there are no known first premises on which the demonstration depends, and thus there is no scientific knowledge.[2] Some, Aristotle says, argued that demonstrations may proceed in a circle: a premise *p* used in a demonstration of conclusion *c* may appear as the conclusion of another demonstration in which *c* is a premise (or perhaps as the conclusion of a series of demonstrations where *c* is a premise at the first stage). Thus, on this account, all that is scientifically known is demonstrated: in one sense, the regress of premises never ends, but there are nevertheless only finitely many propositions in the regress since it wraps around in a circle. Aristotle replies to such a challenge by denying that scientific knowledge can only be deduced from other scientific knowledge: there must in addition be premises that are known but not known through demonstration.

There is much more to Aristotle's reply, however, than the simple assertion that such premises must exist. Interpreters have often seen Aristotle's position as a form of foundationalism. Briefly, Aristotle would then be arguing as follows:

1. If we have any knowledge, then there must be knowledge that does not rest on demonstration.
2. We do have some knowledge.
3. Therefore, there is knowledge that does not depend on demonstration.

Such a way of understanding Aristotle's procedure leaves open what the role of the syllogistic is. Presumably such 'foundationalists' simply regard it as no more than an example of 'accurate reasoning'. What I want to suggest in this chapter is that Aristotle's notion of demonstration is intimately and explicitly tied to the technical details of his syllogistic. That is to say, Aristotle's way of identifying the principles of scientific knowledge cannot be understood apart from certain facts about his logical theory.

Consider, for example, the set Σ containing the two propositions 'Every A is a B' and 'Every B is a C' plus all the consequences that can

[2] In fact, Aristotle makes it clear that the objectors' argument is a dilemma: *either* the regress continues forever (in which case there is no knowledge), *or* it 'comes to a stop' (in which case there is no knowledge).

be derived from these two syllogistically. There are, then, nine propositions in the set Σ:

1.	Every A is a B	
2.	Every B is a C	
3.	Every A is a C	[from 1 and 2 by Barbara]
4.	Some C is a B	[from 1 and 3 by Darapti]
5.	Some B is a C	[from 3 and 1 by Darapti]
6.	Some C is an A	[from 3 by conversion *per accidens*]
7.	Some A is a C	[from 6 by conversion]
8.	Some B is an A	[from 1 by conversion *per accidens*]
9.	Some A is a B	[from 8 by conversion]

This set is deductively closed under the syllogistic,[3] and in fact the deductive closure of any finite set is itself finite. Moreover, the set Σ has further peculiarities: if we delete either (1) or (2) from it, the resulting set remains deductively closed. Thus, although Σ is the deductive closure of {(1), (2)}, both (1) and (2) are deductively independent in Σ. Still further: if we replace (1) with its denial 'Some A is not a B' the resulting set is still deductively closed (the same holds for (2)).

We may take Σ as a picture of the truths of a toy science, that is, all the truths about a particular subject matter. What these results then show is that in that science, there are two propositions such that (a) they cannot be *deduced* from other truths and (b) all other truths *can* be deduced from them. It follows in addition that if either (1) or (2) is known, it cannot be on the basis of other knowledge, since neither proposition can ever be deduced from other true propositions. On the other hand, if both (1) and (2) are known, then every other proposition in Σ can be deduced from known propositions. Consequently, with respect to Σ, we can say that there is knowledge by demonstration of its truths if and only if *precisely these two propositions (1) and (2)* are known without dependence on demonstration. Let us say, then, that (1) and (2) are the only possible principles for this science: on the one hand, they must be known as principles if they are known at all, and on the other hand, if they are known, then every other truth of Σ

[3] Of course, it is not deductively closed in any system that incorporates propositional logic in its deductive theory, since in propositional logic any proposition p is logically equivalent to infinitely many other propositions ($p\&p$, $p\vee p$, $p\&p\&p$, etc.) and in addition any tautology is a deductive consequence of any set of propositions.

can (we may reasonably suppose) be known by deduction from them. This status of (1) and (2) does not depend on any fact about human epistemic abilities.

Given these facts about the syllogistic, Aristotle's next task is to see how to use them in such a way as to show that a (necessary) proposition can always either be demonstrated or be known without demonstration. In *Prior Analytics* I.27–28, Aristotle presents a method (his term is *methodos*) for discovering premises from which a conclusion follows. This may be called a method of "analysis", that is, the inverse of the procedure of deduction (Aristotle uses the verb *analuein* in this sense, e.g., at 78a6–8). Here, Aristotle relies on the claim that all proofs are syllogistic proofs to spell out a method exactly corresponding to the premise regresses of *Posterior Analytics* I.19–22. The object of this procedure is to find the principles (*archai*) of a science, the indemonstrable first premises on which that science's proofs rest. Aristotle claims that this procedure, applied to the totality of truths concerning any subject matter, will lead to the discovery of which of those truths are the indemonstrable principles:

> Consequently, if the facts concerning any subject have been grasped, we are already prepared to bring the demonstrations readily to light. For if nothing that truly belongs to the subjects has been left out of our collection of facts, then covering every fact, if a demonstration of it exists, we will be able to find that demonstration and demonstrate it, while if it does not naturally have a demonstration, we will be able to make that evident. (*An. Pr.*, 46a22–27)

The simplest version of Aristotle's method may be illustrated by looking at the most basic of Aristotle's syllogisms, that is, the case of the one called Barbara, which Aristotle discusses in *Posterior Analytics* I.15,19–23. Suppose that a true universal affirmative proposition AaB can be demonstrated. Since Barbara is the only syllogism with a universal affirmative conclusion, there must then be true premises AaC, CaB from which it can be deduced. If AaC can in turn be demonstrated, then there must be further true premises AaD, DaC from which AaC can be deduced; similarly, there must be true CaE, EaB from which CaB can be deduced, if it can be demonstrated. This process of finding further true premises carries with it the construction of a series of terms A, D, C, E, B such that AaB, BaD, DaC, CaE, EaB.

Aristotle calls such a series of terms $A_1, \ldots, A_n$ such that for each A_i, $A_i a A_{i+1}$, a *chain* (*sustoichia*), and he says that each term in such a chain is *above* its successor (and therefore above all subsequent terms). Thus, the continuation of the regress of premises for a universal affirmative proposition AaB involves the insertion, at each step, of another term, which Aristotle calls the 'middle term', in a chain between A and B. The process of filling in this chain by the insertion of new (middle) terms (a process which Aristotle at least once calls *katapuknoun*, "packing full") can continue so long as there are adjacent terms A_i, A_{i+1} in the chain for which there is a new term T such that $A_i aT$ and TaA_{i+1} are both true. If no middle term can be found, then the regress "comes to a stop" (*histatai*), and each interval in the chain is "unmiddled" (*amesos*), that is, lacks a middle term (and we may also say that the proposition corresponding to each interval is unmiddled). If, on the other hand, the process of finding further premises by inserting more middle terms never comes to a stop, then there must be infinitely many terms in the chain between A and B. For a universal affirmative proposition AaB, then, the regress of premises comes to a stop if and only if there are only finitely many terms that can be inserted into the chain between A and B, and the fully packed chain at which that regress comes to a stop corresponds to a series of unmiddled propositions none of which can be established deductively and from the set of which all other propositions can be deduced.

This gives a concrete picture of how the process works, with the additional detail that it will, in various cases, "come to a stop" just exactly when we find that there is no term that occurs in two sets of terms we have generated in accordance with Aristotle's method. Thus, for a proposition AaB, the regress "comes to a stop" if there is no term both in the set of terms below A and in the set of terms above B. What Aristotle's method does is to *make this evident*, that is, make it evident that AaB is an unmiddled, and therefore indemonstrable, proposition.

What Aristotle has in mind is the kind of "coming to a stop" of terms in a chain as described previously. We find explicit confirmation of this in *An. Post.* I.19–22, the goal of which is to prove that all regresses of premises inevitably "come to a stop" in just this way. He first argues that any regress of premises for a universal affirmative proposition must proceed by inserting terms into a chain between its subject and its predicate. Thus, if there is an infinite regress of premises for AaB, then there must be a chain containing infinitely many terms between A and B. Next, he argues for the following claims. An *ascending* chain above term A is a series of terms $A_1 \ldots A_n$ such that for each A_i, $A_{i+1}aA_i$ and $A_1 = A$. A *descending* chain below A is a series of terms $A_1 \ldots A_n$ such that for each A_i, A_iaA_{i+1} and $A_n = A$. Aristotle argues that if every ascending chain is finite and every descending chain is finite, then there cannot be infinitely many terms between any two terms. Thus, if every ascending or descending chain is finite, then no premise regress for a universal affirmative proposition can continue forever.

All that Aristotle needs therefore is a way of setting limits on the possible middle terms. And if the recipe he sets out in *An. Post.* I.13 is to make sense, it would seem that this is just the kind of guarantee that he thinks is available. For if we are trying to demonstrate that every A is a B, then we have to choose a C which has a necessary connection with both A and B. That is why there could be no science of accidents, since, as Aristotle tells us at *Met.* Γ4, 1007a14–15, there are infinitely many accidental features of anything, and so you could have an infinity of middle terms. If accidents are ruled out, Aristotle clearly thinks that the possible terms will be finite – and this in turn means that a demonstration of 'every A is a B' will involve only terms which have a necessary connection with A and B. And, as we have seen in our discussion of deductive closure in the syllogistic, the deductive closure of any set of propositions made up from a fixed finite set of terms will be finite.

The case of Barbara is particularly simple, since the propositions involving the new middle terms are all universal affirmatives. But the principle is the same for all syllogistic propositions relating A and B. In fact, Aristotle argues in *An. Post.* I.21 for a lemma: if every ascending or descending chain is finite, then every regress for a universal *negative* proposition must terminate. Finally, he argues in I.22 that every ascending or descending chain is finite.[4] In proving these results,

[4] I have passed over several issues concerning these arguments. First, Aristotle does not need the finiteness of *both* ascending *and* descending claims to prove the finiteness of universal affirmative regresses: since every chain between two terms is both an ascending chain over its bottom term and a descending chain below its top term, the finiteness of *either* all ascending *or* all descending chains would be sufficient. (Jonathan Lear 1979) thinks that Aristotle might need both types of finiteness because there might be a chain between two terms that is infinite in one direction but not in the other, but this notion is incoherent: for a decisive refutation, see Scanlan 1983.) Next, the argument in I.21 treats a "third way" of universal negative proof that appears erroneous, but it is easy

Aristotle relies on results concerning his syllogistic theory: specifically, that any proof of a universal affirmative proposition must be in the form Barbara and any proof of a universal negative proposition in one of the forms Celarent, Cesare, Camestres.[5] In effect, then, he takes it as established that these are the only possible forms of argument with universal conclusions. That result is argued for in *Prior Analytics* I, where Aristotle says that he has shown that all deductions whatever can be "reduced to" arguments "in the figures", i.e., in his syllogistic theory.

The syllogistic is presented as a mathematical theory. Aristotle's concerns are primarily with what we would call metalogic. He first establishes all the valid forms of argument in the syllogistic, in the process showing that every syllogism can be reduced to one of the four first-figure syllogisms. Next, he shows that an even further reduction is possible: the particular first-figure forms (Darii and Ferio) can be reduced to the universal ones (Barbara and Celarent) by way of intermediate reductions (using arguments through impossibility) to second-figure Camestres, which in turn is reduced to first-figure Celarent. This is not at all the kind of result we would expect from a simple exposition of valid forms of proof to be used in demonstrative sciences, but it is exactly the kind of result Aristotle needs in order to establish such things as the finiteness of all premise regresses. He must know all the forms a regress of premises might take in order to determine the ways a regress for any type of conclusion might continue. If all deductions can be reduced to deductions using only the two universal first-figure forms, then that problem is greatly simplified: all that he needs to show is that all universal regresses terminate. This is in fact what he does in *An. Post.* I.19–21.

Aristotle takes pains to argue that the syllogistic is not simply a preferred style of valid arguments for demonstrations but in fact the

to reconstruct Aristotle's argument coherently. Third, only the finiteness of ascending chains is required even in the universal negative case, but it may be significant that if we extend Aristotle's arguments to particular (i.e., non-universal) propositions, the termination of regresses requires that all *descending* as well as ascending chains be finite.

[5] Aristotle's proof for universal negative regresses, as presented, is incoherent, since it seems to appeal to a third-figure syllogistic form in a regress for a universal negative proposition (which is impossible since any third-figure form must have a non-universal conclusion). Commentators have noted this problem since ancient times.

full and complete story of valid inference: this is a crucial element in his arguments. Aristotle's argument for the finiteness of all premise regresses is closely linked with the argument for the discovery of principles that can be seen in *An. Pr.* I.27–28. In *An. Pr.* I.30, Aristotle proudly proclaims that he has presented the one true scientific method (method for the construction of demonstrative sciences and the discovery of principles that this requires): we can hardly understand the point of this method without keeping the *Posterior Analytics* before us.

In summary, then, Aristotle's *Posterior Analytics* relies essentially on results Aristotle thinks he establishes in the *Prior Analytics* for one of its most central claims, and from this claim we can reconstruct not simply a picture of the structure of a demonstrative science but also an answer to a vital question: what are the principles of a science? A critical part of that reliance is the claim that all deductions whatever are, or can be reduced to, syllogistic deductions. Aristotle undertakes to prove this in the *Prior Analytics*, and whether his arguments are successful or not, it is clear that he thinks they are.

What my argument so far shows is that the connections between the *Posterior Analytics* and the syllogistic theory of the *Prior Analytics* are deep and even, one might say, essential. A major argument of the *Posterior Analytics* depends crucially on results established in the *Prior* and nowhere else. Conversely, the exposition of the syllogistic in the *Prior Analytics* appears to have been expressly designed with these uses in mind. Against this background, what can we say about the modal syllogistic presented in *An. Pr.* I.8–22? That is, what do these chapters contribute to Aristotle's theory of demonstration? In fact, no result about modal syllogisms is used for any argument in the *Posterior Analytics*. While Aristotle does say that the principles of demonstrations must be necessary, and while it may be true that the finiteness of demonstrations requires essential predication (to prevent the possibility of infinitely many middle terms), Aristotle does not say that they must be propositions asserting necessary predication (that is, explicit necessity propositions). Moreover, the most that might be claimed is that the *Posterior Analytics* agrees with the *Prior* that a necessary conclusion follows from necessary premises. We cannot even find a recognition of the claim, defended in the *Prior*, that in some cases a necessary conclusion follows from premises not all of which are necessary. There is no trace in the *Posterior Analytics* of the modal syllogistic concerning possible premises or mixed possible and necessary premises at all. We thus have no indication what use Aristotle's modal syllogistic might have been for his theory of demonstration.

This is not to deny that we can learn much about the content of the modal syllogistic by studying what Aristotle has to say in the *Posterior Analytics*, and elsewhere, about necessity and possibility. However, nothing we find either in the modal syllogistic itself or in the *Posterior Analytics* points to any issue about which a modal syllogistic might have made any difference for the theory of demonstrative science.

2. The Modal Syllogistic and Possibility

If we do not find a motivation for the modal syllogistic in Aristotle's theory of demonstrative science, where might we turn? One answer is to look at the *Prior Analytics* itself. As we know, Aristotle claims in a famous passage at the end of the *Sophistical Refutations* that he had created logic from nothing. Whatever he may have meant precisely it seems a plausible conclusion that he was discovering features of formal reasoning that go beyond the scientific uses he wants to put it to. We know that he recognized that the (assertoric) formal logic that he was developing works equally when applied to premises which are not necessary, and even to premises which are not even true. (See, for instance, Rini, Chapter 1 of this volume.) For that reason one could argue that he has motivation enough to develop a modal logic.

However, I would like to pursue a different direction. Previous interpreters of Aristotle's modal syllogistic have usually begun with necessity syllogisms and tried to give an account of those independently of the syllogisms involving possibility (which include mixed-modal cases involving combinations of necessary and possible premises). This is perhaps a natural approach given two points:

(a) Aristotle treats necessity syllogisms first;
(b) the theory of necessity syllogisms is much simpler.

A further reason appears if the place we look to for possible applications of the modal syllogistic is the theory of demonstration, where only necessary premises play a role. I propose that it might be more fruitful to suppose instead that it is possibility premises that are at the focus of Aristotle's attention. First, we may note that while there are 4 chapters of the assertoric syllogistic (24b30–29b28) and 5 chapters of apodeictic (29b29–32a14), there are 10 chapters of problematic

(32a15–40b16). But more important for my purposes is that we can find an explicit textual link between two passages which discuss the issue of unactualized possibilities. In contrast to the silence of the *Posterior Analytics* about modal arguments, there is a strong similarity between a passage in the modal syllogistic and a passage in *Metaphysics* Θ: *Met.* Θ4, 1047b14–30, echoes an argument in *An. Pr.* I.15, 34a5–24. Moreover, the issue that arises in the former passage is one to which modal logic is highly relevant. Aristotle attacks a position he attributes to "Megarians" according to which nothing is possible except what is actual, from which it readily follows that whatever is necessarily as it is. This is a direct challenge to Aristotle's own notion of potentiality (the subject of *Met.* Θ) and thereby to the foundations of his physics and metaphysics. In *De Int.* 9, Aristotle responds to a related position according to which the necessity of true statements about the past leads to the necessity of true statements about the future, and thereby to the necessity of whatever is the case. While Aristotle does not ascribe this position to anyone, it is obviously close to the 'Master Argument' of Diodorus Cronus, and whether or not Diodorus was in some sense affiliated with the Megarian philosophers, there is a line of continuity between Megarian positions and this argument.

The Megarian challenge is that there are no possibilities except those that are actualized. For instance, there is nothing that is possibly *A* except what is actually *A*. Megarian universal necessitation, as Aristotle sketches it in *Met.* Θ and in *De Int.* 9, can be argued for as follows:

> You cannot now be both seated and not seated
> You are now seated.
> Therefore, you cannot now not be seated.

This is easily generalized to any proposition whatever to show that if it is now the case that *P*, then it is now necessary that *P*, so that whatever is true is necessary and whatever is false is impossible. More specifically, a Megarian objection might go as follows: if we suppose that *x* is not *A* and is at that very time also possibly *A*, then we are supposing that *x* is possibly (*A* and not *A*). However, nothing can be *A* and not *A* at the same time: that would be a straightforward contradiction.

It appears to depend only on the principle of contradiction. There is an interesting relationship of this argument to the argument Aristotle wants to counter in *De Int.* 9. That latter argument depends in some way on the law of excluded middle. The principles of contradiction and excluded middle are the two principles given special treatment in *Metaphysics* Γ: while Aristotle argues that the former is the most secure of all principles, excluded middle does clearly have an important place for him. This is a little evidence that *Metaphysics* Θ3 and *De Int.* 9 are parts of a general response to Megarianism.

The principle on which Aristotle's reasoning depends is articulated in *Metaphysics* Θ4, and in *An. Pr.* I.15. Both passages may be translated as follows:

If when *A* is, *B* necessarily is, then when *A* is possible, *B* will be possible of necessity. (*An. Pr.* 34a5–7) (*Met* 1047b14–16)[6]

While the interpretation of this test is controversial, Aristotle's use of it in the modal syllogistic is generally associated with proofs through impossibility. In the course of such an argument, if we have a possibility premise – which must be a genuine possibility, such as that a coat is cut up or a log is burned, not that the diagonal is measured – then we may assume the corresponding assertoric premise; if, having made this assumption, we reach an *impossible* result, then we may reject the original *reductio* hypothesis. I shall call this rule 'R', and show how it functions in the syllogistic.

What the Megarian position denies is that there are any unactualized possibilities. To respond to this, Aristotle needs a principled way to distinguish between what is false but possible and what is false and impossible. How might Aristotle reply? He might first say that while the property of being possibly (*A* and not *A*) is self-contradictory, the property of being (possibly *A* and possibly not *A*) need not be. Let us further suppose that this latter property is a single occurrent property and that it is consistent both with the property of being *A* and with the property of not being *A*. Such a property is intimately connected with the notion of "two-sided" possibility that he gives as his standard

[6] The *Metaphysics* passage reads:πρῶτον δὲ λεκτέον ὅτι εἰ τοῦ Α ὄντοςἀνάγκη τὸ Β εἶναι, καὶ δυνατοῦ ὄντος τοῦ Α δυνατὸν ἔσταικαὶ τὸ Β ἐξ ἀνάγκης. (34a5–7) ἅμα δὲ δῆλον καὶ ὅτι, εἰτοῦ Α ὄντος ἀνάγκη τὸ Β εἶναι, καὶ δυνατοῦ ὄντος εἶναι τοῦΑ καὶ τὸ Β ἀνάγκη εἶναι δυνατόν· (1047b14–16)The language here is almost identical with the passage from *Prior Analytics* (the differences are δυνατὸν ἔσται … ἐξ ἀνάγκης versus ἀνάγκη εἶναι δυνατόν and minor variations in word order).

account in the *Prior Analytics*. In fact one can say that the property he seeks is simply the property of being (two-sided) possibly *A*. In these terms we can say that being (two-sided) possibly *A* is logically independent of the property of being *A*, so that it is equally possible for something that is *A* either to be (two-sided) possibly *A* or not to be (two-sided) possibly *A*. The possession of such a property will then be what differentiates between that which is not *A* but could be *A* and that which is not *A* and could not be *A*. His definition of 'possible' may be designed to provide just that: that which leads to nothing impossible if it is assumed to be true. This definition appears in *Met.* Θ in simple form: "that is possible [for something] which, if the actuality of which it is said to have the potentiality obtained, there will be nothing impossible" (1047a24–26). In *An. Pr.* I.13, it appears as "that which is not necessary but, when it is supposed to obtain, there will be nothing impossible because of this" (32a19–21).

We might think of this as an unnecessarily prolix definition of 'possible', if it simply amounts to 'the possible is whatever does not entail anything impossible.' A first point to note is that, as Marko Malink argues (this volume) 'impossible' here has to be understood as one-sided impossibility. A second point is that Aristotle's 'definition' is one which already assumes a modal notion – one-sided (im) possibility – and we can therefore think of Aristotle's definition of two-sided possibility as presupposing the following principle of one-way possibility:

(I) If every *B* which follows from *A* is possible, then *A* itself is possible

What is added to this by the further principle that if *A* entails *B*, then if *A* is possible *B* is possible? It allows us to infer one possibility from another. If whatever is walking is moving, and if *X* is possibly walking, then *X* is possibly moving. This principle is, in a sense, the converse of (I). That is to say, the converse of (I) is

(II) If *A* is possible then every *B* which follows from *A* is possible.

It is precisely principle II which is the one stated in Met. Θ4 and *An. Pr.* I.13.

If we were to express (I) and (II) as principles of current modal logic using □ for necessity and ◊ for possibility they become

(Ia) If, for every B, $\Box (A \supset B) \supset \Diamond B$, then $\Diamond A$

(IIa) If $\Diamond A$ then, for every B, $\Box (A \supset B) \supset \Diamond B$

(IIa is simply a standard principle of all (normal) modal logics.)

(IIb) $\Box (A \supset B) \supset (\Diamond A \supset \Diamond B)$

(Ia) is in fact derivable using only the principle $\Box(A \supset A)$. For if the antecedent of (Ia) is true then, by (Ia) we have, in particular, $\Box(A \supset A) \supset \Diamond A$, from which $\Diamond A$ follows.

Still, the fact that (I) and (II) are converses may well make it plausible that they are linked in Aristotle's mind, and perhaps that he did not clearly distinguish between them. What Aristotle needs to show in *Metaphysics* Θ is that there are possibilities that never are actualized. Aristotle is arguing against a criticism of his own criterion for possibility according to which that is possible which leads to nothing impossible if it is assumed to be actual. The objector undertakes to show that, by this criterion, it is possible that the diagonal of the square will be measured even though it never will actually be measured, just as it is possible that this cloak will wear out, even though it will never actually be worn out (since it will be cut up first), and just as it is possible that this log will be burned, even though it will never actually be burned (since it is at the bottom of the sea). To reply to this, Aristotle must draw a reasonable distinction between the case of the diagonal and the cases of the cloak and the log. The relevant difference is that in the case of the diagonal, an impossibility would result that is not merely a matter of an inconsistency between the supposition itself and the actual state of affairs (in the case in point, that 'odd numbers are equal to even numbers').[7]

The differences between the passage in the *Metaphysics* and that in the *Analytics* indicate, I think, that the latter is an attempt to develop the idea found in the former in terms of the syllogistic. In the *Metaphysics*, both principles are stated in terms of the language of actuality and potentiality. In the *Analytics*, Aristotle seems to me to be

[7] Section 1 stressed that syllogistic proofs are finite. It is therefore of some significance to note the connection between the case of the diagonal and the finitude of proofs. For the proof of incommensurability proceeds by showing that if the diagonal is measured by any n/m then it is more accurately measured by some n'/m' so that its value cannot be established in a finite number of steps.

making an explicit effort to expand this to a generic notion of possibility. I see two bits of evidence for this.

First, after stating the principle that $\Box (A \supset B) \supset (\Diamond A \supset \Diamond B)$ in 34a5–12, Aristotle adds a qualification (34a12–15): possible must not be understood as only applying to 'becoming' (genesis) but also in all the other ways that 'possible' is said. I propose that this is a deliberate move away from *Met.* Θ's focus on potentiality and actuality, which is indeed a focus on becoming. It also perhaps prepares the way for the next point (34a16–24), to which I will return in a moment.

Second, in the earlier presentation of the definition of 'possible' in 32a18–20, Aristotle follows with a passage Ross brackets because he sees in it a logical error (32a21–29). The passage concludes with "the possible will therefore be not necessary and the not necessary possible". But I suggest that this is not an error but instead an endorsement of a twofold distinction between things which cannot be otherwise (which includes both what is necessary and what is impossible) and things which can be otherwise (which exactly coincides with "possible in accordance with our definition" in Aristotle's usage, i.e., neither necessary nor impossible). Since what is impossible is what is necessarily not the case, Aristotle is here using 'necessary' in the sense 'necessarily as it is' (and thus its negation is 'not necessarily as it is'). What Aristotle is doing here is, I suggest, arguing again for a generic sense of 'possible' not limited to the sphere of actuality and potentiality.

And now to return to 34a16–24: in this second explanatory note to the principle $\Box(A \supset B) \supset (\Diamond A \supset \Diamond B)$, Aristotle adds a fundamental result he believes he has established in the *Analytics*: nothing follows from just one premise. Consequently, in 'if *A* is then *B* must be', *A* must be (at least) a pair of premises. He has, then, carried the entire doctrine of *Met.* Θ over to the syllogistic environment, and in so doing he has both generalized it (to any notion of possibility, not simply what can be expressed in terms of actuality and potentiality) and also in a sense narrowed it (because of his belief that the syllogistic is the full story of inference).

What I have just said does imply that the modal syllogistic is chronologically later than *Met.* Θ. But I think that is a desirable result. The modal syllogistic is (or so I think) clearly an unpolished work, an extension of the plain syllogistic attempted by Aristotle in hopes that it would give him a more powerful tool for responding to the Megarian position.

3. The Necessity Syllogistic

In Section 2 I have argued that the possibility syllogistic should be taken as basic. In this final section I will be concerned to show that the necessity syllogistic can based on possibility. If we take (two-sided) possibility as primary, it is a straightforward matter to define necessity and impossibility. That which is necessarily *A* is that which is *A* but is *not* (possibly *A* and possibly not *A*); that which is necessarily not *A* is that which is not *A* and in addition is not (possibly *A* and possibly not *A*). In what follows, I will write '*P*(*A*)' for the (two-sided) possibility of being *A* – the property of being possibly *A* and possibly not *A*. As I am interpreting him, then, Aristotle treats *A* and *P*(*A*) as distinct properties. The property of being necessarily *A* is compounded from these: what is necessarily *A* is what is *A* and not *P*(*A*). This reflects Aristotle's assertion in several places that what is necessary is what cannot be otherwise. We may also define 'necessarily not *A*' as 'not *A* and not *P*(not *A*)', but since *P*(*A*) = *P*(not *A*), that is equivalently '*A* and not *P*(*A*)'. The property of being necessarily *A* is a matter of having the first property without the second.

With this in place, a plausible translation scheme for the propositions of the apodeictic syllogistic into standard first-order predicate logic, using the term letters *A*, *B* etc., as (non-empty) one-place predicates, with modal operators treated as predicate qualifiers, can be provided as follows:

Necessarily *A*: $N(Ax) = Ax \wedge \neg P(A)x$
Impossibly *A*: $N(\neg Ax) = \neg Ax \wedge \neg P(A)x$

We can extend this to Aristotelian categorical sentences:

'Every *A* is necessarily *B*' = $N(BaA) = \forall x(Ax \supset (Bx \wedge \neg P(B)x)) = BaA \wedge P(B)eA$
'Every *A* is necessarily not *B*' = $N(BeA) = \forall x(Ax \supset (\neg Bx \wedge \neg P(B)x)) = BeA \wedge P(B)eA$

This can be extended to particular propositions using the principles of 'ekthesis':

'Some *A* is *B*' (= BiA) iff: for some C, $BaC \wedge AaC$
$= \exists C(\forall x(Cx \supset Ax) \wedge \forall x(Cx \supset Bx))$
$= \exists x(Ax \wedge Bx)$
'Some *A* is not *B*' (= BoA) iff: for some C, $BeC \wedge AaC$

$$= \exists C(\forall x(Cx \supset Ax) \land \forall x(Cx \supset \neg Bx))$$
$$= \exists x(Ax \land \neg Bx)$$

We then have:

'Some A is necessarily B' (= N(BiA)) iff, for some C, N(BaC) ∧ AaC
$= \text{for some } C, BaC \land P(B)eC \land AaC$
$= \exists C(\forall x(Cx \supset Ax) \land \forall x(Cx \supset (Bx \land \neg P(B)x)))$
$= \exists x(Ax \land Bx \land \neg P(B)x))$

'Some A is necessarily not B' (= N(BoA)) iff, for some C, N(BeC) ∧ AaC
$= \text{for some } C, BeC \land P(B)eC \land AaC$
$= \exists C(\forall x(Cx \supset Ax) \land \forall x(Cx \supset (\neg Bx \land \neg P(B)x)))$
$= \exists x(Ax \land \neg Bx \land \neg P(B)x))$

These translation schemes give a good account of the necessity syllogistic except in one crucial respect. That is, they cause trouble for rules of modal conversion, and for those modal syllogisms which depend on them. Aristotle holds that $N(AeB)$ is convertible, that is, that it entails $N(BeA)$. The translations given previously do not immediately yield this, however. $N(AeB) = AeB \land \neg P(A)eB$, while $N(BeA) = BeA \land \neg P(B)eA$. AeB entails BeA, but the inference from $P(A)eB$ to $P(B)eA$ requires further resources. In brief, we need some rule giving a logical relationship between the predicate A and the predicate $P(A)$, and as presented earlier these two predicates are logically independent. In fact the rule required is the rule that I called R, which is based on the principle that Aristotle has defended in the doublet passage from *Met.* Θ.4 and *An. Pr.* I.15:

If when A is, B necessarily is, then when A is possible, B will be possible of necessity.

Using this principle, then, a proof of the convertibility of $N(AeB)$ can be constructed as follows:

1. $N(AeB) = AeB \land \neg P(A)eB$	assumption
2 AeB	1, simplification
3. BeA	2, conversion
4. $P(BiA)$	*reductio* hypothesis
5. $P(B)iA$	4, definition
6. BiA	5, assumption by R

7. AiB	6, conversion
8. $P(A)iB$	If AiB is possible but false, then $P(A)iB$
9. $P(A)eB$	1, simplification
10. $P(B)eA$	5, rule R: assuming BiA leads to the impossible $P(A)iB$, by contradicting 9
11. $N(BeA)$	3,10, definition

What then should be said about this proof? While it does seem to reflect the way Aristotle argues when discussing this conversion in *An. Pr.* I.3 (see especially 25a31–32), it does not appear to respect the translation scheme we have been using. Suppose, to use the kind of examples Aristotle often supplies, that A means 'horse' and B means 'moving', and suppose that only humans are moving. Then it would seem that $N(AeB)$, for no moving thing is a horse, and it is not possible that any moving thing, i.e., any human, be a horse. But then, although no horse may be moving, it is certainly possible that a horse be moving. For this reason the following formula is not valid:

$$(\forall x(Bx \supset \neg Ax) \wedge \forall x(Bx \supset \neg P(A)x)) \supset (\forall x(Ax \supset \neg Bx) \wedge \forall x(Ax \supset \neg P(B)x))$$

If we look at the proof (1)–(11), just given, and replace its steps by their 'translations' what we get is

1. $\forall x(Bx \supset \neg Ax) \wedge \forall x(Bx \supset \neg (P)Ax)$	assumption
2 $\forall x(Bx \supset \neg Ax)$	1, simplification
3. $\forall x(Ax \supset \neg Bx)$	2, conversion
4. $\exists x(Ax \wedge P(B)x)$	*reductio* hypothesis
5. = 4	
6. $\exists x(Ax \wedge Bx)$	5, assumption by R
7. $\exists x(Bx \wedge Ax)$	6, conversion
8. $\exists x(Bx \wedge P(A)x)$	If AiB is possible but false, then $P(A)iB$
9. $\forall x(Bx \supset \neg P(A)x)$	1, simplification
10. $\forall x(Ax \supset \neg P(B)x)$	5, rule R (assuming BiA leads to the impossible $P(A)iB$), by contradicting 9
11. $\forall x(Ax \supset \neg Bx) \wedge \forall x(Ax \supset \neg (P)Bx)$ 3,10	

To see what has gone wrong look at how the rule R has been applied: at line 4 we assume, for *reductio*, that there is an A which is a possible B.

So we then, according to R, are entitled to suppose that some A really *is* a B. With A as 'horse' and B as 'moving' this means that we are supposing that some horse is moving. This is a consistent supposition; but notice what it does. What it does is change the things which are moving so that one of them is a horse. It does *not* envisage a circumstance in which one of the actually moving things – one of the humans – has become a horse. But at line 8 we conclude that some moving thing *is* a possible horse.

The explanation of what has gone wrong is not easy, and for that reason it is not clear whether Aristotle was aware of it, or how he might have responded to it. Perhaps he would say that in applying the rule R you have to be careful how you choose your terms.[8] Or he might say that we must 'ampliate' the movers so that it covers the possible movers,[9] and certainly if you ampliate you get

$$(\forall x(Bx \supset \neg Ax) \land \forall x(P(B)x \supset \neg)\ P(B)x) \supset (\forall x(Ax \supset \neg Bx) \land \forall x((P)Ax \supset \neg P(B)x))$$

which is valid.

Alternatively one might argue somewhat as follows. Starting from Striker 1994, one might argue for the view that the modal syllogistic is, in the form we have it, unsalvageable because of its multiple serious inconsistencies. But saying this is perhaps not particularly helpful in diagnosing just *why* Aristotle might not have seen these inconsistencies. If we can assume a principle of reasoning which seems, on the surface, a plausible one, but yet a principle which, when looked at closely, might be seen to be fallacious, then we can explain why such an astute logician as Aristotle might nevertheless have not discerned where the trouble is. One might then argue that a rule like R is one of these places. We have evidence that he uses such a rule, and the explanation of why it may not be justified is one which someone who had just invented logic might easily have missed.

What I have tried to do in this chapter is address the problem of why Aristotle needs a peculiarly *modal* syllogistic, given that, while his theory of scientific demonstration demands essential use of the *assertoric* syllogistic, it makes no use of the *modal* syllogistic. I have

[8] That seems to be the line taken in Rini 2011 (pp. 151f).
[9] Something like this seems to be the view of Nortmann 1996.

suggested that the answer lies in his theory of possibility, and I have used a principle set out in *Metaphysics* Θ to show how he puts modal principles to work in answering an objection to his claim against the Megarians that there can be unactualized possibilities. We have, then, identified not only a textual connection between the modal syllogistic and another part of Aristotle's work but also an issue in which his modal logic can be seen to play a crucial role.

4 Necessity, Possibility, and Determinism in Stoic Thought

VANESSA DE HARVEN

1. Introduction

Necessity and possibility in Stoic thought run deep, through logic, physics, and ethics together. There is a wealth of scholarship just on Stoic responses to Diodorus Cronus's Master Argument that everything possible either is or will be true, revealing the Stoics' sophisticated propositional logic and modal theory. Likewise, scholars have long mined the Stoics' innovative theory of causation and their unique commitment to an everlasting recurrence of the cosmos, elucidating the scope of necessity and possibility in Stoic physics. And, perhaps most famously, the vexed relationship between determinism and responsibility testifies to another vein of modality running through Stoic ethics. This chapter will argue that underwriting the Stoics' moves in all these areas is a distinction among *logical*, *metaphysical*, and what I will call *providential* necessity and possibility.

Complicating matters, the Stoics are known to embrace apparent paradox, a trace of Eleatic influence through the Megarians.[1] It will become apparent as we proceed that understanding Stoic thought is doubly challenging: in addition to the exegetical and interpretative challenge of dealing with ancient, self-consciously paradoxical views, there is also, inextricably, the basic reconstruction of Stoic views through often hostile lenses. The textual evidence of Stoic thought is fragmentary, scant, and subject to widespread interpretative debate.

[1] To name just a few: only bodies exist, but not everything that is Something exists (Alexander, *In Ar. Top.* 301,19–25 (27B); Democritus's cone split in half yields surfaces that are unequal, and neither equal nor unequal (Plutarch, *Comm. not.* 1079E (50C5); we consist of a determinate number of parts (head, trunk, and limbs) and yet no determinate number of parts (in virtue of infinite divisibility of their continuum physics) (Plutarch, *Comm. not.* 1079B–D (50C3). Parenthetical citations like 38D refer to the chapter and order of passage in Long and Sedley 1987, hereon LS; unless otherwise noted, translations are LS with modifications.

The method of this chapter will be to undertake a close textual analysis of a single piece of testimony from Diogenes Laërtius on Stoic modality, bringing other passages to bear as needed.[2] I will show that we can find these three senses of *necessary* and *possible* in Diogenes' report, and that the Stoics can exploit this systematic ambiguity to embrace the air of paradox while crafting a unique compatibilism in their ethics, making room for counterfactual truth in a predetermined physics, and upholding bivalence amid genuinely possible futures.[3]

By *logically necessary*, I will mean what is true a priori (e.g., *If it is day, it is day*).[4] By *metaphysically necessary* I will mean what is true a posteriori, as a function of the active principle (*archē*), i.e., god or divine logos, working on the passive principle, matter (*hulē*): a world order according to certain metaphysical principles and natures.[5] By *providentially necessary* I will still mean what is true a posteriori as a function of the divine active principle (god), but this time god conceived as the immanent chain of efficient causes, acting through the natures and principles established as metaphysically necessary, to bring about the same world order time and time again. The Stoics are committed to an everlasting recurrence of the cosmos, punctuated by periods of conflagration when the world turns into fire and then starts over. Crucially, however, while the laws of nature are metaphysically necessary, the fated (or determined) chain of causes is not. So what is providentially necessary is not defined in terms of what will be, but only in terms of what is or has been. A particular event only becomes providentially necessary once it is under way or in the past; until then

[2] I have chosen Diogenes Laërtius's testimony (hereon DL) at 7.75 for its completeness and greater detail over Boethius, *In Ar. De Int.* 234.27–235.3, which only gives the first three definitions and is otherwise in general agreement with DL.

[3] I am going to operate on the hypothesis that it is legitimate to speak of *the Stoics* as a whole in reference to orthodox views that unify the school. I do not thereby deny that there was internal debate or that successive heads of school and other individual Stoics left their mark on Stoic doctrine. Rather, I take there to be core doctrine against which, as an agreed-upon background, there were more fine-grained internal debates. In this case, the relevant orthodoxy is the recognition of these three senses of necessity and possibility (Cleanthes and Chrysippus on the Master Argument notwithstanding, cf n. 59).

[4] I have avoided characterizing logical truths as analytic because I do not think the analytic/synthetic distinction is entirely applicable; see Kneale 1937, for an effort to apply this distinction to Stoic modality.

[5] Kripke 1972.

the future remains providentially contingent – even though bivalence for propositions about the future holds. Hence for the Stoics there are future truths that are not necessary.[6]

Corresponding to these three senses of necessity, according to my reading, what is *providentially possible* is what the circumstances allow; this will include ordinary circumstances such as not being in jail so it is possible to go to the agora, and what one might characterize as technical circumstances, such as being in future, so it is not yet settled, given the unwinding of the rope of fate so far.[7] *Metaphysically possible* is what the metaphysical principles and natures allow regardless of the external circumstances; and *logically possible* is what is not self-contradictory. So what is providentially possible is a subset of what is metaphysically possible, which is itself a subset of the logically possible. I hope this suffices as a preliminary account of the three senses of necessity and possibility I have in mind. I will fill in the details of this picture in Section 2, as I argue from the textual evidence.

2. Textual Evidence

As I have alluded, at the heart of the Stoic theory of modality is a strict commitment to bivalence, even for future contingents. Naturally, this commitment to both future truth and contingency has often been thought paradoxical, and the hostile commentators from whom we get much of our scant testimony make the most of it. So we will begin with testimony from Diogenes Laërtius, a neutral and generally reliable (if sometimes superficial) doxographer. I will provide the entire passage A here, then undertake a close reading of the numbered segments.

A. (1) Further, there are those [*axiōmata*] on the one hand that are possible (*dunata*), as well as those on the other that are impossible (*adunata*). And there are those on the one hand that are necessary (*anankaia*), and those on the other that are not necessary (*ouk anankaia*). (2) Now, possible (*dunaton*) is what (a) admits of being true (*to epidektikon tou alēthes einai*), (b) not being opposed externally (*tōn ektos mē enantioumenōn*) in being true, e.g., "Diocles is alive." (3) Impossible (*adunaton*) is that which (a) does not admit

[6] As well as false futures that are not impossible; see Cicero, passage B, later.

[7] For the image of the rope see Cicero, *Div.* 1.127 (55O), Gellius 7.2.

of being true, <or (b) what does admit of being true but is opposed by things external to it from being true>, e.g., "Earth flies;" and (4) necessary (*anankaion*) is (a) that which being true, does not admit of being false, or (b) what on the one hand does admit of being false but is opposed by things external to it from being false, e.g., "Virtue benefits;" and (5) not necessary is that which both is true and can be false, being contradicted by nothing external, e.g., "Dion is walking."[8]

Note first, from A1, that the bearers of modal properties are strictly speaking *axiōmata*,[9] which we can translate loosely as *propositions*; they are incorporeal entities distinct from the corporeal words uttered and world spoken about, a species of what the Stoics call *lekta*, or *sayables*, which also include questions, imperative, oaths, et al.[10] *Axiōmata* are introduced by the Stoics to be the bearers of truth value, enter logical relations, and ground causal analysis, as well as to be shared in communication and to stand as the objects of our assent as rational agents. The Stoics' correspondence theory of truth is well attested, so it will be uncontroversial to take the modal properties of propositions as evidence about what is necessary and possible in the world itself.[11] Note, too, that I take it as given that the Stoics' modal notions are interdefinable in the customary way, with *necessarily* contradicting *possibly not*, and so on.[12] I will thus move freely between speaking of what is possible and not necessary, impossible and necessarily not.

Another interpretative question about this passage concerns the connection between (2a) and (2b): whether they are two distinct conditions for possibility, or whether (2b) merely explains (2a). Michael Frede has argued and most have agreed that the genitive absolute *not being contradicted externally in being true* should be read as giving *further* conditions for possibility, rather than as an explanation of what it means to admit of truth.[13] I will also take this as given, so that

[8] 7.75 (38D).

[9] Thus, as Bobzien 2003: 117 puts it, not a logic of modal propositions but of non-modalized propositions insofar as they themselves are possible, necessary, etc. Cf. Bobzien 1999: 99-102.

[10] DL 7.66–7.

[11] DL 7.65 (34A); for arguments to this effect see D. Frede 1990, Bobzien 1993.

[12] This is largely agreed but not settled. See Paolo Crivelli's thorough accounting of the current state of this debate in the Complements to Mignucci 1978 in Brunschwig 2006.

[13] M. Frede 1974. NB: this debate is related to the question of interdefinability; on the conjunctive interpretation of the genitive absolute the necessary and

the possible is what both (a) admits of truth (can be true) *and* (b) is not opposed, or hindered, externally.

But what does it mean to be *opposed externally*, and what does that tell us about *admitting of truth*? Here too there is some debate, and for my purposes, an opportunity for some added clarity on what the relevant externals are. Susanne Bobzien, embracing the conjunctive reading, has suggested that admitting of truth is a matter of being internally consistent, i.e., not self-contradictory; the Stoics are concerned to exclude items such as round squares.[14] Thus the relevant externals are those things external to the logical subject of the *axiōma*; the first part of the definition (2a) embraces Philo's modal definition of the possible: that which is capable of being true, by the proposition's own nature, i.e., intrinsically or internally as a matter of logical consistency.[15] I agree with this reading and take *what admits of truth* (2a) to capture the Stoic notion of what is logically possible, their widest modal notion and minimal constraint on contingency. However, using the external opposition of (2b) as a contrast case to understand the possibility of (2a) as internal to the concept says nothing yet about what it means to be hindered externally.

Bobzien and most other commentators have understood external hindrances, quite sensibly, as circumstantial barriers to something's happening; "ordinary, physical hindrances: for example, a storm or a wall or chains that prevent you from getting somewhere; the surrounding ocean that prevents some wood from burning."[16] So Bobzien defines the possible as (a) what is logically possible and (b) is not in fact prevented by the circumstances from happening. If wood is not (or will not be) at the bottom of the sea, then it is possible for it to burn. If you do not spend the rest of your life in chains, it will be possible for you to go to the agora.

However, one might worry, as Margaret Reesor does, that such a reading conflicts with the well-attested Stoic response to Diodorus Cronus's Master Argument, namely, their commitment to there being

the impossible are disjunctive definitions, while possible and not necessary are conjunctive, as required for interdefinability. Cf. Mignucci 1978, Sorabji 1980a.

[14] 1993: 76–77; Hankinson 1999 concurs on this point.

[15] Boethius 234.10-21: *possibile esse, quod natura propria enuntiationis suscipiat veritatem, ut cum dico me hodie esse Theocriti Bucolica relecturum.*

[16] 1999: 119; cf. also 1993: 77, and 1998: 112–16.

something possible that neither is nor will be.[17] Briefly, the Master Argument consists of three mutually inconsistent premises designed to restrict the possible to what is or will be, i.e., to the actual. They are that 1) Every past truth is necessary; 2) An impossible cannot follow from a possible; and 3) There is something possible that neither is nor will be.[18] Diodorus is committed to the first two in exclusion of the third, so that what is possible is what is actual. Although Cleanthes and Chrysippus may disagree about whether the first or the second premise is to be rejected, the Stoics are in any case committed to the truth of the third. So here again we see the Stoic commitment to future truth and contingency together. Reesor's worry, as I understand it, is that casting external hindrances in terms of opportunity conflicts with the Stoic commitment to things that are in fact hindered being nonetheless possible; the conflict arises because *ex hypothesi* they will not take place, so they will be hindered, and thus will not have the opportunity. Likewise, Reesor worries that the following testimony from Cicero about a jewel that is in fact hindered but nonetheless possible to be broken suggests that possibility is not restricted to opportunity.

B. For he [Diodorus] says that only that is possible which either is true or will be true; that whatever will be is necessary; and that whatever will not be is impossible. You [Chrysippus] say that even things that will not be are possible. For example, that this jewel be broken, even if that will never be the case. And you say that it had not been necessary that Cypselus should rule in Corinth, even though the oracle of Apollo had foretold it a thousand years earlier. But if you are going to endorse these divine predictions you will also hold things falsely said about the future to be in such a class that they cannot happen, so that if it be said that Scipio will take Carthage, and if that is truly said about the future and it will be thus, you must say that it is necessary. And that is precisely Diodorus's anti-Stoic view.[19]

[17] Reesor 1965: 292, n. 17.

[18] Epictetus, *Diss.* 2.19.1–5 (38A). There has been a tremendous amount of scholarship devoted just to reconstructing the argument so that any two premises entail the contrary of the third; see, for example, Barreau 1978, Denyer 1981, Kneale 1937, and Prior 1955 (who argues we can find Lewis's S4 system in Diodorus). We can bypass these issues because it is uncontroversial that the Stoics agreed in accepting the third premise of the argument, i.e., in holding that there is a possible that neither is nor will be. Therefore, I will also largely bypass the details of Chrysippus's and Cleanthes' diverging strategies with respect to the other two premises (cf. n. 59).

[19] *Fat.* 13 (38E3).

Now, one might yet reply that the Stoic commitment to the third premise of the Master Argument is compatible with the reading that external circumstances must provide the opportunity for something to take place if it is to be considered possible, even if it will not take place as a matter of fact. One could maintain that as a matter of fate (and thus fact) a certain piece of wood will not burn, but that what makes it nonetheless *possible* is that it is not (or will not be) at the bottom of the sea; and that what makes it possible that the jewel will break is that someone has the opportunity to break it, and can thus be praised for not doing so.[20] What I would like to say at this juncture is that, while it is true that moral responsibility requires opportunity (i.e., that you cannot be praised for not breaking a jewel if you were nowhere near it), this cannot be the whole story of possibility in Stoic thought.

Indeed, I think it likely that Zeno's agreement with Philo extends beyond logical possibility as conceptual coherence to possibility as metaphysical coherence, i.e., from the first part of the definition (2a) to the second (2b). Philonian possibility is described as follows by Alexander of Aphrodisias.

C. "That which is said in accordance with the bare fitness of the substrate (*to kata psilēn legomenon tēn epitēdeiotēta tou hupokeimenou*), even when having been prevented (*kekōlumenon*) from coming about by some necessary external factor." Accordingly, he said that it was possible for chaff in atomic dissolution to be burnt, and likewise chaff at the bottom of the sea, while it was there, even though being prevented of necessity by its surroundings.[21]

According to this passage it is possible for wood (or chaff) at the bottom of the sea to burn because it is of a nature to burn, regardless of whether circumstances prevent its burning. Wood is combustible according to its *bare fitness* or *suitability* to burn, i.e., according to its intrinsic nature, so it is not just logically but metaphysically coherent to call it combustible even when surrounded by water (in a way that it would not be to call water itself or iron combustible). Similarly, a shell at the bottom of the sea is perceptible, even if circumstances prevent its being perceived. To locate possibility in the bare fitness of a subject

[20] As Sorabji 1980a: 272 insists, and LS: 235 agree.
[21] Alexander, *In Ar. An. Pr.* 184.7–9 (38B2); cf. Sharples 1983.

is thus to move away from possibility as opportunity and toward possibility as intrinsic capacity, placing it in the nature inherent to the entity. The issue of how far the Stoics agree with Philo is fraught because of conflicting evidence that points both ways.[22] My suggestion is that the Stoics agree with Philo in locating possibility in something's bare fitness, or intrinsic nature; but disagree with Philo insofar as they also recognize possibility as opportunity. This would explain testimony in both directions, and solve what Bobzien calls the Philonian Paradox, that possibility defined in terms of bare fitness leaves no reasonable room for responsibility.[23]

Reesor has argued persuasively that possibility for the Stoics is "found in the principal cause (e.g., breakable), which was a quality inseparable from its substratum (e.g., gem) and the cause of a predicate (e.g., is broken) which may or may not be realized; and a possible event is one in which the predicate is derived from the principal cause but which may or may not be realized (e.g., The gem is broken)."[24] The principal cause is what I have been calling the intrinsic capacity, or nature inherent to the entity. A gem is of a nature to be broken, whether it is broken or not, so that opportunity is not the salient issue but rather what things are true in virtue of (what predicates can be derived from) the principal cause or nature of the entity, i.e., what it can do given the kind of thing that it is.[25] Another well-known example of a principal (alternatively, perfect or complete) cause is the shape of a cylinder.[26] Although an external (alternatively, initiating or auxiliary) cause may be required to set a cylinder in motion, say,

[22] Both in the ancient evidence: Plut., *St. Rep.* 1055E (for), Boethius 234 (against); and in contemporary commentators: Bochenski 1951: 87, Mates 1953: 41, Rist 1969: 115 (for); Bobzien 1993: 75–76, Kneale 1962: 122–123, Salles 2005: 82-89, Sorabji 1980a: 271 (against).

[23] 1993: 75.

[24] 1965: 296; see also Reesor 1978 and Stough 1978 in the same spirit.

[25] I prefer not to speak in terms of predicates (*kategorēmata*), which represent an interpretative challenge of their own that we can put aside without damage to the present thesis. Predicates are considered incomplete *lekta* (sayables) and they figure prominently in the Stoic account of causation, in the role of effects. D. Frede 1990 argues that *lekta* are the laws of nature themselves; and Reesor in effect makes the same point; cf. also Striker 1987. I agree that the Stoic theory of modality is concerned to underwrite counterfactual truths about nature, but I disagree that the *lekta* are the laws themselves, or that *lekta* should be cast as events, facts, or states of affairs in any case.

[26] e.g., Cicero, *Fat.* 42 (62C5).

someone pushing it, the responsibility and complete explanation for that motion lie in the shape of the cylinder. Crucially, this intrinsic capacity is where the Stoics locate moral responsibility – someone may have pushed the cylinder, but it rolls because of its shape; analogously, a thief may be pushed to steal by seeing a jewel, but a thief steals because of character.

Reesor argues also that the Stoics recognized two distinct senses of fate: that of the internal, principal cause (the nature with which you are born and its habituated states developed in reaction to the environment, i.e., the nurture it receives), and that of the series of external, initiating causes, i.e., the sequence of events to which principal causes react. The unwinding of the rope of fate is thus the interaction of external causes with internal natures.[27] So, for Reesor, a possible event is a predicate derivable from the principal cause, which may or may not be realized due to external circumstances. Or, as I would prefer to put it, what is possible is what something's nature (in the first sense of fate) allows, independent of what in fact goes on once fate in the second sense, as the chain of external causes, is in motion. Thus a piece of wood at the bottom of the sea is combustible because the predicate *is burnt* resides in the principal cause (i.e., it is of a nature to burn), even if external causes prevent that predicate from becoming true of it. By locating possibility in the principal cause, internally, Reesor rightly separates possibility from opportunity.

R. J. Hankinson also recognizes something like this in the Stoics, with two distinct types of modality he calls *species possibility* and *actual possibility*. Species possibility corresponds to what is "possibly applicable to an individual of natural kind K just in case K's can, other things being equal, be P's;" while actual possibility is defined by what actually obtaining circumstances do not prevent; he adds, "if this is right, it is false to say that the only type of possibility available to the Stoics is epistemic."[28] My purpose in limning this debate in the literature over possibility as

[27] The image of the rope unwinding is not entirely felicitous since the idea is more like the winding of a rope with internal and external causes coming together to intertwine; perhaps we might now prefer the image of a zipper.

[28] 1999: 528. Given more space, I would have liked to engage with these and other scholars whose analyses are in the neighborhood, so to speak; e.g., D. Frede 2003 as an alternative to equating laws with *lekta;* the Kneales 1962 on absolute vs. relative necessity; Kneale 1937 for logical vs. metaphysical vs. empirical facts; LS as an alternative or addition to possible for vs. possible that; Rist 1969: 122, 132 for the role of natures in the Stoic retreat from

opportunity as opposed to intrinsic capacity is to show that intuitions on both sides are not only correct but in fact complementary once we see that the Stoics recognized both: what I am calling metaphysical possibility, located in the intrinsic nature of each thing, and providential possibility, located in the chain of external causes.

With our central passage from Diogenes in mind, then, and in particular the interpretative question of what it means to be hindered externally to the concept, let us turn now to testimony from the late Stoic Epictetus. The following passage gives further reason to think that the Stoics employed a notion of metaphysical possibility, and in particular that they thought of external hindrances this way.

D. Fittingly enough, the one thing the gods have placed in our power (*eph' hēmin epoiēsan*) is the one of supreme importance, the correct use of impressions. The other things they have not placed in our power. Is this because they didn't want to? My belief is that they would have entrusted them to us too, had they been able to, but that they simply weren't able. For we are on the Earth, bound by an Earthly body and Earthly partners. How then could we have failed to be hampered (*empodizesthai*) by externals with regard to these things? What does Zeus say? "Epictetus, if it had been possible, I would have made your wretched body and trappings free and unhindered. But as it is, please note, this body is not your own, but a subtle mixture of clay."[29]

Here I take Epictetus to be expressing a sense of *being hampered by externals* that means being hampered by those things external to the mind that makes use of impressions, i.e., those things external to what is conceivable, or logically possible. In particular, we find that God is constrained by the matter he has to work with; being on Earth, bound by earthly partners is to be bound by the properties of matter. Crucially, even God is subject to these constraints in crafting the cosmos: *if it had been possible, I would have made your wretched body free and unhindered*. Otherwise, the sphere of possibility would be constrained only by logical consistency; as it is, however, God is constrained by matter and so what is cosmically, or metaphysically, possible is narrower than what is logically possible.

necessity; Sorabji 1980a for his rejection of empirical vs. logical; and Stough 1978 for the room she finds in *cannot* for moral responsibility.

[29] *Diss*. 1.1.7–11 (62K, part).

Thus, what is metaphysically possible is what is logically possible *and* not hindered externally to the concept by God's decisions within the given material constraints; and to define it this way, as what is *not* hindered, is of course to work around the external hindrances that there *are*. What is possible is what is left over after what is necessary (and thus impossible) has been established. According to this picture, then, God cannot make just any possible world – only those that can be crafted out of the matter he is given. Moreover, God does not make just any of those possible worlds, but rather the best one.[30] God takes the matter, and crafts the best world he can; in so doing he establishes what is metaphysically necessary and possible. My next suggestion is that in addition to the various *natures* he creates within the constraints of matter, God establishes certain metaphysically necessary *principles*, most crucially the principle of bivalence and its correlate that there is no motion without a cause.[31]

Dorothea Frede has argued that the Stoics treated bivalence (future truth) not only as a consequence of Stoic fatalism (if everything is predetermined, then there must be future truths), but also as an argument *for* it.[32] If bivalence does not hold, the causal order of the cosmos is threatened: anything could happen at any time (as indeed the Epicureans held in introducing the swerve of an atom), contrary to Stoic determinism. Frede writes:

> This coherence and stability can be maintained only if there is an eternal uninterrupted causal series. But not only that: there must be a fixed order within the causal series, and that can be guaranteed only if the same cause is always followed by the same effect. Take for example the rule: "When someone puts a burning match to a dry haystack, the haystack will burn." To guarantee the stability of the physical order the Stoics had to assume that there are such eternally valid causal principles that obtain even if nothing actually corresponds to the hypothetical state of affairs at a particular moment.

Thus to question future truth, i.e., to consider suspending bivalence, is to question the very laws of nature; future truth is a necessary condition for causality.

30 This claim is not to be confused with a teleological account of nature; it is rather a story of providential creation; cf. Hankinson 1999: 539 for agreement on "limitations on creative possibility supplied by material recalcitrance."

31 Cicero, *Fat.* 20–21 (62G), Alexander, *Fat.* 185,7–11 (62H), 191,30–192,28 (55N).

32 1990: 210; cf. Hankinson 1999: 519–520.

For all that, however, it is perfectly conceivable, i.e., logically possible, that bivalence does not hold and that the world is not determined – the Stoics were only too aware of this fact, surrounded as they were by lively criticism and debate over fatalism and future truth in a competitive intellectual era. Thus we would not want to say that the principles and natures that God creates hold a priori as logically necessary. Rather, the necessity of metaphysical principles is a posteriori, according to the decisions God makes in crafting the cosmos within the constraints of matter. So, what is metaphysically possible is what is unhindered external to the concept, i.e., unhindered by the metaphysical principles and natures that God establishes. It is metaphysically possible for wood on the ocean floor to burn even if it will always be surrounded by water, because it remains combustible by nature.

We can also discern this gap between what is logically and metaphysically possible from the Stoic defense of infinite, extracosmic void. One of the many criticisms of the Stoic notion of void is that they cannot simultaneously say that the cosmos is at the center of void and yet claim that the void is isotropic. The Stoic response is that the cosmos is not in fact at the center of independently subsisting void, so they are not committed to an identifiable center of something infinite and isotropic.[33] Nonetheless, they can speak as though there were such a middle, and even consider the possibility of a cosmos not at the center of void. Hence Plutarch quotes Chrysippus as saying that "if it [the cosmos] should be imagined elsewhere [than in the middle], destruction would almost certainly be attached to it"; and that "it has also in some such way been an accident of substance (*ousia sunteteuchen*), from the very fact that it is the kind of thing it is, to have occupied everlastingly the middle place, so that otherwise but also accidentally it does not admit of destruction."[34] In imagining the cosmos elsewhere, in a void that is in fact self-subsistent, Chrysippus evinces a recognition that what is logically possible is distinct from what is metaphysically possible. It is also clear that substance (*ousia*), which for the Stoics picks out the material substrate, is what accounts for that distinction.[35] Thus what is metaphysically possible is distinct from what is logically possible, as a function of God's decisions within the

[33] As I argue in de Harven 2015.

[34] Plut., *St. Rep.* 1054C–D, trns. Cherniss; see also 1055D.

[35] That *ousia* refers to the material substrate is largely agreed, e.g., by LS, Sorabji 1988, and Sedley 1982; *pace* Menn 1999.

constraints of matter, establishing the principles, natures, and causes of the one and only cosmos.

What is metaphysically possible is also distinct from what is providentially possible, which captures the notion of possibility as opportunity that commentators have favored. In this sphere, as in the last, there are things that remain possible even though they are not and will not be – herein lies moral responsibility. It is true enough that one cannot be held responsible for actions one did not have the opportunity to commit: a thief who never steals again because locked up is not thereby reformed and cannot be given credit for not stealing anymore. Only if the opportunity to steal is presented and passed up can there be credit for not stealing. What is providentially possible is therefore what your circumstances allow, including the absence of ordinary hindrances such as being locked up and what one might think of as technical hindrances such as being too late, i.e., in the past. What is possible is thus what is not yet settled, what is possible prior to the external, initiating causes that set the principal cause in motion. It is thus providentially possible for the thief who is not locked up and has the opportunity to steal not to steal, even if as a matter of fact he will.

On this basis even what is fated is not necessitated, and the Stoics can coherently stand by the third premise of the Master Argument, that there is a possible that neither is nor will be (at both the metaphysical and providential levels). Now, it is true that from the cosmic perspective what is providentially possible must in some sense be considered necessary, since it is impossible that the world order turn out otherwise.[36] However, this cosmic perspective notwithstanding, there is an important difference between what is providentially possible and necessary: namely, that what is providentially possible is not yet settled insofar as the relevant initiating causes that will set the principal cause in motion are not yet present to it; the rope has not yet unwound that far. Thus it is providentially possible that Socrates will escape from jail, until he drinks the hemlock – only then does his death become necessary (more on this below, with the Stoic account of necessity).[37]

In summary, I submit that Diogenes' definition of Stoic possibility in passage A should be understood as what admits of truth (logical

[36] Cf. the Stoics' "bifocals" in Long 1971, also Rist 1969.

[37] I take this to address Sorabji's concerns about responsibility amid necessity, since I deny that the chain of fate is necessary except once an event is past.

possibility) and is not hindered externally to the concept, either as a matter of metaphysics (metaphysical possibility) or as a matter of external circumstances (providential possibility). The example Diogenes provides, *Diocles is alive*, does not settle matters in one direction or the other. This proposition could be considered possible either as a matter of metaphysics – Dion is of an animate nature – or as a matter of providence: Dion is not dead (he either is or will be alive). Boethius's example is similarly neutral, and thus capable of illustrating both senses: *that I will reread Theocritus' Bucolica today*. Our testimony is extremely compressed; although we cannot be certain, it is at least possible that these examples were carefully chosen to illustrate both kinds of possibility external to the concept that I describe.

I turn next to Diogenes' definition of the impossible (A3): that which (a) does not admit of being true, <or (b) what does admit of being true but is opposed by things external to it from being true>, e.g., *Earth flies*. Accepting Michael Frede's emendation (in brackets), this disjunctive definition is the contradictory of what is possible.[38] It is (3a) what is logically impossible, e.g., *It is day* and *Not: it is day*; or (3b) what is logically possible (i.e., internally consistent), but opposed externally. Again, we ask what it means to be opposed externally. Here the example, *Earth flies*, is of help, although intuitions do diverge. Bobzien takes the example to illustrate the first part of the definition, (2a), a logical impossibility.[39] But the reason the Earth cannot fly (as far as the Stoics are concerned) is that the Earth does not have wings.[40] So I would argue that the reason *Earth flies* is impossible is that as a matter of metaphysical fact the Earth does not have wings and so cannot fly. It is at least conceivable that the Earth have wings, so it does not seem like a logical impossibility that is being illustrated.[41] Further, it is clear that the relevant hindrance to Earth's flying is no external circumstance since there are no such circumstances for God, who is identical with everything there is.[42] So the relevant impossibility must

38 With LS and Mignucci 1978, contra Inwood & Gerson 1988; cf. earlier remarks in n. 13 concerning the conjunctive reading of possibility.

39 1999: 118.

40 SE, *PH* 2.104–106 (35C3), *M*. 8.113.

41 And to reply that it is logically impossible in the context of this wingless Earth that it fly, would be to concede the point that it is not true as a matter of logic that Earth does not fly.

42 Plut., *St. Rep*. 1050C–D (54T).

be metaphysical, and knowable only a posteriori, having established that as a matter of fact the Earth has no wings.

Now to Diogenes' definition of Stoic necessity, passage A4: (a) that which, being true, does not admit of being false, or (b) what, on the one hand, does admit of being false but is opposed by things external to it from being false, e.g., "Virtue benefits." It should be familiar by now that the first part of the definition (4a) corresponds to what is logically necessary, i.e., impossible not; and that the second part (4b) corresponds to what is metaphysically necessary. Here too scholars have thought that the example *Virtue benefits* illustrates a logical necessity for, as the Kneales put it, "a Stoic could scarcely have conceived that in certain circumstances 'Virtue is beneficial' might be false."[43] Here again I would argue that while the Stoics are certainly committed to the necessity of this proposition, they would not have seen it as logically necessary. Following in the footsteps of Socrates, they would have been all too aware that few take this position to be true, let alone logically true, and concerned to argue for the necessity of this principle naturalistically. Indeed, it is uncharitable to think the Stoics would conflate their metaphysical principles with logical truths. Rather, *Virtue benefits* is necessarily true because our natures are such that virtue cannot fail to be good for us (the predicate is derived from the principal cause, as Reesor would put it); and it is not logically true since, as Glaucon so well illustrates in Plato's *Republic*, it is certainly conceivable that humans are *pleonectic* in nature so that justice is only instrumentally good, and never beneficial in itself.

The necessity of Stoic metaphysical principles and natures is confirmed by ample textual evidence. For instance, Alexander testifies that for the Stoics "what each does is in accordance with its proper nature.... Hence, they say, none of the things each of them does in accordance with its proper nature can be otherwise: everything they do is done of necessity.... For the stone, if released from a height, and not prevented, cannot fail to travel downwards."[44] Note here the idea that natures are as they are as a matter of necessity (as a function of its nature, the stone necessarily falls when unhindered by external circumstances), and that this corresponds to Reesor's first sense of fate

[43] 1962: 123; cf. Reesor 1978, Gourinat 2000: 206.

[44] *Fat.*, 181,18–27 (62G2–4); cf. Long 1970 for a detailed exegesis of Alexander's *Fat.*

as principal cause (and to the Philonian bare fitness extended to the second part of Stoic possibility).[45] Alexander then adds: "the very fate, nature and rationale in accordance with which the all is governed is god. It is present in all things that exist and happen, and in this way uses the proper nature of all existing things for the government of the all."[46] Having crafted the principles and natures that constitute fate in the first sense, God then relies on the necessity of their natures (*uses the proper nature of all existing things*) in unwinding the rope of fate by the chain of initiating causes.

The necessity of these natures is further emphasized when the Epicurean Philodemus tells us that for the Stoics the inference "Since the humans familiar to us are mortal in that and in so far as they are human, humans everywhere are mortal too" is correct (*orthōs*).[47] The reason the inference is correct is that the conditional "If something is human, then it is mortal" is sound; i.e., it meets the Stoic criterion of implication called *cohesion*, which says that "a conditional is sound whenever the contradictory of its consequent conflicts with its antecedent."[48] Stoic implication is stronger than Philonian, which is merely truth-functional, and stronger again than Diodorian implication, which states that a conditional is sound when it is impossible for there to be a true antecedent and false consequent. But the criterion of Stoic implication is unclear to the extent that the nature of the conflict between affirmed antecedent and negated consequent is unclear.

It has been suggested by S. Sambursky and by M. Frede that we understand the conflict as logical, in contrast to empirical.[49] Perhaps retreating from the force of logical necessity, Long and Sedley suggest that the relevant sense of conflict is a *conceptual* rather than empirical incompatibility, such that "when the consequent is hypothetically eliminated the antecedent is *thereby* be 'co-eliminated' – a formulation which appears to outlaw any considerations which cannot be extracted from our understanding of the antecedent and consequent themselves."[50] I agree that the consequent and antecedent will be coeliminated, but would like to give a finer point to the discussion,

[45] Cf. Stough 1978: 223 about the necessity of acting according to our natures.
[46] 192,25–28 (55N4).
[47] *Sign.*, fr. 6 (42G4).
[48] SE, *PH* 2.111 (35B4); NB: *sound* for the Stoics, here, is captured by our *true.*
[49] 1959 and 1974, respectively; see also Sorabji 1980a: 267 on this.
[50] LS: 211.

because to say that the relevant necessity behind cohesion is conceptual rather than empirical elides the difference between logical and metaphysical necessity.

The example Sextus gives of a sound conditional according to the criterion of cohesion is *If it is day, it is day*, which is logically necessary and thus clearly knowable a priori. Certainly, analytic truths make for sound conditionals and that is why Sextus cites it as an example; but it is clear the Stoics do not want to restrict their inferences to logical or analytic truths. They are with Diodorus in aiming "to convert the sound conditional into the kind of necessary truth which might ground scientific or dialectical inferences."[51] Recognizing this, Long and Sedley say it is our understanding of the antecedent and consequent themselves that makes the connection conceptual rather than empirical. Here I interject to suggest that the connection between antecedent and consequent is better thought of as metaphysical: necessary but a posteriori, and therefore already empirical, not logical. At the same time one would not want to deny that there is a certain conceptual connection between antecedent and consequent in a sound conditional. *What Earth is* makes it metaphysically necessary that Earth does not fly, and our concepts are correspondingly related: the concept of Earth does not admit flying. Likewise, human nature makes it necessary that virtue is to our benefit, and this is reflected in our concepts so that it is scarcely conceivable that matters be otherwise; nevertheless, no formal contradiction emerges from the supposition.[52]

In case doubts remain that these necessary truths are a posteriori and conceivably otherwise, I offer this additional evidence that shows even such correlative concepts as good and evil, justice and injustice, pain and pleasure, virtue and vice, and even true and false are a posteriori in the way I have been describing, as a function of material constraints and God's decisions for the best.

E. In his judgment it was not nature's principal intention to make people liable to disease: that would not have been fitting for nature, the creator and mother of all good things. But, he [Chrysippus] adds, while she was bringing about many great works and perfecting their fitness and utility, many

[51] LS: 210.

[52] Cf. Kneale 1937, for whom what is metaphysically necessary is that whose consequent is inconceivable but not self-contradictory.

disadvantageous things accrued as inseparable from her actual products. These, he says, were created in accordance with nature, but through certain necessary "concomitances" (which he calls *kata parakolouthēsin*). Just as, he says, when nature was creating people's bodies, it was required for the enhancement of our rationality and for the very utility of the product that she should construct the head of very thin and tiny portions of bone, but this utility in the principal enterprise has as a further, extraneous consequence the inconvenience that the head became thinly protected and fragile to small blows and knocks – so too, illnesses and diseases were created while health was being created.[53]

It is clear that the necessity of illness alongside health is not a logical necessity, since nature would have preferred to do otherwise, and thus can at least conceive of health without illness. It is also clear that the concomitances result from material constraints, and that the relevant necessity in play is material necessity. Even such correlative concepts as good and evil, health and illness, virtue and vice are connected conceptually only because they are connected metaphysically, as a function of matter; and such necessity is entirely a posteriori.

Thus I suggest that what underwrites the Stoic criterion of cohesion is not logical but metaphysical necessity; and although it is inconceivable that necessary truths fail to hold, it is perfectly conceivable that what is metaphysically necessary be otherwise. This is why the Mowing Argument is invalid: 1) if you will mow, you will mow; and, if you will not mow, you will not mow; 2) but necessarily, either you will mow or you will not mow; 3) therefore, if you will mow, you will mow necessarily; and, if you will not mow, it is impossible that you mow; the contingent is eliminated.[54] The principle of bivalence expressed in premise 2 is itself metaphysically necessary, but the proposition that must now be true or false according that principle, as expressed in premise 3, is not itself necessary or impossible; it is merely predetermined, as expressed in premise 1.

We can now see why the Stoics retain the Philonian conditional (material implication), which they express as a negated conjunction, *Not both: p and not-q*. Philonian implication expresses what is true as a matter of fate, i.e., what is determined (and thus subject to divination)

[53] Gellius 7.1.1–13 (54Q); examples cited in introducing the text occur earlier in the passage, not quoted here, but are subject to the same account.

[54] Ammonius, *In Ar. De Int.* 131, 20–32 (28G).

but not, for all that, necessary. As Long and Sedley observe, a negated conjunction asserts no direct logical connection between *p* and *q*, and "since laws of divination assert empirical rather than logical connections between past and future truths, the negated conjunction is indeed the appropriate means of formulating them."[55] Here again I interject to suggest we say instead that the laws of divination assert *providential* rather than *metaphysical* connections – what is determined rather than necessitated. This is why the Stoics can maintain that what is determined not to be is nonetheless possible: no necessity attaches to the future truth, which is reflected in the use of the conjunction to express future truths (while the conditional is reserved for metaphysical and logical truths that do hold of necessity). Future truths are providentially possible, not necessary. Thus the Stoics avoid the charge that appeared above in passage B (Cicero's testimony with the jewel), that bivalence commits them to false futures as impossible, and true futures as necessary.[56]

This leaves open the question, what is providential necessity? One might still worry, as I did above, that the entire world order is providentially necessary, since it cannot turn out otherwise. But providential necessity is not just the ineluctable hand of fate. Rather, it corresponds to what is already settled, how far the rope has unwound. Alexander again:

> F. Whenever the external causes that encourage the stone's natural motion are also present, the stone of necessity moves with its natural motion. And these causes of its moving are, at the time, unconditionally and necessarily present. Not only is it incapable of not moving when they are present, but it also moves of necessity at that time, and such a movement is brought about by fate through the stone.[57]

This passage expresses two levels of necessity: the metaphysical necessity that the stone act according to its nature upon encountering the relevant external causes (that the stone will fall *if* dropped), and the providential necessity of the moving causes once they are present or actual (the providential necessity of the stone's falling *when* dropped).

[55] 236.

[56] Cf. Bobzien 1993: 82 for a denial that future truths are contingent, Reesor 1978 and Rist 1969 for an endorsement.

[57] 181, 28–30 (62G4, part).

Hence, the stone is counterfactually incapable of not falling when dropped (as a result of the metaphysical necessity of its nature), and actually incapable of not moving once it is in fact dropped (because of the providential necessity of what is current or past). Once the stone is dropped, it is providentially necessary, i.e., settled, that it falls. Until then it is merely possible, albeit predetermined. Hankinson says something similar: "Consider an example of Aristotle's: a new cloak might perish as a result of ordinary wear, or it might be cut. For the Stoics, *sub specie aeternitatis* there is only one thing that can happen to it – the unraveling of fate will see to that. However, there is nothing now in the world that prevents either outcome, for no causally efficient state of affairs is now making it the case that it will (or will not) be cut."[58] What is providentially necessary is what a causally efficient state of affairs makes to be the case, what is happening now and what has been.[59]

Finally, I turn to Diogenes' definition of what is not necessary: that which both is true and can be false, being opposed by nothing external, e.g., *Dion is walking*. Here again the central interpretative question concerns what it means to be opposed externally. As with the case of *Diocles is alive* to illustrate possibility I suggest that we hear the example as illustrating what is not necessary either as a matter of metaphysics or as a matter of providence. Dion's nature does not dictate that he be walking all the time (the way, say, fish nature might dictate that Nemo be swimming all the time); nor do his life circumstances dictate that he walk all the time. Note that even if Dion's fate were to

58 1999: 528.

59 Thus it is clear how Chrysippus can affirm the first premise of the Master Argument, that all past truths are necessary. Although Cleanthes denies this premise, we need not take this as a deep divide in Stoic modal theory (as do, e.g., Rist 1969: 118, 1978: 197, Barreau 1976: 30–40). Cleanthes could deny that past truths are necessary in the sense that they remain contingent events, not metaphysically necessary. It would then be open for him to embrace past events as providentially necessary, along with Chrysippus, while denying they are metaphysically necessary in a way that meets the dialectical requirements of the Master Argument. Alternatively, Cleanthes could deny providential necessity altogether, holding that fate is always determined but never necessary, so even past events do not become necessary; he could argue from everlasting recurrence that past contingents are future contingents, and thus never necessary. Both of these, however, constitute a legitimate retreat from necessity, obviating the need to see possibility as merely epistemic, on which see Long 1971: 189, Reesor 1978: 194–197, Sorabji 1980a: 276–278.

spend his whole life walking, it would not thereby be providentially necessary that he walk. This is the point of making the distinction between what is providentially possible and necessary: namely, to preserve contingency even for future truths. Thus even if as a matter of fate *Dion is walking* were always true (while he lives), it will not be a future truth expressible by a sound conditional (since that is reserved for logical and metaphysical truths) but by the negated conjunction: *Not: Dion lives and Not: Dion walks*. What is not necessary is thus the sphere of things that can be otherwise, as a matter of either metaphysics or providence.

3. Conclusion

In conclusion, if I am right that the Stoics recognized these three senses of necessity and possibility, the Stoic retreat from necessity is accomplished across logic, physics, and ethics. In the domain of logic, we see that the Stoic response to the Master Argument is underwritten by the fact that providential necessity applies only to present and past truths, so future truth and contingency are compatible. Also, that we have reason to see metaphysical necessity as the principle of cohesion. In physics we see that metaphysical necessity underwrites the counterfactual truths that guarantee the stability of the cosmos. And in ethics, the sphere of responsibility is secured by what is providentially possible. Thus the Stoics embrace the apparent paradox of future truth and contingency without falling into contradiction.

5 Necessity in Avicenna and the Arabic Tradition

PAUL THOM

1. Introduction

The expression 'Arabic logic' is used by scholars to cover a tradition of writing over a period stretching from the eighth century to the present day. The treatises and handbooks that belong to this tradition were written in various languages, including Arabic, Persian, Turkish, Hebrew, and Urdu. The central figures are Alfarabi (d. 950), Avicenna (d. 1037) and Averroes (d. 1198) (Street 2013a, introduction). Because of limitations of space, this chapter will omit Alfarabi, discussing only Avicenna and some of the more significant thinkers (including Averroes) whose work is based on, or reacts against, the latter's logical theories.

2. Avicenna

Avicenna uses the term 'necessary' to mark out (a) a class of beings, and (b) a class of propositions. His doctrine of the unique Necessary Being from which all contingent beings are derived is well known. Contingent beings are described as 'possible'; but 'possible' here means two-sided possibility. A contingent being is one that can either exist or not exist. Whether it exists depends on a cause external to it. A necessary existent is one that cannot not exist, because its existence does not depend on anything external:

TEXT 1. And so we say that the things that fall under 'being' can be divided by the intellect into two groups. One of these comprises things which, when considered by themselves, do not possess necessary being; and it's clear also that their being is not impossible, otherwise they would not fall under 'being', and these are contained in the term 'possibility'. The other comprises

I want to extend my warm thanks to Tony Street and Riccardo Strobino for giving me the opportunity to present an earlier version of this paper to the senior research seminar in the Faculty of Divinity, The University of Cambridge, and for their insightful and helpful comments.

> those which when considered by themselves possess necessary being. And so we say that a being which is necessary by itself does not have a cause, and that what is possible by itself has a cause. (Avicenna 1977, I, 6, 8 ff)

What is the connection between the usage of 'necessary' as a predicate of beings and its usage as a predicate of propositions? There is some evidence that, in Avicenna's view, the first usage is a particular case of the second – that to say a being is necessary is to say that the proposition expressing the being's existence is a necessary proposition. Avicenna's standard example of a proposition which is absolutely necessary is a proposition about the Necessary Being, namely, the proposition 'God exists' (Street 2004, 551).

This, however, does not imply that Avicenna's exposition of the sense in which the Necessary Being is necessary can be applied to the necessity of necessary propositions. The Necessary Being is necessary in the sense that it does not depend on anything other than itself for its being. But this is not the sense in which a necessary proposition is necessary. There does not seem to be any textual evidence for thinking that when Avicenna ascribes necessity to a proposition he means that it does not depend on any other proposition. Such a conception might be applied to *indemonstrable* propositions; but the concept of necessity, as applied to propositions, is much broader than that of indemonstrability. So it seems that, for Avicenna, even though propositions expressing the existence of the Necessary Being are themselves necessary propositions, the necessity ascribed to it is more specific than the necessity ascribed to them.

Avicenna knows that the generic necessity which is ascribed to all necessary propositions can be defined as the impossibility of the contradictory proposition; and he also knows that such a definition is circular (Avicenna 1977, I, 5, 65ff).

According to Avicenna, a necessity-proposition may be either predicative or hypothetical. Both types of proposition contain temporal operators. A predicative proposition affirms or denies that its predicate applies to its subject in some temporally qualified way: either always, or at some specified or unspecified time (Street 2002, 132). Similarly, according to him, 'if-then' propositions take forms such as:

Always: if every A is B then every J is D
Never: if every A is B then not every J is D
Once: if every A is B then not every J is D

Khaled El-Rouayheb comments:

> The prefixing of modal operators to conditionals may strike scholars familiar with the Latin logical tradition as odd. It is in fact distinctive of the Avicennian tradition of Arabic logic. Avicenna explicitly drew an analogy between these modal operators prefixed to conditional propositions and the quantifiers in categorical propositions. (El-Rouayheb 2009, 210)

However, Avicenna's discussion of propositional necessity is mainly concerned with predicative propositions:

> Avicenna does not discuss modal conditional propositions except very sketchily .... What we describe as necessary, possible ... etc. is the judgement that a certain consequent follows from a certain antecedent. Saying this he goes on to explain what is meant by the modal notion 'necessity'. A universal connective, according to him, is necessary if the consequent follows (in any of the senses of following) the antecedent every time we posit the antecedent and so long as the antecedent remained posited. (Shehaby 1973, 253)

Avicenna was familiar with Aristotle's definition of the syllogism, and thus with the notion that the conclusion of a syllogism follows *with necessity* from the premises. But he does not seem to have devoted much thought to a specific consideration of this type of necessity.

In some of his works Avicenna seems to tie the concept of necessity to that of perpetuity. For example, in the logic section of his *Deliverance* (*Najāt*) he portrays the idea of perpetuity as the central element in the concept of necessity: every necessary proposition 'has some share in perpetuity'. In his logical writings Avicenna shows little interest in the concept of *absolute* necessity. As mentioned earlier, his standard example of an absolutely necessary proposition concerns the Necessary Being. The concept that interests him is the concept of relative necessity. For beings other than the Necessary Being, all necessities are relative. Avicenna distinguishes different types of relativity. In his classification of six types of necessity-proposition he notes that (i) the proposition 'God is living' is true in perpetuity because God exists eternally. (ii) The proposition 'Every man is by necessity an animal' is true in the sense that its predicate applies to its subject at all times when there are men; but since men do not exist eternally, this proposition expresses a perpetuity only in a qualified sense. (iii) The proposition 'By necessity every white thing has a colour which dilates sight' is true 'as long as the substance of the subject is described by a description that is posited along with this substance'. (iv) 'Zayd is walking' is necessary 'as long as the predication exists'; i.e. Zayd is by necessity walking for as long as he is walking. (v) The proposition 'The moon eclipses' 'is not a necessity that persists perpetually, but for some exact and specified time'. (vi) The proposition 'By necessity every man breathes' holds 'for some unspecified span of time' (Avicenna 2011, §48.ii–vii).

Avicenna distinguishes a proposition's matter (*mawādd*) from its mode (*jihāt*). The matter is the objective relationship of necessity, possibility or impossibility linking the proposition's predicate and subject; the mode is what is expressed in the proposition regarding that relationship. In identifying a proposition's matter we speak of a *necessary* proposition, and so on for the other sorts of matter; in identifying its mode we speak of a *necessity*-proposition, and so on. His six types of necessity-propositions also happen to be necessary in matter (given the appropriate relativities). But Avicenna knows that a proposition's matter may not be the same as its mode. For example, in the proposition 'It is possible for Zayd to be an animal' the matter is necessary but the mode is one of possibility (Avicenna 2011, §41 and 45).

Among the types of necessity-proposition Avicenna gives priority to type (ii):

TEXT 2. Our statement, 'Every *B* is *A* by Necessity' means 'Everything that is described in the intellect as *B* – whether [it is so described] perpetually or non-perpetually – is perpetually described as *A*, as long as its individual substance exists'. [An example is] your statement, 'By Necessity, every mover is a body'. (Avicenna 2011, §53)

In addition to the reduction of necessity to perpetuity, two points stand out in this treatment. First, propositions of type (ii) are treated as propositions about every *individual* falling under the subject:

TEXT 3. 'Every man is by necessity an animal', i.e. each one among humans is always an animal as long as his substance exists. (Avicenna 2011, §48.iii)

Second, these propositions are understood as making a predication about their subject relative to the continued existence of the *substance* underlying that subject.

Both these points are also evident in the parallel passage from Avicenna's magnum opus, *The Cure* (*Shīfā*):

TEXT 4. But we say: 'Every whiteness is a colour' and 'Every human is alive' meaning not that every single thing which is white is a colour which always was and will be [a colour], or that every human is alive and always was and always will be [alive]. Rather, we are just saying that everything that fits the description [WHITENESS], and that is [properly] said to be a whiteness, is a colour so long as its essence continues to be satisfied. And likewise everything (properly) said to be human (is not alive in the sense) that it always was and always will be an animal; but rather so long as its essence and substance continue to be satisfied. (Hodges 2010, 11–12)

The example of whiteness and colour gives rise to a terminological question. At *Topics* I.9 (See Aristotle 2009) Aristotle uses 'essence' (*ti esti*) in two different ways in the same paragraph. First (103b20), he applies the expression to the category of substance as distinct from the categories of quantity, quality, etc. Then (103b25) he says that the expression 'essence' applies not only to substance but also to quality and the other categories. His point is that in all the categories we find the genus-species relation – a relation that depends on the notion of essence, since the genus is of the essence of the species (Topics I.8, 103b15). If we were to follow Aristotle's usage in the first passage, we would use 'essence' and 'substance' interchangeably. If we were to follow the second usage, we would say that both man and whiteness are essences, but of these two only man is a substance. I shall follow the second usage, and thus I call type (ii) predications 'essential'. For type (iii) predications I will use Tony Street's term 'descriptional' (Street 2004, 551).

Where in this classification can we fit the necessity with which a consequent follows logically from its antecedent? Earlier I quoted Avicenna's phrase 'the consequent follows (in any of the senses of following) the antecedent every time we posit the antecedent and so long as the antecedent remained posited'. This phrase suggests that Avicenna thinks of the necessity of a consequence as a kind of *descriptional* necessity. What he has in mind is that the mental act of positing the consequent is relative to the act of positing the antecedent: the latter remains in force only so long as the former remains in force.

But this does not get to the question of the necessary logical relation between the *propositions* which are antecedent and consequent. If Avicenna had considered the question whether this necessity was essential or descriptional, I think he could have come to the conclusion that it is an essential necessity, as distinct from the necessity of statements like 'Khaled's latest statement follows with necessity from

Khaled's latest statement'. But in order to develop this thought, he would have had to articulate an ontology of propositions, analogous to the Aristotelian ontology of substances and accidents. So far as I know, he did not do this. His account of the necessity of logical consequence remains sketchy.

Avicenna held that first-figure syllogisms whose minor premise states a possibility are valid. For example, he held that if every *J* is possibly *B*, and every *B* is necessarily *A*, then it follows that every *J* is necessarily *A*. Now, if '*B*' in the second premise applies only to what is actually *B*, it seems that the inference will be invalid, because if we consider a *J* that is possibly *B* but not actually so, the premises will not imply that this *J* is necessarily *A*. And yet, in *The Cure* Avicenna states that the subjects of predicative propositions (including necessity-propositions) should *not* be understood as applying to whatever possibly falls under the subject-term:

TEXT 5. When we say 'Every white thing', its sense is definitely not 'everything that could legitimately be white'. Rather it means 'everything that in actuality fits the description [WHITE], where besides being actual, it can be so for some time which is indeterminate or determinate or permanent'. (Hodges 2010, 11)

It is not clear how Avicenna can maintain the validity of the syllogism mentioned, at the same time as denying that the subject of a true necessity-proposition applies to whatever possibly falls under it. Some later Arabic logicians thought that Avicenna could avoid inconsistency if he distinguished what can *be B* from what can *become B*:

By excluding ampliation to the possible, Ṭūsī took Avicenna to be excluding the ampliation of a term like 'men' to include things that could become men, like sperm. I would therefore tend to take the *J*s "in mental supposition" *(fī l-fardi l-dhihni)* as licensing a kind of ampliation to the possible. (Street 2002, 134)

In his different logical works Avicenna seems to espouse conflicting views on (a) the nature of necessity, (b) the application of the subject-term, and (c) what is quantified over in necessity-propositions.

For example, in his later *Pointers and Reminders* (*Ishārāt*), perpetuity is not identified with necessity, and some of his remarks explicitly distinguish the two concepts. In this work, Avicenna gives separate truth-conditions for necessity and perpetuity propositions:

TEXT 6. An example of a statement with such an addition is, "Necessarily every C is *B*," as if saying, "Every one of that which is qualified as *C*, always or not always, is, as long as its essence exists, *B* by necessity" ....

Another example is the statement, "Every *C* is always *B*," as if saying "Every one of the things which is *C* according to the previously mentioned manner is found to have *B* always, as long as its essence exists, yet without necessity." (Avicenna 1984, Fourth Method chapter 5)

Furthermore, he gives examples of non-necessary perpetuities (relative to the continued existence of the subject):

TEXT 7. An example of that which endures and is non-necessary is something like the affirmation or negation, applicable to an individual, [of a quality] accompanying him in a non-necessary manner as long as he exists; as you may correctly say that some human beings have white complexions as long as their essence exists, even though that is not necessary.....

You must know that what endures is other than the necessary. Thus, writing may always be negated of a certain individual at the time of his existence, let alone at the time of his non-existence, without that negation being necessary. (Avicenna 1984, Fourth Method chapter 3)

So, a singular negative proposition might hold perpetually without holding necessarily. Avicenna adds that, even if there are reasons for holding that there are no non-necessary *universal* perpetuities (i.e. if there are reasons for accepting the Principle of Plenitude), these will be non-logical reasons:

TEXT 8. Is it possible for that which is not necessary to be always present in every individual, or is it always negated of every individual? Or is this not possible, and that which is not necessary must unavoidably be negated of some others? <This> is a matter concerning which the logician need not make any judgment. (Avicenna, Fourth Method chapter 3)

His view appears to be that if logicians have to refrain from making judgments on that issue, then they have to allow for a distinction between perpetuity and necessity even in universal propositions; and accordingly they have to develop separate logics for propositions of these two types.

We may add that, in the case of conceptual truths such as 'Horses are animals', which are both perpetual and necessary, their necessity is primary, and their perpetuity derivative, both epistemically and ontologically. We know these truths are perpetual only because we know they are necessary. And this implies that what we know when we know that a proposition is necessary cannot be just that every individual of a given sort is always such as the proposition describes; rather, what we know is something concerning a relationship between two conceptions. Avicenna does not make this clear; but we shall see presently that this is the way he was understood by later Arabic logicians.

3. Formalisation

Since Avicenna presented different ideas about necessity in different works, and since there are sometimes questions about the internal consistency of what he presents in a single work, there is no straightforward answer to the question of how his logic of necessity should be formalised. Two formalisations have been published in recent times.

Wilfrid Hodges takes *The Cure* as his main text. He reads Avicenna as identifying necessity with perpetuity, and as holding that necessity-propositions quantify over individuals. He explains Avicenna's references to an underlying essence as follows:

TEXT 9. Ibn Sīnā often uses a strange phrase: 'for as long as its essence (d̲āt) is satisfied'. From his examples it's clear that he means 'for as long as it exists'. (Hodges 2014, 160)

He proposes the following formalisation, which he styles 'two-dimensional semantics' because it involves quantification over individuals and over times. Singular predications are relativised to a time: '$Bx\tau$' says that the individual x is B at time τ. There is an existence predicate: '$E!x\tau$' says that the individual x exists at time τ. Thus, the proposition 'Every B is necessarily A', understood as being relative to the essence underlying B, is true iff

(1) $\exists x \exists \tau\, Bx\tau \wedge \forall x\, [\exists \tau\, Bx\tau \supset \forall \tau\, (E!x\tau \supset Ax\tau)]$

The same sentence, understood as being relative to the description 'B', is true iff

(2) $\exists x \exists \tau\, Bx\tau \wedge \forall x\, [\exists \tau\, Bx\tau \supset \forall \tau\, (Bx\tau \supset Ax\tau)]$

(Hodges 2014, 157; notation altered). Possibility-statements are understood in an analogous way.

The present author takes *Pointers and Reminders* as his main text. Necessity-propositions involve quantification over individuals, but necessity is not identified with perpetuity. Two different necessity-operators are taken as primitive: a *de dicto* operator 'L' forming propositions from propositions, and a *de re* operator '$_L$' forming predicates from predicates: 'LBx' says that Bx is necessary, 'B_Lx' says that x is necessarily B' in the sense that B is the essence of x. 'Every B is necessarily A', understood as expressing an essential relationship, is true iff

(3) $L\forall x\, (Bx \supset A_Lx)$.

Understood as expressing a descriptional necessity, it is true ff

(4) $L\forall x\, (Bx \supset Ax)$

(Thom 2008, 286; notation altered). Possibility-statements are treated analogously. Temporal statements can be represented using term-forming operators for 'always' and 'sometimes' that are notationally distinct from the necessity-operators (Thom 2008, 284).

Both these formalisations accept the validity of the necessity/possibility syllogism mentioned earlier. Hodges does so by effectively ampliating the subject-term of a modal proposition to apply to what possibly falls under it; for the possibly-B is the same for him as the sometime-B. Thom does not directly invoke such an ampliation; instead, he requires that *de dicto* necessity obeys the principles of S5 – which indirectly achieves the same result (Thom 2008, 288).

In Thom's formalisation the notion of essence appears under the name of *de re* necessity. Hodges' formalisations deal with Avicenna's notion of an underlying essence in an indirect way, through a proxy, namely, the notion of existence. Of course, there is nothing wrong with using proxies in logical analysis, if the purpose of the analysis is to construct a formalism that makes valid just those inferences that are supposed to be valid. But achieving this purpose is not the same as giving a philosophical analysis that captures the thought of the text being analysed. Neither formalisation does that.

4. Averroes

Most Arabic logicians post Avicenna took Avicenna rather than Aristotle as the springboard for their logical reflections. Averroes (d. 1198), who lived in Muslim Spain, was an exception. He wrote numerous commentaries on Aristotle, and was harshly critical of Avicenna's logic, including the latter's treatment of necessity. Here is my translation of a passage from Averroes' *Logical Questions* (Quaesita) as printed in the sixteenth-century Latin edition:

TEXT 10. Now Avicenna was in error about this, and his error was a big one. [Averroes gives a summary of Avicenna's sixfold division of necessity-propositions.] But all this is error and perversion. For 'Man is an animal' never fails to be the case, nor will such a predication ever fail to be the case, regardless of whether or not there exists a single man necessarily and permanently. For universals do not go in and out of existence – and I mean those things from which a universal proposition is composed, as when we say 'Man is an animal'..... But I think that this man [Avicenna] held the opinions worked out under these headings because certain men had defined the necessary as that which never fails to be the case and will never fail to be the case, and they said that the absolute is what holds when something is in a subject for so long as its being is preserved, either essential (if it is not in a subject) or in a subject (if its being is in a subject). And this man said that the name 'necessary' is said equivocally of this meaning and of the absolute, since this man thought that the claimed varieties of absolute being were varieties of universal necessary propositions, and that the first definition of the necessary is 'what is always the case in the singulars', and this definition contains species of the necessary in the singulars insofar as it is universal..... And this, I believe, is the source of this man's error. But God knows the truth. (Averroes 1562, 79vb–80rb)

It is not completely clear what error Averroes is ascribing to Avicenna. What is clear is that Averroes takes essential predications such as 'Man is an animal' to concern a relationship between two universals – a relationship that holds omnitemporally, and therefore independently of the existence of individuals falling under the proposition's subject-term. By contrast, he accuses Avicenna of thinking that affirmative necessity-propositions are about the individuals instantiating the subject-term, and therefore fall short of the pure necessity which is found in God, precisely because they do *not* hold at all times.

Averroes is right about this. Avicenna does require that necessity-statements are false when their subject-terms are not instantiated:

TEXT 11. The condition is either [1] perpetual [relative] to the existence of the substance [of the subject] (*ḏāt*), as in '*Man is necessarily a rational body*'; by which we do not mean to say that man is and always will be a rational body, because this is false taken for each human individual. Rather we mean that while he exists as a substance (*mā dāma mawjūda ḏ-ḏāt*) as a human, he is a rational body. (Avicenna 1984, §264–266, Street translation; Street 2013a, §2.3.1)

But neither Hodges' nor Thom's formalisation represents Avicenna's necessity-statements in this way. In both cases the formalisation of an essential necessity comes out as a proposition that does not change its truth-value over time; if true at all, such propositions are eternally true.

Hodges' formalisation could be adapted so as to capture this feature of Avicenna's analysis, if we introduce a time-constant ν ('now'), governed by the assumptions

(5) $Bx\nu \supset \exists\tau\,(Bx\tau)$

(6) $\forall\tau\,(Bx\tau) \supset Bx\nu$.

'Every B is necessarily A' will then be true as an essential predication iff

(7) $\exists x\, Bx\nu \wedge \forall x\, [\exists\tau\, Bx\tau \supset \forall\tau\, (E!x\tau \supset Ax\tau)]$.

Thus, 'Every man is necessarily an animal' is not now true if there are now no men.

Similarly, Thom's formalisation could also be adapted so as to be false if there are no men, if we stipulate that 'Every B is necessarily A' is true iff

(8) $\exists x\, Bx \wedge L\forall x\, (Bx \supset A_L x)$.

The fundamental criticism that Averroes makes of Avicenna is that his analysis of necessity-statements runs together two features that should be kept apart: he wants essential necessity-propositions to express omnitemporal relations between essences, and he also wants them to express intermittent truths about singular things.

5. Rāzī

Working within a basically Avicennian framework, but attempting to improve what they saw as weaknesses in Avicenna's logic, a number of thirteenth-century Arabic logicians arrived at a similar criticism of Avicenna's ideas about necessity. The first of these logicians was the Persian Fakhraddīn al-Rāzī (d. 1210). Rāzī proposed a distinction between two types of predication. One of these is extensional, and is based on what exists in the external world (the *khāriji* reading); the other is intensional and is based on judgments about what would be true were certain kinds of thing to exist (the *ḥaqīqī* reading). In his *Compendium* (*Mantiq al-Mulakhkhas*) Rāzī writes:

TEXT 12. [When we say 'Every *J* is *B*'] we don't mean by it what is described as *J* externally, but rather something more general, which is: were it to exist externally it would be true of it that it is *J*, whether it exists externally or not. For we can say "every triangle is a figure" even if there are no triangles existent externally. The meaning is rather that everything which would be a triangle were it to exist would be – in so far as it existed – a figure.....

By the second reading, we mean by "every *J*" every single thing which exists externally among the individual *J*s... On this hypothesis, were there no septagons existent externally, it wouldn't be correct to say "every septagon is a figure"; if the only figures existent externally were triangles, it would be correct to say "every figure is a triangle." (Translation in Street 2013a, §2.3.2)

Intensional (*ḥaqīqī*) propositions are about relationships between conceptions; extensional (*khāriji*) propositions are about all the individuals under the subject-term.

The distinction between intensional and extensional predications was widely taken up by logicians after Rāzī.

6. Formalisation

The extensional predications can presumably be formalised in the same way as standard Aristotelian categorical propositions. 'Every *B* is an *A*' would be true iff

(9) $\exists x\, Bx \wedge \forall x\, (Bx \supset Ax)$.

It is not obvious how an extensional necessity-proposition should be represented. We will return to this question.

In order to represent the intensional predications we need a language in which we can name abstract entities. (I have been calling these entities 'conceptions'. Averroes talks instead of 'universals'.) An intensional proposition then affirms or denies a logical relation between two conceptions. There are two approaches to this.

One could require that the logical relation linking conceptions is that of inseparability ($\Leftarrow$), taken as primitive. The properties of this relation include:

(10) $\beta \Leftarrow \beta$

(11) $\alpha \Leftarrow \beta \wedge \beta \Leftarrow \gamma \supset \alpha \Leftarrow \gamma$.

The intensional proposition 'Every *B* is *A*' then says that

(12) $\alpha \Leftarrow \beta$;

the intensional 'No *B* is *A*' says that

(13) $\forall \delta\, [\beta \Leftarrow \delta \supset \neg\, (\alpha \Leftarrow \delta)]$.

The particular propositions are the contradictories of the universals of opposite quality. Given that, by (10) and (11), the universal affirmative can equivalently be expressed

(14) $\forall \delta\, [\beta \Leftarrow \delta \supset \alpha \Leftarrow \delta]$,

the universal negative denies of the subject exactly what the universal affirmative affirms of the same subject. Thus, if the two particular propositions are defined as the contradictories of the universals, the four propositions will form a square of opposition.

Now, Rāzī states that a universal negative necessity-proposition expresses the predicate's incompatibility with the subject:

TEXT 13. The meaning of 'no J is possibly B' is that J and B are mutually incompatible due to their essences (*li-dhātayhimā*). (Street 2014, 495)

But in our formalisation the universal negative proposition does not say that the predicate is incompatible with the subject. In order to say that, it would have to say that the opposite of the predicate is inseparable from the subject. In order even to raise the question whether the intensional affirmation entails that the opposite of the predicate is inseparable from the subject, our formal language would need to be expanded to include, for every conception α, a corresponding negative conception $\bar{\alpha}$. The question would then be whether (13) implies

(15) $\forall\delta\,[\beta \Leftarrow \delta \supset \bar{\alpha} \Leftarrow \delta]$.

We would have an affirmative answer if

(16) $\neg\,(\alpha \Leftarrow \delta) \supset (\bar{\alpha} \Leftarrow \delta)$.

This is a very strong requirement, stating that a pair of conceptions must be related either by inseparability or by incompatibility. The pair *round*, *white* provides an obvious counter-example. (But see Thom 2012, where principle A7 defines a class of conceptions for which (16) is true.)

It seems preferable therefore to find an alternative formalisation. Fortunately, one springs readily to mind. We can introduce a primitive relation of incompatibility ($\Downarrow$) alongside that of inseparability. The formal properties of incompatibility include:

(17) $\beta \Downarrow \alpha \supset \alpha \Downarrow \beta$

(18) $\alpha \Downarrow \beta \wedge \beta \Leftarrow \gamma \supset \alpha \Downarrow \gamma$.

The intensional universal negative necessity-proposition then says that

(19) $\beta \Downarrow \alpha$.

The particular propositions will be, as usual, the contradictories of the universals (12) and (19). However, these four propositions do not form a square of opposition: (12) and (19) are not incompatible with one another. In order to guarantee their incompatibility we have to make another assumption:

(20) $\beta \Leftarrow \alpha \supset \neg (\alpha \Downarrow \beta)$.

If we view the relations of conceptual inseparability and incompatibility as analogous to the relations that hold when one proposition, or its negation, is implied by another proposition, then we can say that the inseparability relation required by (20) is analogous to a *connexive* implication relation, i.e. an implication relation satisfying the principle known as Boethius' Thesis:

(21) If $A \rightarrow B$ then $\neg (A \rightarrow \neg B)$.

(21) is not a thesis of classical logic; but there are consistent connexive logics containing it as a thesis (Wansing 2014). Analogously, it should be possible to develop a consistent logic of the relations of inseparability and incompatibility between conceptions which contains (20).

Nonetheless, the acceptance of (21) places certain limitations on such a logic. Claudio Pizzi points out that a logic containing (21) cannot contain

(22) $(A \wedge B) \rightarrow A$

if it is to be consistent:

In fact, two instances of [(22)] would be $(A \wedge \neg A) \rightarrow A$ [and] $(A \wedge \neg A) \rightarrow \neg A$, which are jointly incompatible with [(21)] (Pizzi 2008, 143).

Adapting this point to our conceptual logic, it becomes clear that if our language includes a notation for conjunctive conceptions (say, 'β & α' for the conception of being both *B* and *A*), and if we allow the substitution of signs denoting negated conceptions (e.g. '$\overline{\beta}$') for the letters α, β, then we should not adopt as axioms the principles

(23) $\beta \Leftarrow \beta \,\&\, \gamma$

(24) $\beta \Downarrow \gamma$ iff $\overline{\beta} \Leftarrow \gamma$.

Otherwise, both

(25) $\beta \Leftarrow \beta \,\&\, \overline{\beta}$

and

(26) $\overline{\beta} \Leftarrow \beta \,\&\, \overline{\beta}$

will be theses (by (23)), where (26) implies

(27) $\beta \Downarrow \beta \,\&\, \overline{\beta}$.

(by (24)). But (25) and (27) jointly conflict with (20).

So, the adoption of (20) cannot be combined with the admission of conjunctive and negated conceptions governed by (23) and (24), on pain of inconsistency.

6.1 *Impossible Subjects*

In fact, some thirteenth-century Arabic logicians, wishing to develop a logic applicable to impossible conjunction of conceptions, adopted (20) in combination with (23). Afdaladdîn al-Khūnajī (d. 1248) and Athīraddīn al-Abharī (d. 1265) were among these. I will say something about Khūnajī. (For a discussion of Abharī, see Thom 2010.) In his *Disclosure of Secrets as to the Obscurities of Thought* (*Kashf*) Khūnajī discusses the convertibility of predicative propositional forms containing various temporal and alethic modalities (Street 2014, 454 ff). He argued that 'nothing that is at some time *J* is at all times *B*' implies 'Something that is at some time *B* is at no time *J*'. His reasoning was as follows. First, in place of the antecedent we can substitute 'Nothing that is at all times *B* is ever *J*'. Then:

TEXT 14.... If ["no always-*B* is ever *J*"] is true ..., then "some [sometime] *B* is never *J*" is true; because in this case "every always-*B* is [at some time] *B*" is true, and "no always-*B* is ever *J*", which produces "some [sometime] *B* is never *J*." (Street 2014, 469)

His reasoning takes the form:

(28) Everything that is at all times *B* is at some time *B*; nothing that is at all times *B* is at some time *J*; so, something that is at some time *B* is not at any time *J*.

But this syllogism (Felapton) deduces a particular conclusion from two universal premises, and thus, if its premises and conclusion are read intensionally, the inference is valid only if (20) is accepted.

Furthermore, Khūnajī understood the conception *moon* to be inseparable from the conception *eclipsed*; and in discussing the truth-value of 'Something that is eclipsed is not a moon', he introduced the conjunctive conception *eclipsed-which-is-not-a-moon* – a conception which (it seems) is governed by (23):

TEXT 15. So the eclipsed-which-is-not-a-moon, even though it is impossible, is among those individuals which would be eclipsed, were they to come to exist. (Street 2014, 470)

On this basis he arrives at the conclusion that, even though *moon* is inseparable from *eclipsed*, something is both eclipsed and not a moon, namely, the eclipsed-which-is-not-a-moon. Since (24) seems to be merely notational, Khūnajī seems to be committed to the combination of (20), (23) and (24) – and thus to the inconsistency presented a moment ago. My aim here is not to explore possible ways out of this inconsistency; it is only to show that some of the Arabic logicians wanted to extend their logic of conceptions to impossible conceptions.

6.2 *The Logical Relations between Intensional and Extensional Propositions*

The logical relations between intensional and extensional propositions can be determined on the basis of the following assumptions:

(29) If $\alpha \Leftarrow \beta$ then $\forall x[Bx \supset Ax]$

(30) If $\alpha \Downarrow \beta$ then $\forall x[Bx \supset \neg Ax]$

(29) provides us with a way of understanding Rāzī's idea (taken up by Khūnajī and others) that the (possible/impossible) *B*s would be *A* were they to exist. For (29) entails that if 'Every *B* is an *A*' is true (on its intensional reading) and something is *B* then 'Every *B* is an *A*' is true (on its extensional reading). This, I suggest, is what it means to say that the *B*s would be *A* were they to exist. Thus, the relations between intensional and extensional propositions is as in Figure 5.1.

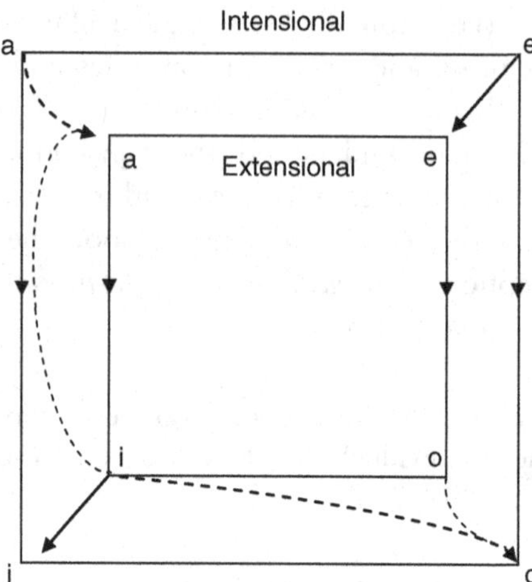

Figure 5.1 Squares of opposition for intensional and extensional predications.

6.3 The Relation between the Intensional/Extensional and the Essential/Descriptional Distinctions

On the face of it, the distinction between intensional and extensional necessities seems to cut across the distinction between essential and descriptional necessities. In order to test whether this is in fact the case, we need a language in which both distinctions are expressible. Such a language can be developed if we include a concept of essence in our formal language of conceptions. But in doing so we need to be clear that an 'underlying' essence of the proposition's subject is not evident on the surface of the proposition. It does *not* underlie the subject in the sense that it is a logical subject of the proposition's subject. That would imply that the essence is less general than the proposition's subject, whereas the essence should be *more* general than that subject. The essence underlying the conception *slave* is *human*: this is not because a human is essentially a slave, but because a slave is essentially human.

The proposition 'Men are necessarily animals', understood as being essential and intensional, states that men are necessarily animals by their essence; and the essence here is a middle term by which *man* falls under *animal*. Thus, the proposition states that *animal* is inseparable from *man*, and there is an essence intermediate between them, from which *animal* is inseparable, and that is inseparable from *man*. What the proposition states is true according to Aristotelian ontology: the intermediate essence can be *animal* itself, since *animal* is an essence and inseparability is reflexive.

The proposition 'What walks necessarily moves' can be understood as being descriptional and intensional, in which case it states that *moving* is inseparable from *walking*. There is no reference to essences here. There is no essence intermediate between *moving* and *walking*. There is an essence that is inseparable from *walking*, namely, *animal*: only animals walk, and this is a necessary truth about the Aristotelian genus *animal* according to Aristotle. But *moving* is not inseparable from *animal*: it is possible for animals to be stationary.

The two propositions just considered can also be understood extensionally rather than intensionally, in which case they make statements about the class of men or the class of walkers, rather than about the conceptions *man* and *walking*.

With these points in view, let 'E' be a primitive predicate, applicable to conceptions. The sentence '$E\alpha$' will be read as 'α is an essence'. By an essence I mean a conception which is of a kind of thing that belongs, as such, in an Aristotelian category.

Then the distinction between universal intensional predications that are essential and those that are purely descriptional can be represented as the distinction between, on the one hand,

$$(31)\quad \exists\delta\,[E\delta \wedge \alpha \Leftarrow \delta \wedge \delta \Leftarrow \beta]$$

or

$$(32)\quad \exists\delta\,[E\delta \wedge \alpha \Downarrow \delta \wedge \delta \Leftarrow \beta],$$

and (12) or (19), on the other.

(If a symmetrical semantics is required for the universal negative proposition, then it should be represented as

$$(33)\quad \exists\delta,\varepsilon\,[E\delta \wedge E\varepsilon \wedge \varepsilon \Downarrow \delta \wedge \delta \Leftarrow \beta \wedge \varepsilon \Leftarrow \alpha].)$$

The distinction between universal extensional predications that are essential and those that are descriptional can be represented as the distinction between, on the one hand,

$$(34)\quad \exists x\, Bx \wedge \forall x\,[Bx \supset \exists\delta\,(E\delta \wedge \alpha \Leftarrow \delta \wedge \delta \Leftarrow \beta)]$$

or

$$(35)\ \forall x\ [Bx \supset \exists\delta\ (E\delta \land \alpha \Downarrow \delta \land \delta \Leftarrow \beta)]$$

and

$$(36)\ \exists x\ Bx \land \forall x\ [Bx \supset \exists\delta\ (\alpha \Leftarrow \delta \land \delta \Leftarrow \beta)]$$

or

$$(37)\ \forall x\ [Bx \supset \exists\delta\ (\alpha \Downarrow \delta \land \delta \Leftarrow \beta)].$$

Given the properties of inseparability and incompatibility stated in (10), (11), (17), (18), we can see what relations of entailment hold among our eight types of universal predication. Among universal propositions, extensionals entail the corresponding intensionals; thus, (34), (35), (36), (37), respectively, entail (31), (32), (12), (19). Essential predications entail the corresponding descriptionals; thus, (31), (32), (34), (35) entail (12), (19), (36), (37). These relationships are shown in Figure 5.2.

Putting all this together, the entailment-relations among the four types of universal affirmative necessity-proposition are as follows: the extensional essential (33) entails both the extensional descriptional (34) and the intensional essential (31), and both of these entail the intensional descriptional (12). These relations are shown are shown in Figure 5.2.

7. Summary

The Arabic logicians analysed necessity-propositions by considering various sorts of things to which the necessity which these propositions express is relative. The two most important relativities are to the subject-term's underlying essence, and to the description explicitly given by the subject-term.

The post-Avicennian logicians also distinguished those necessity-propositions whose subject is picked out purely extensionally from those where the subject is identified intensionally. Some of the thirteenth-century logicians extended the application of intensional propositions to those having impossible subjects. The distinction between

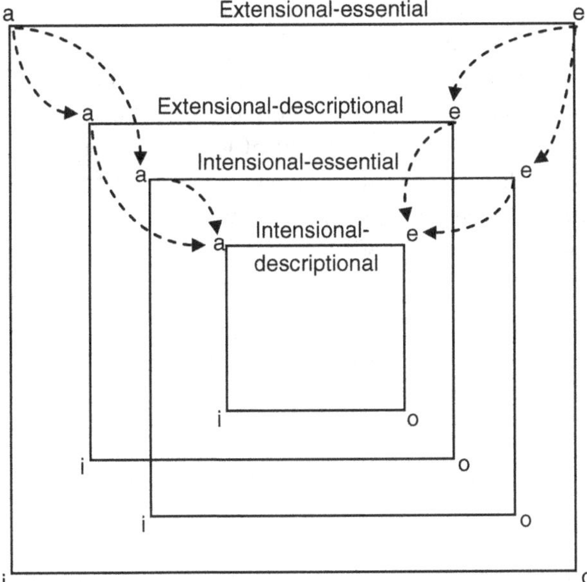

Figure 5.2 Logical relations among the four types of universal necessity-proposition.

extensional and intensional necessities can be combined with the essential/descriptional distinction.

There is no concept of purely *logical* necessity among the Arabic logicians surveyed. The necessity of the Necessary Being's existence is ontological. The necessity of essentially necessary propositions like *Man is an animal* depends on a metaphysical concept of essence. Even the necessity of a descriptionally necessary proposition like 'Whatever walks moves' derives its necessity from the essential necessity 'Walking is a species of movement'.

Telling though these points are, we should remember that *logic*, for the Arabic logicians, included the material in Porphyry's *Introduction to the Categories*, and thus included the study of essential predications such as those that predicate a genus of its species. The concept of essence, for them, was a logical concept, and they frequently included a discussion of essence and the essential early on in their logic books (Avicenna 1984, §151; Avicenna 2011, §7). The fact that this never happens in modern logic textbooks is an indication of how the idea of logic, and that of logical necessity, have changed since the time of Avicenna, Averroes, Abharī, and Khūnajī. The Arabic logicians may not have developed a theory of logical necessity as it is specifically exemplified by valid logical consequences and more generally by logical truths; but they certainly had the concept of logical consequence, and they knew how to demonstrate that a given consequent followed with necessity from a given set of antecedents – as any reader can see by inspecting the working of their logical proofs (Street 2002; Thom 2012).

6 *Modality without the* Prior Analytics: *Early Twelfth Century Accounts of Modal Propositions*

CHRISTOPHER J. MARTIN

Peter Abaelard's birth around 1079 coincided with the promulgation by Pope Gregory VII of a decree instructing his bishops to ensure that all clerics were trained in the liberal arts, and Abaelard's life was shaped by the consequent increase in the number and importance of teaching masters. These early twelfth century teachers commented on the works of what would later be called the *logica vetus*, including from Aristotle only the *Categories* and *De Interpretatione*. Abaelard also had some access to the *Sophistical Refutations* and, indeed, seems to have known a little about the *Prior Analytics*. The latter work, however, played no role in the development of thinking about modal propositions in the period to be considered here. Rather it was Aristotle's remarks on the proper placing of the negative particle in forming the contradictory of a given modal proposition in *de Interpretatione* 12, and 13, and a distinction made by Boethius between different forms of modality in his *De Syllogismis Hypotheticis* that prompted early twelfth century philosophers to investigate the nature of modal propositions.

1. Ancient Sources

The general problem for the *De Interpretatione* is to characterise the logical relationship of opposition in truth-value. Aristotle has no notion of negation as a propositional operator and proceeds, rather, syntactically, to consider where the negative particle is to be inserted into each form of categorical proposition to produce the desired opposite. This leads him in chapters 12 and 13 to the question of the proper negations of, and logical relations between, propositions involving nominal

Work on this study was supported by an award from the Marsden Fund Council administered by the Royal Society of New Zealand.

modes. For the Latin writers we are to consider here these are propositions such as '(For) Socrates to be a man is possible' ('*Possibile est Socratem esse hominem*') and '(For) Socrates not to be a stone is necessary' ('*Necesse est Socratem non esse lapidem*').[1] Aristotle holds that negation is an operation on terms which takes the narrowest possible scope. For modalities this entails, he concludes, that it must be applied to the modal adjective to yield, for example, 'not necessary to be' as the negation of 'necessary to be'. In chapter 13 of *De Interpretatione* he summarises his results in a set of four tables of equipollent combinations of mode and quality which were the starting point of early twelfth century discussions of modal sentences. Aristotle's presentation is almost entirely schematic with no indication of whether there is any difference between the application of modal determinations to singular and to quantified propositions.

Before turning to Aristotle's text Boethius begins his longer commentary on *De Interpretatione* 12 with an extensive discussion of the negation of modal propositions.[2] He observes that modal adverbs such as 'rapidly' in 'Socrates walks rapidly' apply directly to the verb as predicate to indicate its 'quality' but notices no difference between them and the application of nominal modes to produce propositions such as '(For) Socrates to be a man is possible'. In the case of the latter he follows Aristotle and takes the mode to be the predicate and the infinitive in the subordinate clause to be the subject of the modal proposition. Like Aristotle, too, Boethius has no notion of propositional negation.[3] Just as the negative particle is applied to the mode in the case of adverbial qualifications, so, too, he insists, it is applied to a nominal mode.

While Boethius does little more than rehearse Aristotle's construction of schematic tables of equipollent modals in his commentary on *De Interpretatione* 13, in his own treatise on the hypothetical syllogism[4] he introduces a form of modal proposition not mentioned

[1] In what follows I will where possible give English versions of the sentences under consideration, but the original Latin should be easily recoverable from these examples. 'For' is needed to yield something that looks like colloquial English but perhaps has a tendency to suggest that the sentences are to be construed *de re*. They are not; rather the formulation is intended to leave the range of possible readings quite open.

[2] Boethius 1877.

[3] See Martin 2009: 56–84.

[4] Boethius, 1969.

by Aristotle and an important distinction. Modal propositions may, he claims, take three forms, two in which the modality is temporally qualified and a third in which it is predicated without qualification. The temporal qualification is indicated with the conjunction '*dum*,' which I will translate as 'while'. It may mean simply 'now', as in English present progressive constructions, such as 'Socrates is whistling while he is working', or habitual association, as in the English gnomic present 'Socrates whistles while he works'. The discussion in the twelfth century of such constructions anticipates the recent development of what has been characterised as hybrid logic with indexicals – the combination of tense and modality with so-called nominals such as 'now' to fix temporal reference rigidly.[5]

Boethius distinguishes the three 'grades' of necessity and the corresponding possibility, using nominal rather than adverbial modes:

(**Mod1**) The temporally qualified clause merely repeats the main clause as in '(For) Socrates to sit is necessary while he sits' and '(For) Socrates to sit is possible while he sits'. Since Boethius holds that such modals are equivalent to the simple categorical 'Socrates sits' the sense of 'while' here must include now.

(**Mod2**) The temporally qualified clause is different from the main clause, as in '(For) a man to have a heart is necessary, while he is alive' and '(For) Socrates to read is possible while he exists'. Boethius ambiguously glosses such sentences as asserting that the necessity or possibility holds 'just so long as'[6] the condition is satisfied, leaving it open whether it is presently satisfied.

(**Mod 3**) Finally, modal claims may involve no temporal determination – as '(For) God to be immortal is necessary'.

2. Eleventh and Twelfth Century Developments

It is difficult to date texts from this period precisely but we do have two closely related discussions of modal propositions, designated by Yukio Iwakuma as *M1* and *M3*,[7] preserved in manuscript Orléans

[5] For a survey of hybrid logic see Braüner 2006. On the addition of indexicals to hybrid logic see Blackburn and Jørgensen 2012: 144–160.

[6] '*quamdiu*'.

[7] These are not the only texts which included discussions of modal propositions from this period, but they are the most extensive.

266, with *M1* also found in Paris BN Lat 13368.[8] These must report something of the debates about such propositions, which took place in the first decade of the twelfth century since they refer to the opinions of Masters W., Gos., and Gosl, presumably William of Champeaux, who had left Paris by 1113, and certainly Goselin of Soissons, who remained there until around 1115. *M3* seems to be the continuation of the commentary on *De Interpretatione* which immediately precedes it in the manuscript and in which there are also references to Master W., Master G., and Master Gos., who at one point is dignified as *Magister noster*. Likewise *M1* is found in both of the manuscripts in which it is preserved immediately after the *de Generibus et Speciebus*,[9] which also contains references to both Master G. and Master W. In addition to these texts we have the *Dialectica* of Garlandus Compotista: a slightly earlier work, which develops logic in the anti-realist fashion associated with Abaelard's first master, Roscelin. Together they give us some idea of the theories of modal propositions current at the beginning of the twelfth century to which Abaelard is responding in his *Dialectica*.[10]

Garlandus's *Dialectica* contains only a brief discussion of modal propositions but it goes well beyond Boethius. Nominal modes are contrasted with the corresponding adverbs and Garlandus notes that modality lies properly in the modification of a predicate. Nominal modes, he tells us, may be 'resolved' into the corresponding adverbial forms with '(For) Socrates to read is necessary', for example, equivalent to 'Socrates necessarily reads'.

While Boethius employs only schemas in discussing modality Garlandus applies modes to quantified categorical propositions and provides lists of modal equipollents for both singular and quantified modal propositions. He goes on to list the various forms of temporally qualified modality repeating what Boethius had said about them but then goes beyond him with a new classification of the varieties of possibility:

(1) Possibility which is actual – *it is possible for a man to be an animal*
(2) Possibility which is not actual

[8] *M1* = Orléans, Bibl. Municipale, 266: 166a–169a; Paris, Bibl. Nationale, lat. 13368, 175va–177ra. M3 = Orléans, Bibl. Municipale, 266: 252b–257b.
[9] See King 2015.
[10] Garland 1959.

(a) which actually occurs after the time at which it is possible
  (i) follows necessarily – *it is possible for the sun to set*(ii) does not follow necessarily – *it is possible for Socrates to become pale*
  (b) which does not occur after the time at which it is possible – *it is possible for Garland to become a bishop.*

Here we meet for the first time a possibility which will in some cases never be actualised but which will attract many twelfth century logicians, the possibility of becoming a bishop, which Garlandus characterises as that 'to which nature is not repugnant but which will never come about'.

The Orléans treatises, like a number of texts from this period, are very closely related though *M3* is considerably longer than and contains much material not found in *M1*. When they do deal with the same issues the two texts often have the same structure, the same arguments, and the same wording although they also contain different arguments and formulate their claims in different ways.

In these texts we soon meet again the possibility of being a bishop, this time attributed to a peasant. We are first introduced to nominal and adverbial modes and told that propositions containing the latter differ from the corresponding simple categoricals only in that they cannot be converted by the non-Aristotelian procedure of term contraposition introduced by Boethius,[11] to convert '*S is P*' into '*not P is not S*'. This we are told is because, as Aristotle insists in *De Interpretatione* 12, we cannot negate, or 'infinitize,' a phrase.

Both *M1* and *M3* go on to note, following Boethius, that an adverbially modified term stands to the unmodified term as a part to a whole. We may thus appeal to the topical *locus a toto* warranting an inference from part to whole, to conclude that a predicate holds without modification of a subject if the corresponding modified predicate holds of it: to infer, for example, from the fact that Socrates is running rapidly that he is running. The solution to the puzzle that we may thus prove that our peasant is a bishop is, we are told, to distinguish between adverbs which modify both wording and sense and those which modify only the wording of the simple propositions to which they are added. To modify both word and sense the application of an adverb to

[11] In Boethius 2008.

a term must *restrict* its extension to only part of that of the unmodified term. It must modify it, that is, in just the way required for the *locus a toto* to be applicable. The kind of adverb, on the other hand, which *M1* and *M3* characterise as a modifier 'only of a word' does in fact, according to them, also modify meaning. Adverbs used in this way may, like 'possibly', *ampliate* the terms to which they are applied, or transform their meanings into the opposite, as in the true proposition 'a stone impossibly lives'[12] where the extension of 'lives' is modified to include only those things which cannot possibly be alive.

The inference from 'a peasant is possibly made a bishop' to 'a peasant is made a bishop' thus fails because '*possibly*' ampliates the predicate verb to produce an expression which applies to both bishops and non-bishops. This solution according to *M1* was not accepted by Master Goselin, referred to in the present tense, who argued rather that adverbial modals should be expounded using the corresponding nominal modes as having the *figurative* sense of '(For) a peasant to be made a bishop is possible' from which it does not follow that the peasant is a bishop – presumably, though we are not told this, because the possible truth of a proposition does not entail its actual truth.

This solution is apparently the one accepted by *M3*. As an alternative, *M1* tells us, we may follow Master W., who retained the adverbial formulation of the claim, but proposed that here the mode functions to form an expression that signifies a single concept applying equally to actual and potential bishops, but for which no name has been introduced. Taking the expression in this sense we may thus infer that someone who is a bishop is possibly a bishop but not conversely.

Neither *M1* nor *M3* has any more to say about the logic of adverbial modals, which, as we saw, reduces for them essentially to that of simple categoricals. Nominal modes, they agree, are quite different and they suggest that it was the problem of establishing their logical relations that led Aristotle to pay them special attention. The crucial question here is how we are to understand properly the various ways in which quality, quantity, and mode may be combined in forming such propositions. Both *M1* and *M3* give the schematic tables of equipollence for nominal modes and hold that they are satisfied for all permissible substitutions of quantifiers. The discussion of simple modals in *M1* is limited to the question of the proper interpretation of the interaction

[12] '*lapis impossibiliter vivit*'.

of negation and quantity in such sentences and we will return to that in a moment. *M3*, on the other hand, has much more to say about the nature of such propositions.

First there is the general question of the grammatical structure of a categorical modal proposition such as '(For) every man to be an animal is possible'. According to Aristotle the subject (**SUBJ**) of such a sentence is 'to be an animal' and the predicate (**PRED**) is the mode 'is possible'. The quantified term 'every man' is taken in our two texts to function as a determination (**DET**) of the grammatical subject. The proposition thus has the grammatical structure:[13]

(For) [(*every man*)**DET**(*to be an animal*)**SUBJ**,(*is possible*)**PRED**].

According to *M3* all such propositions are grammatically indefinite but since Boethius assigns them different quantities we must do so too and *M3* and *M1* agree that the quantity to be assigned to a nominal modal is that of the unmodalised simple categorical of whose nominalisation the modal adjective is predicated. '(For) every man to be an animal is possible' is thus universal and, since its quality is that of the copula, affirmative.

M1 and *M3* agree that, save for the failure of term contraposition, adverbial modals convert in just the same way as simple categoricals, but according to *M3* every form of conversion fails for nominal modals. It first gives a series of arguments for these failures for each form of conversion, which it then rejects but which we will meet again in Abaelard's discussion of nominal modes. For example, against the simple conversion of nominal modals *M3* tells that '(For) no dead (thing) to be a man is possible' is true but '(For) no man to be dead is possible' is false, apparently reading the propositions *de re*. Later, however, it denies this and claims that the second proposition, like the first, is true, now apparently reading them *de dicto*. While these pairs of propositions are thus not counter-instances to the principles of conversion, according to *M3* there are others which are. For example, against simple conversion '(For) no old man to be a young man is possible', which *M3* says is true, and which is so read either *de re* or *de dicto*, and '(For) no young man to be an old man is possible', which *M3* says is false, which it is if it is read *de re* but not if it is read *de dicto*.

It is thus not clear from its examples whether *M3* understands the logical relations between nominal modals to be those between

[13] [(*omnem hominem*)DET(*esse animal*)SUBJ,(*est possibile*)PRED].

the corresponding adverbial modals, and so reads them *de re*, or whether it reads them *de dicto*. The only explicit remarks on the sense of nominal modals in the text indicate the latter, but these present us with a serious puzzle about one of Abaelard's references to his master.

Having dismissed the proposal that with nominal modal propositions a modality is attributed to individual things, to words, or to the corresponding understandings, *M3* turns to the suggestion that such modals are equivalent in meaning to the corresponding adverbial forms, a position which it attributes to Master W., referring to him in the imperfect tense:

> Master W. in fact said that the common sense of all modals with a nominal mode is that they should be expounded by those with an adverbial mode, just as this: '(For) Socrates to be a man is possible', that is 'Socrates is possibly a man'.

M3 goes on to offer a number of objections to this position, which make it clear that it really is the proposal to explicate the sense of nominal modals as that of the corresponding adverbials. Against it *M3* sets down its own position:

> It seems to us that the sense of modals should be expounded in another way, when we say '(For) Socrates to be a man is possible' or '(For) Socrates to read is possible', we are not saying that Socrates is possible, or that Socrates' reading is possible, or that the words or understandings are possible, but rather that the sense *Socrates to read* is possible, that is it is possible that Socrates reads.

The position attributed here to Master W. is that insisted on by Abaelard in his *Dialectica* in opposition to the theory that he attributes to his master ('*magister noster*'), which is, if not identical with, very close to the position which *M3* says is its own and perhaps is also that mentioned previously by *M1* as held by Goselin of Soissons. One further complication, though not so embarrassing for the standard identification of references by Abaelard to *magister noster* with William of Champeaux, is found in *M3*'s discussion of the signification of 'possibile'. *M3* rejects the suggestion that 'possible' signifies a property, *possibleness*, which inheres in its bearer, apparently understood to be a state of affairs. In response to the arguments against this account of modality, it tells us:

> Master W. (referred to in the imperfect tense again) expounded these propositions in a negative way: as in the case of '(For) Socrates to be an animal is possible', that is, it is not repugnant to the nature of the thing that Socrates be an animal.

We have been taught to associate this account of possibility in terms of non-repugnance with nature with Abaelard[14] but as we have seen was already employed by Garlandus and indeed it seems to be the standard view at the beginning of the twelfth century.

As I noted above, most of the treatment of nominal modals in *M1* and a substantial part of that in *M3* is concerned with the question of what quantity and quality they should be assigned. These discussions are particularly interesting because there seems to be no awareness at all in them that negation may be propositional, or as Abaelard says 'extinctive',[15] yielding for any proposition another with the opposite truth-value, rather than predicative, Abaelard's 'separative' negation, serving to indicate the separation of the predicate from the subject of a categorical proposition.

Both *M1* and *M3* maintain that the negative particle 'not' apposed to the nominalised quantifier must modify the quantity of the subordinate clause, and so of the whole proposition, but also that it must transform either the infinitive or the finite verb into the opposite quality. Such propositions, it follows, may be assigned different qualities depending on how they are construed.

Here in order to see the point being made it is better to leave our sentence in the regimented Latin of our texts. In the case, for example, of '*non omnem hominem esse animal possibile est*'[16] the application of '*non*' to the universal quantifier '*omne*' results in a proposition whose quantity is particular. If the negative particle is construed as determining the infinitive, the resulting proposition is affirmative '*non omnem hominem esse animal est possibile*'. If it is construed as determining the copula as '*non omnem hominem esse animal est possibile*', it is

[14] See Knuuttila 1980: 163–257.
[15] See Martin 2004: 158–199.
[16] Literally 'Not every man to be an animal is possible'.

negative. A distinction of scope, then, but one according to which the negative particle must be construed in addition to modifying the quantity as modifying either the nominalised expression which is the subject of the modal proposition or the main verb of the modal proposition as a whole. *M1* goes on to note that the application of the negative particle to the main verb alone also modifies the meaning of the nominalised quantifier. Thus '*omnem hominem esse animal non est possibile*' is negative because the negative particle is applied to the main verb and particular because it must also affect the quantity:

> We say that it is rendered particular in sense because just as we say that when '*non*' is preposed to the sign it is referred to a verb, so conversely, the negative particle conjoined with '*est*' is referred back to the sign to be cancelled.

Applying these principles our two texts go on to discuss the quality and quantity of quantified modals with two negative particles but no more since according to *M3*:

> To appose three negations would make no sense and have no reason, as in the case of: '*non omnem hominem non esse animal non est possibile*'.

There is thus again no hint that negation may be propositional.

M1 and *M3* complete their discussions of modal sentences by considering the temporal qualification of modality, and for this they now have a technical vocabulary. Composite modal propositions are formed with qualifying terms and the one of most interest to our authors is '*while*' as used by Boethius to distinguish the two forms of conditioned modality mentioned above as (**Mod1**) and (**Mod2**). Such propositions are referred to as *determinate modal propositions* and have the general form **mod**(*A*)**det**(*B*). Where '*A*' and '*B*' are the same, as they are in propositions of type **Mod1**, the determination is said to be *intrinsic*. Where they differ in any way, as in type **Mod2**, the determination is *extrinsic*. For any determinate modal let us call '*A*' the modalised proposition, '*B*' the determining proposition, '**mod**(*A*)' the modal proposition, and '**det**(*B*)' the determination.

Prior to any further analysis determinate modal propositions are characterised as categorical since modality is a modification of predication. Their basic logical properties according to *M1* and *M3* are:

(**det 1**) so long as the modal proposition does not involve negation a determinate modal is true only if its determining proposition is true.

(**det 2**) intrinsically determinate modalisations of the same simple proposition which do not include negation, for example '(For) Socrates to sit is possible while he sits', '(For) Socrates to sit is necessary while he sits' and 'Socrates sits', are equipollent, i.e. identical in truth-value.

We will return to the problem of negation in a moment.

Having introduced determinate modals as categorical propositions our two texts note that they may in fact be expounded in two different ways. The basic structure of such sentences is determined by Aristotle's remark that in nominal modals, the infinitive is the subject and the mode the predicate. Determinate nominal modals thus have a grammatical structure that we may represent as follows, for example, for '(For) Socrates to be a man is necessary while he is a man':

(For) $[(\textit{Socrates})_{\mathrm{DET}}(\textit{to be a man})_{\mathrm{SUBJ}},(\textit{is necessary})_{\mathrm{PRED}}]$ $(\textit{while he is a man})_{\mathrm{det}}$

Remembering that their basic logical property is that where the modality is affirmative the truth of the determination follows from the truth of the whole proposition, there are, according to *M1* and *M3*, two possible readings of determinate modals, depending on whether the determination is construed as applying to the subject or to the predicate. If it is to the subject, the result is what *M1* calls a ***modal categorical***:

(For)$[(\textit{Socrates})_{\mathrm{DET}}(\textit{to be a man})_{\mathrm{SUBJ}}(\textit{while he is a man})_{\mathrm{DET}},(\textit{is necessary})_{\mathrm{PRED}}]$,

which is construed as the claim that Socrates is such that it is necessary that now, while he is man, he is a man; i.e. he is such that it is not possible that he is now both a man and not a man.

Alternatively the determination may be construed as applying to the predicate and expounded as what our texts call a ***temporal hypothetical***,[17] where 'hypothetical' means no more than compound:

(For) $[(\textit{Socrates})_{\mathrm{DET}}(\textit{to be a man})_{\mathrm{SUBJ}},(\textit{while he is a man})_{\mathrm{DET}}(\textit{is necessary})_{\mathrm{PRED}}]$

[17] For the treatment of temporal hypotheticals in later mediaeval logic by Ockham, Burleigh, and Buridan see Øhrstrøm 1982: 166–179.

In this case '*while*' is treated as a propositional conjunction and the resulting compound proposition might be more clearly expressed with 'now, while Socrates is a man, it is necessary that Socrates is a man'.

The modal categorical is true as long as Socrates is a man but the temporal hypothetical is false since while Socrates is a man we may detach the consequent to conclude that without qualification he is necessarily a man, a claim about all future times according to our texts and so false.

M3 but not *M1* insists that the proper way to read determinate modals is as categorical propositions with the determination applying to the subject and goes on to discuss the sense of '*while*' in such propositions. Given that we can infer the truth of the determining proposition it must include the sense of 'now'. *M3* notes, however, that if we are to agree with *The Book*, i.e. Boethius, we should include in the sense of '*while*' the temporal coincidence at any time expressed by 'whenever', which he employs in expounding the second type of conditioned necessity. *M3* argues, however, that if we include such temporal coincidence in the sense of '*while*', true claims about possibility will turn out to be false, for example, '(For) today to be is possible while the sun shines', where 'today' is, as we would say, rigid. The solution is to distinguish the use of '*while*' in formulating determinate possibilities, where its meaning is simply that of '*now*', and its use in determinate necessities, where its sense includes that of '*whenever*', and, since *M3* insists, '*whenever*' does not include the present instant, also the sense of '*now*'.

The problem with determinate modals involving negation according to *M1* is that if the relations of equipollence which hold for simple categoricals also hold for determinate modals, then what we may call the **Equipollence Argument** could be employed to prove the impossible, that Socrates is a stone.

As it appears in *M3* the proof is:

(**E*1**) '(For) Socrates to be stone is possible while he is a stone' is false, because the determination is false. So
(**E*2**) '(For) Socrates to be stone is impossible while he is a stone' is true since what is not possible is impossible. So
(**E*3**) '(For) Socrates not to be a stone is necessary while he is a stone,' is true if the modalisations are equipollent.

It follows that Socrates is a stone.

M3 notes that because of this argument some people conclude that equipollence fails for determinate modals. Others attempt to maintain that equipollence remains but since they construe the determination as applying to the predicate of the determined proposition, and so treat the determinate modal as a temporal hypothetical, they cannot avoid the same conclusion. *M3* itself holds that the argument fails because the determination must be taken to apply to the subject of the modal proposition. The negation understood predicatively thus, as it says, serves to remove 'to be a stone while he is a stone' from Socrates.

3. Abaelard on Simple and Determinate Modalities

Abaelard's views on modal propositions are found in the *Dialectica*[18] written around 1112 and reflecting the debates in which he had been involved in Paris in the preceding decade and in the commentary on *De Interpretatione* in his *Logica 'Ingredientibus'*[19] written some years later.

Aside from the reference to Master Goselin's position in *M1* the Orléans texts do not discuss the reduction of nominal to adverbial modes or indicate whether they think that the adverbial formulation is in some way prior to the nominal one. This claim is, however, central to Abaelard's account of simple modals in his *Dialectica*, where much of the discussion is concerned with the refutation of the alternative position, which he tells us was held by his master. Abaelard begins by distinguishing between the *grammatical*, or *surface structure*, of modal propositions and their sense (*sensus*), what we can call their *semantical structure*. He insists that only adverbial constructions are properly modal and that nominal modals simply employ an alternative grammatical structure to express the same semantical structure, the same qualification, that is, of the same semantical subject and predicate. Thus the meaning of '(For) Socrates to be a bishop is possible' is that Socrates is possibly a bishop, or, better, that Socrates is able to be a bishop, a claim about him which Abaelard expounds negatively in terms of its non-repugnance to his species nature:[20]

[18] Abaelard, *Dialectica*, Tract. II.2: 191–210.
[19] Abaelard, *Glossae*.
[20] Abaelard, *Dialectica*, p. 193.

'When we say '(For) Socrates to be a bishop is possible', even if this is never so, it is nevertheless true since his nature is not repugnant to <his> being a bishop; which we conclude from other individuals of the same species, who we see to actually participate in the property of <being a> bishop'.

This is precisely the same exposition of the nominal modals as that attributed in *M3* to Master W.

Given Abaelard's exposition, the truth-values of nominal modals and their logical relations are just those of the adverbial modals with the same semantical structure. The logical properties of '(For) every man to be white is possible' are thus exactly the same as those of 'Every man is possibly white'. The grammatical structure of adverbial modals does not, however, according to Abaelard, necessarily mirror the true semantical structure of the intended claim. If it did, we would be able to infer from Socrates' possibly being a bishop that he actually is one. The inference, however, fails, he tells us, for just the reason reported by *M1* to have been given by Master W.

Immediately after defusing the objectionable inference from part to whole, Abaelard goes on to report the alternative account of nominal modals, which he tells us was held by his master, referring to him in the present tense:[21]

It is our master's theory that modals descend from simple categoricals in such a way as to be concerned with the sense (*de sensu*) of <these categoricals>, so that when we say '(For) Socrates to run is possible, or necessary' what we are saying is that what the proposition 'Socrates runs' says to be so is possible or necessary.

Abaelard characterises his master's theory here as holding that modality applies to the sense of the nominalised categorical and although he refers on a couple of occasions in the discussion of modals in the *Logica 'Ingredientibus'* to modal claims understood '*de dicto*', his preferred formulation is that this view construes modality '*de sensu*'. According to it, what is possible or necessary is the state of affairs which the non-modalised categorical proposition assets to be so, apparently the position which we saw *M1* attribute to Master Goselin in contrast to that held by Master W.!

21 Abaelard, *Dialectica*, p. 195.

Abaelard criticises his master for assigning the wrong quality, the wrong truth-values, and the wrong logical relations to nominal modals. '(For) no man to be white is possible', for example, according to the *de sensu* theory is, Abaelard insists, an affirmative non-modal proposition rather than a negative modal one, and '(For) a man to be dead is possible', expounded as 'what the proposition "a man is dead" says is possible' is false rather than, as he claims it should be, true. This is so, of course, only if one assumes that the *de re* reading of nominal modals is the correct one.

It is not at all clear in fact that Abaelard's master had a *theory* of modal sentences since Abaelard's most powerful objection is that his treatment of conversion is inconsistent because he argues for his conclusions on the basis of the *de re* reading of modals, ignoring the fact that these do not agree with his official *de sensu* readings. In effect Abaelard charges his master with failing to distinguish the two readings, a shortcoming of which, as we have seen, *M3* also appears to be guilty.

The conversions considered by Abaelard are those initially given by *M3* to illustrate the failure of its various forms but which *M3* eventually rejects. According to Abaelard the advocates of the *de sensu* theory did indeed maintain that all forms of conversion fail but according to their own account they should have accepted some at least of those which they rejected. They claim, for example, he tells us, that '(For) no dead (thing) to be a man is possible' is true but what they take to be its simple converse '(For) no man to be dead is possible' is false, just as the *de re* reading of such sentences has them. Read *de sensu*, however, since 'No man is dead' and 'No dead (thing) is a man' are logically equivalent the two modal propositions must have the same truth-value.

The advocates of the *de sensu* theory, according to Abaelard, also had an argument to show that term-contraposition fails for nominal modals, an argument which is given in *M3*, although again it is rejected there. Abaelard insists, however, that according to the *de sensu* theory equivalence must be accepted here since it holds for the simple categorical from which the modals descend. The objection to it, according to the *de sensu* theorists, is that while '(For) every non-stone to be a non-man is possible' is true, its converse by term-contraposition '(For) every man to be a stone is possible' is false. Their argument for the truth of the former is interesting and in the *Dialectica* Abaelard does not indicate what he thinks might be wrong with it. Let us call it the **Contraposition Argument**. Both

(1) '(For) every man to be a non-man is possible' and
(2) '(For) every non-man to be a non-man is possible' are true.

(2) is obviously so and (1) is so, Abaelard tells us, they claim, because every man will be a non-man, and 'what is future is possible'. This relies on an appeal to the *de re* reading, but exactly the same justification seems to be given by *M3*. Since a non-stone is either a man or a non-man it follows that (3) '(For) every non-stone to be a non-man is possible' is true. For Abaelard's account of the flaw in the argument we have to turn to the *Logica 'Ingredientibus'*.

After working through the orders of equipollence and entailment for quantified modals, and discussing the signification of modal adjectives, where he again rejects the *de sensu* theory that modal claims are simple categorical statements concerning states-of-affairs, Abaelard moves on to discuss determinate modals, as in the Orléans texts he distinguishes between their exposition as modal categoricals and as temporal hypotheticals. His solution to the Equipollence Argument in the *Dialectica* is the same as that in *M3* save that he explicitly construes the negation as what he elsewhere calls *extinctive*, that is to say, propositionally. Here again the point being made is clearer if the example is left in Latin:[22]

> When we say that if '*possibile est Socratem esse lapidem dum est lapis*' is false, then '*non possibile est Socratem esse lapidem dum est lapis*' is true, with the negative particle we extinguish the whole sense of the proposition.

Rather than considering what he has to say about determinate modals in the *Dialectica*, however, let me turn to the *Logica 'Ingredientibus'*, where Abaelard shows what is wrong with the Contraposition Argument and what is right with the Equipollence Argument.

Abaelard's discussion of modal propositions in the *Logica 'Ingredientibus"* takes the form of short treatise on the subject placed before his glosses on *De Interpretatione*, 12, in which the material is organised in the same way as it is in the *Dialectica* and the Orléans treatises. Abaelard retains here the distinction between grammatical and semantical structure, continuing to insist that only adverbial modals properly represent the semantical structure of modal claims,

[22] Abaelard, *Dialectica*, p. 208.

and indeed only some of them, since 'possibly', for example, is not semantically a mode, and the sentence 'Socrates possibly runs' must be expounded as 'Socrates is able to run'. New in the '*Ingredientibus*' is an account of the proper way to form the various converses of adverbial and nominal modals with a 'that' clause (*quod*). '(For) no body to be a man is necessary', for example, converts simply with 'Nothing that is (such that it is) necessary (for it) to be a man is a body'. With this understood the principles for the conversion of modals are just those for the conversion of simple propositions.

Abaelard no longer refers to his master in the discussion of modals in the '*Ingredientibus*' and the Contraposition Argument[23] which was associated with him in the *Dialectica* is not in the '*Ingredientibus*' invoked as part of the criticism of the *de sensu* exposition of such propositions, but rather analysed as involving a quite general mistake. The argument, Abaelard presumably realised, applies equally well against the *de re* account of nominal modals and so must be refuted rather than invoked to criticise the *de sensu* theorists' exposition of them. It is an instance of a fallacy which may be equally well committed in converting simple categoricals. Whereas in the *Dialectica* Abaelard's complaint was that the advocates of the *de sensu* theory did not recognise that they were committed to the logical equivalence of term-contrapositives, here he complains only that those who appeal to this argument have not properly formulated the contrapositive of '(For) every non-stone to be a non-man is possible', which should rather be 'Every thing that is (such that it is) not possible (for it) to be a non-man is a stone', just what is required by the *de re* theory. In this form, however, it is false, while the argument has apparently shown that the original sentence is true. The problem Abaelard realises is that in Aristotelian syllogistic all quantification carries an existential presupposition. But in this case the predicate of the converted sentence, in *de re* terms 'able to be a non-man', has universal extension and so when contraposed it is empty. The inference fails, he tells us, in just the way that the inference from 'every non-human is a non-rose' to every 'rose is a human' fails in the only circumstance in which the premiss can be true, i.e. when there are no roses.

[23] For a discussion of this argument in the *Logica* '*Ingredientibus*' see Astroh 2001.

Although Abaelard does not mention his master here, he does introduce the theory associated with him in the *Dialectica* in one of the two passages in which he refers to the *Sophistical Refutations*. Having established that modals convert in the way that simple categoricals do, assuming throughout the *de re* exposition of them, he notes that there is an alternative to this. The *de re* exposition corresponds, he proposes, to what Aristotle calls the *per divisionem* reading of a proposition. In the case of '(For) a standing man to sit is possible', for example, attributing the ability to sit to someone who is presently standing. Alternatively the proposition may be read *per compositionem*, or *coniunctionem*, to yield the *de sensu* exposition 'It is possible that what the proposition "a standing man is sitting" says comes about'. As in the *Dialectica* Abaelard rejects this exposition as not properly modal. He goes on, however, in the '*Ingredientibus*' to investigate in some detail the properties of impersonal modal propositions, noting that there are some which cannot be resolved personally, and so *de re*, but must be construed *de sensu*, and when they are, they convert in that we may substitute for the subject proposition the equipollent or implied converted proposition: again just the position which he had insisted on in the *Dialectica*.

The logics of both *de re* and *de sensu* propositions are, Abaelard argues, perfectly straightforward and, *pace* the objections of various modern commentators, he correctly formulates the relations of equipollence and implication for both of them.[24] When he arrives finally at the mixed logic of *de re* and *de sensu* modals Abaelard notes, however, that if we persist in expounding propositions *de sensu* by predicating the mode of the corresponding simple proposition, a general principle that the *de re* exposition of modal proposition follows from its *de sensu* exposition fails. For example, the proposition '(For) my son to be alive is possible' is false, Abaelard says of himself, expounded *de re*, because he has no son, but true *de sensu* because the proposition that his son is living may at some time be true. While noting that it perhaps does not matter that this principle does not hold Abaelard goes on to offer an alternative account of the *de sensu* exposition for which it does. This version of the *per coniunctionem* reading attributes to an existing subject the ability simultaneously to possess the features attributed to it in the unexpounded modal proposition in contrast to the *per divisionem* reading, which attributes to the subject characterised as

[24] See Martin 2001.

possessing one feature the ability to possess another. The difference is that between 'the standing man is able to be seated while remaining standing' and 'the standing man is able to be seated'. Read in this way, Abaelard claims, the *de sensu* exposition of any modal implies the *de re* exposition but not vice versa.

Rather than going into any more detail here or considering Abaelard's account of determinate modals, the final third of his treatise on modality, let me finish by saying something briefly about his resolution there of the Equipollence Argument.

In the version of the argument that Abaelard considers the problematic proposition is

(E**2) '(For) Socrates to be a stone is not possible while he is a stone'.

Its truth is inferred from the falsity of

(E**1) '(For) Socrates to be a stone is possible while he is as stone'.

Depending on whether this latter proposition is construed as a temporal hypothetical or as a categorical modal, Abaelard notes, there are three different ways in which we may construe (E**2). As an affirmative temporal hypothetical the consequent of which is the negation of the consequent of the original falsehood

(E**2a) While (Socrates is a stone), (for Socrates to be a stone is not possible); as a 'negative' temporal hypothetical
(E**2b) Not (while (Socrates is a stone), (for Socrates to be a stone is possible)) which is the propositional negation of the original falsehood; or as negative categorical

modal

(E**2c) (For) Socrates to be a stone while he is a stone is not possible, with the sense that it is not possible for the compound predicate 'is a stone while he is a stone' to apply to Socrates.

Abaelard's comments on these alternatives are very brief[25] but it is clear is that he has grasped perfectly that the validity or invalidity

[25] Abaelard, *Glossae*: 431–432.

of the argument and so the preservation of the principles of equipollence turns upon the way in which negation is understood and in particular its application both predicatively and as a propositional function. (**E**2a**) follows from (**E**1**) if it is a temporal hypothetical and Socrates is a stone but this is false and so the argument is unsound. (**E**2b**) follows from (**E**1**) if it is a temporal hypothetical and (**E**2c**) if it is a categorical modal, but in both cases the determination is included within the scope of the negation, propositional negation in the first case, predicative negation in the second, and so the argument cannot proceed to infer the truth of the determination.

4. Conclusion

Abaelard was notorious in at the beginning of the twelfth century for the novelties which he introduced into logic and for his attack upon the positions held by other Parisian masters, and he has seemed to modern commentators to break new ground in the theory of modal propositions. It is certainly true that his is the first reference to and use of the distinction between the composite and divided reading of propositions made by Aristotle in the *Sophistical Refutations*. Abaelard's considered application of this distinction, however, does not, as we have seen, ultimately identify the composite reading with the *de sensu* theory of nominal modal propositions. In other areas we can see from the Orléans treatises on modality and other early twelfth century texts not discussed here that Abaelard is working within a well-developed tradition and, incidentally, that the identification of the *de sensu* theory with William of Champeaux must be reconsidered. Abaelard is certainly an innovator in the study of modal propositions, but his innovations have more relation to his sophisticated understanding of scope and, for the first time in the Latin tradition, of negation as a propositional operator on propositional contents than they have to the introduction of a new theory of modality.[26] With this new understanding of propositionality, however, the *de sensu* exposition of modal propositions can be clearly formulated and distinguished from the *de re* exposition, providing twelfth century logicians with a sophisticated tool to use in their first attempts to understand the *Prior Analytics*.

[26] For what was new in Abaelard's metaphysics of modality see Martin 2001.

7 Ockham and the Foundations of Modality in the Fourteenth Century

CALVIN G. NORMORE

Prologue

There was a very considerable development of modal logic within the mediaeval Latin tradition both before and rather soon after the translation of Aristotle's *Prior Analytics*. The subject was explored, largely without benefit of the *Prior Analytics*, by Abelard and his generation and with the *Prior Analytics* in hand by Kilwardby and his. By the early fourteenth century we find a good deal of reflection on just what was involved in a modal logic.[1] In particular by this time there are a well-developed propositional logic and a well-developed syllogistic and the question of how modality interacted with each could be and was posed directly. In this chapter I hope first to present a sketch of the semantic and metaphysical foundation of modality as it appeared in the first half of the fourteenth century and then some of the nuts and bolts of modal logic as it was developed by Ockham.

1. Part I. The Foundation

1.1 *Something between a Fable and a Fact*

Once upon a time, perhaps by the mid- to late thirteenth century, the concept of possibility was intimately connected with that of power – some state of affairs was thought possible just in case it was thought there was some agent who could bring it about. For adherents of the Abrahamic religions the chief among these agents was an omnipotent God. Exactly what omnipotence meant is not entirely clear (it does not seem to require that God be able to make it true that Adam freely eats forbidden fruit, for example) but it at least entailed that if any agent could make or destroy a thing then God could make or destroy that thing. For the Latin West this picture had already been given

[1] Cf. Chapters 5 and 6 in this volume.

a vivid and interesting formulation by Anselm of Canterbury and it had been developed in some earlier thirteenth century thinkers, Robert Grosseteste, for example.[2] It remained central until the middle of the fourteenth century when a number of writers, Thomas Bradwardine and Gregory of Rimini notable among them, launched us on a path which broke the connection between possibility and power and eventually replaced it with a conception of possibility as a semantic or logical matter. This is an idea which goes back at least to the early thirteenth century commentaries on Aristotle's Physics and was imported into theology by Richard Rufus of Cornwall and Henry of Ghent, and developed by Duns Scotus.[3] When this process was complete, instead of considering what agents could do we came to consider first which sentences could be true, and, strikingly, we came to see the question of which sentences could be true as a question of the semantic relations among terms. Thus was born the idea that the only necessity is verbal necessity – an idea we still find it hard to shake.

1.2 Reflection

I find it difficult to overstate the magnitude of the shift which this story postulates. It involves a shift in the logical structure of the modalities (from S4-like systems in which what is possible is not necessarily so to S5-like systems in which it is), a complete reconceiving of the relation between time and modality, the separation of causal possibility from possibility *tout court*, and even the repudiation of the axiom that act precedes potency. If, as I claim, the decisive early battles in this struggle occurred in the fourteenth century, one might expect William Ockham to have been involved and indeed he was – on the side of the older rather than the newer modal model. Moreover his siding with the older picture of modality is not just a quirk of his thought. On the contrary, Ockham's view on this question is central both to his conception of valid inference and to his solution to the problem of future contingents. Nor does Ockham argue merely from his theory. He thinks, as we shall see, that the Scotistic device which was to play a crucial

[2] Cf. for Anselm particularly in *De Casu Diaboli* and his so-called *Philosophical Fragment*. For Grosseteste cf. his *De Libero Arbitrio*.

[3] For a quick sketch of some of these developments cf. Normore 1984, especially pp. 374ff.

role in the development of the picture of modality as verbal – the so-called non-evident capacity for opposites – is based on a logical error. Ockham will turn out, on my telling, to be the hero of the piece but like the most interesting heroes to be not entirely without flaw.

1.3 Ockham

Since Prof. Allan Wolter's seminal paper "Ockham and the Textbooks: On the Origin of Possibility" no one has disputed that there is a close connection for Ockham – and indeed for Aquinas, Aureoli, Scotus and Henry of Ghent as well, to name but a few – between possibility and the power of God.[4] Part of the connection Ockham states explicitly when he writes:

> Rather it should be understood that something's being possible sometimes is understood according to the laws ordained (ordinates) and instituted by God and those things God is said to be able to do by his *potentia ordinata*. [But sometimes] 'to be able' (*posse*) is understood in another way for 'to be able to do all that whose doing does not include a contradiction' whether God ordains this to be going to be done or not, for God can do many things which he does not do as the Master says. … And these things are said to be possible by his *potentia absoluta*. (*Quodlibet* VI, Q.1)

Ockham's view about the relation between God's *potentia ordinata* and *potentia absoluta* is still the subject of much controversy and to take it up here would lead us far afield. What seems clear is that there are not, for Ockham, two sorts of possibility, one corresponding to the *potentia ordinata* and the other to the *potentia absoluta*. Rather to say that something is possible *de potentia ordinata* is just to say that it is possible *de potentia absoluta* and that God has in some sense ordained its doing. The modalities correlated with God's absolute power are the basic modalities.

If this is true it already shows that the Diodorean conception of possibility as truth-at-some-time is certainly not Ockham's conception. There are things which God can do *de potentia absoluta* but never will do. These are, nonetheless, genuinely possible. Thus we are given a lower bound on possibilities – there are at least more things

[4] Reprinted in Wolter 2003 ch. 13.

possible than actual. What of an upper bound? We are torn today between two conceptions of the basic modalities – one sometimes called logical and another sometimes called metaphysical. The logical conception of possibility has at least two species. One identifies it with consistency. On this view a sentence describes a possibility if no contradiction is entailed by it taken together with the axioms of whatever logic is being employed. 'Contradiction' looks like a syntactic concept. It is common for elementary logic books to claim that 'It is not the case that *p*' is the contradictory of *p* and that an explicit contradiction is a conjunction of contradictories. Sometimes a sentence is said to be an implicit contradiction or to imply a contradiction if an explicit contradiction can be derived from it. Another form of the logical conception identifies it with truth in a model where models are constrained only by the requirement that those items picked out as the logical constants behave the same way in all of them.

There are certainly evidences of a consistency conception of possibility in the Middle Ages. As heirs to Aristotle's *De Interpretatione* mediaeval logicians had a formal notion of contradictory and hence of contradiction. But it is not this formal notion of contradiction which Ockham has in mind in his discussion of *potentia absoluta*; nor when Ockham says that by his *potentia absoluta* God can do 'anything whose doing does not include a contradiction' does he mean that if there is no semantic inconsistency in the description of God's bringing something about then God can bring it about. He writes:

> Sometimes some consequence or conditional should be conceded although the antecedent can be maintained without the consequent when the antecedent involves a contradiction. Hence if it is asked whether the Holy Spirit would not be distinct from the Son if he did not proceed from the Son, it should be answered that it is so because through this question nothing is asked but the truth of this conditional "If the Holy Spirit did not proceed from the Son, the Holy Spirit would not be distinct from the Son" which is true though it is not evidently true. Whence many consequences are good and many conditionals true although not evident to us. (*Summa Logicae* III-3, c.42 1.47–56)

Ockham uses examples like this to support the general claim that there are sentences that involve a contradiction but cannot be shown to be inconsistent. He writes:

From all this it is evident that many propositions including a contradiction – that is entailing contradictories – can be maintained in impossible positio, nor should contradictories be conceded on account of this because when such a positio is made not everything following from the positum should be conceded. Rather many things which follow should be denied or not conceded. For all those things which do not follow evidently so that such an inference cannot be made evident by natural consequences should not be conceded because of the positum. (*Summa Logicae* III-3, c.42, 1.68)

The context of the text just quoted is Ockham's discussion of a type of Obligatio known as impossible positio. In this type of obligatio the positum is a sentence which is in fact impossible. Since the obligatio would be trivial in such a case if rules like "from the impossible anything follows" could be invoked there are restrictions on the rules which may be employed. It seems that for Ockham the rules which may be employed are those of what he calls simple formal consequence.

In his discussion of consequentia Ockham distinguishes between those which are 'formally' valid and those which are valid but only 'materially'. He divides formally valid consequences into two kinds – those which hold immediately in virtue of what he calls an extrinsic middle and those which hold mediately through an extrinsic middle but immediately through a middle which is intrinsic. Extrinsic middles are very like inference rules in the twentieth century sense; intrinsic middles are (a subset of) true categorical sentences. Thus formal consequences holding immediately in virtue of extrinsic middles seem to be what we would call formally valid arguments, and formal consequences holding immediately in virtue of an-intrinsic middle are a special kind of enthymeme which can be reduced to a formally valid argument by adding the right kind of premiss. If this is correct then the inconsistent sentences would be just those which can be refuted in an obligatio employing impossible positio.

Support for this claim can be found in Ockham's discussion. For example, giving advice to the respondens in an impossible positio about which posita to accept, he writes:

Whence only that impossible proposition from which it is not possible for contradictories to be inferred through rules and propositions per se nota which no understanding can doubt should be accepted in impossible positio. (*Summa Logicae* III-3, c.42 1.24–27, p. 739)

And just a little later he spells out the kinds of rules which should be accepted and rejected in impossible positio:

Hence this rule can be given: everything following from the positio by natural and simple consequences holding by virtue of propositions or rules per se nota should be conceded. From which it follows that in impossible positio what follows from what has been posited or well conceded syllogistically should be conceded. Similarly what follows by virtue of rules about circumstamces such as 'from an affirmative with an infinite predicate to a negative with a finite predicate is a good consequence' and suchlike should be conceded. For if such consequences were denied there could be no disputation. However that which follows by an ut nunc consequence or a material consequence or other consequences like this can be denied however much it may truly follow from the positum (*Summa Logicae* III-3, c.42 1.36-45, p. 740)

In the text quoted, Ockham claims that there are good consequences which do not follow 'evidenter' and which cannot be made evident 'ex naturalibus'. Such consequences are, I claim, not formal, and while they may be good their consequent cannot be shown to follow from their antecedent except in very special cases.

Given these texts it is reasonable to interpret Ockham as asserting that although sentences like that in the Trinitarian case quoted above entail a contradiction in the sense that it is impossible for them to be true unless contradictions are true, there is no formal consequence or chain of formal consequences holding immediately through extrinsic middles which will get us from the sentence to a contradiction, and, moreover, there is no evident or per se nota proposition which, when added as a premise, will facilitate the deduction. In short Ockham is arguing that there are valid arguments whose validity cannot be known a priori and cannot be guaranteed by logic alone. From this it follows that there are propositions which are impossible but cannot be known a priori to be impossible.

The other widely used contemporary notion of possibility, the one frequently called 'metaphysical possibility', seems suggested by this result. It is a concept still under development. In one influential way of conceiving the matter some sentence expresses a metaphysical possibility if and only if what it expresses is compatible with the identities and natures of the things under discussion – something which may well not be graspable a priori (in the modern sense of that phrase).

Thus if Cicero is Tully it is not metaphysically possible that Cicero be a famous orator and Tully not; nor is it metaphysically possible that water contain no hydrogen – though it is metaphysically possible (we suppose) that Aristotle have been a wrestler rather than a philosopher.

This account too employs a modal term 'compatible' and we might be led to think that any satisfactory account of the modalities will employ some modal term – though we can build a *theory* of the modalities. This seems closer to the way Ockham and his contemporaries proceed; Ockham and many of them accepted the view that a consequence is good just in case it is impossible for its antecedent to be true and its consequent false. Ockham also accepted the rules that "from the impossible anything follows" and "the necessary follows from anything". Thus, for Ockham, including a contradiction is a broader notion than formally entailing a contradiction and, hence, the sphere of the possible is narrower than that of the semantically consistent. Is it then the sphere of the metaphysically possible?

There seem to be two ways in which, for Ockham, something semantically consistent might turn out to be impossible. In some cases, like the Trinitarian case mentioned above, language is simply inadequate to reality. The difficulty here is in our capacity to form concepts adequate to express the natures or identities of the things involved, which seems to be very like the issues involved in the notion of metaphysical modality. There are, however, other cases where Ockham seems to recognize necessities which would not be metaphysical necessities in that sense.

Ockham thought that the past as such is necessary. For example, in *Quodlibet* II, q.5 he writes:

> I say that had God produced the world from eternity he would have produced it from eternity by necessity because supposing that case this proposition would now be necessary 'The world was from eternity' since it would be true and never could be false. However it would have been able to be false just as a true proposition about the past is necessary and yet was able to be false.

Why does Ockham think that the past is necessary? Two lines of thought converge here. One is simply Aristotelian and relates to Aristotle's attempt to define temporal order in terms of the order of potency and act. That is another subject. I want to focus here instead on the second line, which is intimately connected with what I call Peter Aureoli's Principle:

If x is A and it is possible that x be not-A, then x can become not-A.

Aureoli's Principle links possibility and mutability and from it the necessity of the past follows almost at once. No one can change the past for that would require that some sentence of the form 'It was the case that *q*' be true before a time *t* and false after *t*. But 'It was the case that *q*' is true at a time only if '*q*' was true at an earlier time. If that condition is met when we consider just the times before *t* it will certainly be met if we consider these plus some times after *t*. So if 'It was the case that *q*' is true at some time, it is true at any later time.

I call the principle Aureoli's because Petrus Aureoli used it and an argument rather like the one I have just sketched to show that either the future is necessary or some sentences about the future are neither true nor false. The future is no more changeable than the past. But if some sentence, say 'The Antichrist will be', is true but can be false, and if Aureoli's Principle holds, then 'The Antichrist will be' can become false. But then the future would change from one in which there is to be an Antichrist to one in which there is not. That is impossible.

Ockham accepts Aureoli's Principle. He also accepts the necessity of the past and the view that every sentence is either true or false. He agrees that if something is actual and immutable then it is necessary. He evades Aureoli's argument by adopting something very close to a disappearance theory of truth. On this view to claim that '*q*' is true is not to ascribe a property to '*q*' at all and so sentences about the future can be true without their truth being part of what is actual. It was commonly conceded during the Middle Ages that God is immutable and, by extension, that the single act by which God creates is not revisable. Given that that act is actual and that whatever is actual and immutable is necessary it follows that whatever God does God does necessarily. Thus God cannot do anything which God does not in fact do. This was apparently accepted by Peter Abelard (inter alia) and was among the propositions condemned at Paris in 1277.

Ockham does not draw the condemned conclusion. Instead he, in effect, claims that God's single creative act is not all actual. He claims, for instance, that 'God will give Peter grace on the last day' while describing part of God's activity does not describe anything which is. By thus suggesting that the divine creative act unfolds in time Ockham seems to suggest that, in some way, God is in time. Thus it is not surprising that his view aroused opposition. Some of that opposition focussed on Aureoli's Principle.

The way in which the past is necessary seems different from other ways of being necessary. Even if one thinks that a sentence like 'Ockham existed' is now as necessary as '2 + 2 = 4', it is nonetheless reasonable to suppose that it was not always as necessary. We can then distinguish between true sentences which could have been false once but can no longer be false and true sentences which never could have been false. Actuality plus immutability characterizes the first kind but seems too weak for the second. It is central to Ockham's view as I understand it that this appearance is illusory – that at bottom necessity just is actuality plus immutability. If this is right Ockham's notions of necessity and possibility are not the contemporary (to us) notions of metaphysical necessity and possibility either.

Of course there is a difference between necessity and immutability *tout court* and there is also a difference between being always necessary and being necessary from some time on, but if I am right this difference is not a difference in the kind of necessity involved. What is different is just *when* the sentences are necessary. Moreover for Ockham there seems to be no difference between the necessity of the past and the necessity involved in valid inference. I think a case can be made that it is a necessary and sufficient condition for a consequence's being simply valid that not even God by his absolute power can make the antecedent true and the consequent false. Ockham is a modal monist.

Some evidence for this can be had by looking at cases where Ockham employs a sentence about the past as a middle to reduce an enthymematic formal consequence to one holding by virtue of an extrinsic middle; for such a reduction to be valid the necessity of the intrinsic middle must be at least as strong as that of the conclusion. This Ockham does on at least one occasion. He writes:

> This should be noted – that some such consequences are only ut nunc and some are simple. For when the predication of the subject of the antecedent of the subject of the consequent is necessary, then the consequence is simple; however when the [resulting] categorical is contingent and not necessary then the consequence is only ut nunc. Hence this consequence is

only ut nunc "Any divine person was from eternity; therefore something which is creating was from eternity" if the subject of the consequent is taken [to stand for] what is, because this is contingent "Something which is creating is God". However if the subject of the consequent is taken for what was, then the consequence is in a way simple, because this is necessary "Something creating was God" taking the subject for what was just as this is necessary "God was creating". (*Summa Logicae* III-3, c.6, 1.243–256)

One might cavil about the 'in a way'. I take it though that the argument in question is this:

"Any divine person was from eternity"
Therefore, "Something which was creating was from eternity"

The antecedent is simply necessary. The argument, Ockham suggests, is simply valid. Hence the consequent must be necessary in whatever sense is required for validity. The consequent was not always true but is now true and indeed necessarily true in the sense in which the past is necessary. Hence that sense must be strong enough to preserve a valid inference from a premiss that is timelessly necessary.

Since which sentences are necessary changes from time to time, it would seem that which arguments are valid changes from time to time. Hence we might expect the full reduction principles characteristic of S5 to fail for Ockham's logic. He himself suggests as much when he writes:

Sometimes, although rarely, a modal universal may be impossible and each singular possible as is evident in the case of "Each of these of necessity is true" pointing out "Socrates was at [time] a" and "Socrates was not at a". This universal is impossible yet each singular is possible for this is possible "This is necessarily true 'Socrates was at a' ", and similarly the other is possible. (*Summa Logicae* II-10, 1.94–99)

The point can be stated more explicitly, putting *L* for the operator 'it is necessary that', and *M* for the operator 'it is possible that'. The claim is then that if we could derive "It is necessary that Socrates was at a" (*LA*) from "It is possible that it is necessary that Socrates was at a" (*MLA*) then by applying rules for de dicto modality which Ockham accepts we could derive the contradictory of "It is possible that it is necessary that Socrates was not at a" (~*ML*~*A*). That is because we can derive ~*ML*~*A* from *LA*. But we are told explicitly that that is true.[5]

Ockham's modal monism and his acceptance of the view that actuality plus immutability yield necessity were controversial in the Middle Ages. We already find the beginnings of a challenge to them in the work of Robert Grosseteste almost a century earlier. In his treatise "On Free Will" Grosseteste claims:

> Of what is necessary there are two sorts. Of one sort is that which has no possibility in any way for being otherwise. 'Two and three are five' is of this sort for it is not able to be untrue either before time or in time. Such a proposition is simply necessary. There is another sort of necessary thing which has no power to be its opposite in the past or in the present or in the future. 'The Antichrist is going to be' is of this sort and so are all those which are about the future and which are such that given that their truth exists if the truth is not able to have non-being after being as has been shown above. However there is a possibility that from eternity and without beginning it will have been false. (*De Libero Arbitrio* c.6, p. 68. 1.24ff.)

One might wonder whether Grosseteste thinks that God now has the power to do that which he neither did nor will do and which is impossible in his second sense. Grosseteste thinks that God does have this power but in a rather peculiar way: God has the power because he has already exercised it. On Grosseteste's view not only is God's power to do *A* numerically the same as God's power to do not-*A*, but for God to actualize his power to do *A* is just for him to actualize his power to do not-*A*. This view makes it difficult to be certain just how far from Aureoli's Principle Grosseteste has strayed.

If it is not clear whether Grosseteste rejected the doctrine that what is actual and immutable is necessary it is clear that John Duns Scotus did. The context is a mediaeval problem closely related to the question

[5] The proof of ~*MLA*~*A* from *LA* holds even in the system T and is as follows. Let *A* be "Socrates was at a":

1)	*LA*	Premiss
2)	*A*	Necessity Elimination from 1
3)	*MA*	Possibility Introduction
4)	*LA* ⊃ *LMA*	Modal distribution (given that 2 implies 3)
5)	*LMA*	1,4 Modus Ponens
6)	~*M*~~*L*~*A*	Modal interchange
7)	~*ML*~*A*	Double negation

of God's freedom, namely, the problem whether a created will which existed only for an instant would be able to act freely. Suppose a will that exists only at instant *t*. Suppose this will does *A*. To do *A* freely at *t* it seems that the will must be able to avoid doing *A*. But when can this will avoid doing *A*? Not before or after *t* since what does not exist has no powers. But not at *t* either. At *t* the will does *A* and, if Aureoli's Principle is correct, then if it does *A* at *t* and can avoid *A* at *t* then it can change from doing *A* to avoiding *A* at *t*. But change takes time and *t* is just an instant. Hence the will does *A* necessarily. One way out of this difficulty is to abandon belief in instants. A second is to abandon Aureoli's Principle and admit that what is actual and immutable might still not be necessary. This is Scotus' approach. For, he says:

> Therefore there is a power of this cause [the will] for the opposite of it [the volition] so that without any change it may cause the contrary. This real power is a power which as first act is naturally prior to the contraries which as second act are naturally posterior. For a first act considered in that instant In which it is naturally prior to a second act contingently produces that 'second act' as its effect so that as something naturally prior It would equally be able to produce a contrary thing. Moreover this real act of power naturally prior to that which is produced accompanies a logical power which is an absence of exclusion in the terms (*quae est non repugnantia terminorum*). (I. *Sent.*, d.39 t. 16)

Here Scotus abandons Aureoli's Principle while apparently accepting an analogue of it for the order of nature. The text is a remarkably rich one and Scotus' claim that the ability which the will has for the time it freely does *A* to do not-*A* instead "accompanies a logical power which is an absence of exclusion in the terms" seems to be the first explicit identification of necessity and analyticity. It entails abandonment of the tag 'omne quod est, quando est, necesse est esse' and with it Aureoli's Principle, but it is not clear that it entails abandonment of the necessity of the past. Curiously, it may have been Ockham himself who, in his zeal to correct what he perceived as Scotus' error, first showed how the non-evident capacity for opposites connected with the necessity of the past. In his *De Praedestinatione* Q.III Ockham attacks Scotus' view, claiming:

> Suppose that 'the will wills this' is true for the instant A but that the capacity to not will this can be actualized by you at that same instant. Then at

the same instant these would be simultaneously true: 'the will wills this', 'the will does not will this' and thus contradictories at the same time. (trans, based on Adams and Kretzmann 1983, p. 73)

This argument is a rather complex enthymeme involving both time and modality. Ockham continues:

> Suppose instead that if this capacity were actualized so that the will does not will this for instant A, then its opposite 'the will wills this in instant A' will be false ... On the contrary every sentence simply about the present has, if it is true, some necessary sentence about the past [corresponding to it]. But 'the will wills this in instant A' is true by hypothesis and is simply about the present, therefore 'the will willed this in instant A' will always be necessary afterward. Therefore, after instant A, 'the will did not will this in instant A' cannot be true. (Ibid., my translation)

Ockham's argument is valid but seems subtly to miss the point. Scotus does not have to show that 'the will did not will this in instant *A*' can be true but only that 'the will does not will this in instant *A*' could have been true. And Scotus need not accept $PMp \supset MPp$, which seems to be the principle Ockham employs.

Christopher J. Martin has pointed out that in formulating his doctrine of the non-evident simultaneous capacity for opposites Scotus uses the framework of the Obligatio, and that his innovation consists in rejecting a rule of Obligatio that corresponds in that framework to Aureoli's Principle. It is within this picture that the subsequent debate about the necessity of the past is carried on. Yet, in a sense, by accepting this framework the Ockhamists have already sold the past. To refute someone in an Obligatio one has to show that the positum and the proposita form a set from which a contradiction can be formally derived. In other words within Obligatio impossibility collapses into semantic inconsistency.

Nevertheless the debate about the necessity of the past was vigorous and in it both sides deployed the immutability of God. Those like Thomas Buckingham who claimed that God cannot make a different past argue that to suppose God can is to suppose an impossible change of the divine mind. Those like Bradwardine and Gregory of Rimini who argue that God can make a different past claim that to suppose he cannot is to suppose an impossible diminution of the divine power over time. Starting from divine immutability Gregory argues that the proper test of whether God can make a sentence to have been true is whether the terms of the sentence exclude one another. Gregory writes:

> The third consequence is that God is able to make any past thing not to have been. I prove this proposition thus: any past thing not to have been does not imply a contradiction, therefore any past thing God can make not to have been. The consequence is evident from the preceding argument. The antecedent I prove because Adam's not having been and similarly Eve's and Abel's and so on for singulars does not imply a contradiction. For in saying that Adam was not it is no more said that Adam was and was not, nor does any other contradiction follow from the same, than would if it were said that the Antichrist or some other thing which is going to be in fact will not be.

The second argument goes thus:

> God did not will to produce Adam, therefore God did not produce Adam and further, therefore Adam was not. The consequences are necessary and the first antecedent is possible, therefore so is the last consequence and further it follows 'it is possible for Adam not to have been, therefore God is able to make it that Adam will not have been.' [If] one asks if the first antecedent is possible, it is proved thus: everything that God was able from eternity to will, he is now able from eternity to have willed and what he was able not to will he is able not to have willed. Therefore, although he will have willed from eternity to produce Adam he is able not to have willed it and he is able to will not to produce him just as he was able from eternity. (I. *Sent.* d. 42 – UU, q. 1, art. 2, f. 162 r. b)

Those, like Buckingham, who argue for the necessity of the past are fond of enumerating implausible consequences of the contrary view – consequences like that it would be possible for a man who in fact had parents to lack them and possible for Christ to be born only yesterday. These are not silly objections. If the origins of things are essential to them, then it is an essential property of me that I had the parents I had, and it may be essential to me that I was born when I was. Careful consideration of just how the parts of the past depend on one another might well show that God's power to make a different past is significantly more limited than was his power to make a different future. Later things may require earlier ones in a way in which earlier ones do not require later ones. But for Ockham and most other theologians of the fourteenth century this type of argument is a non-starter. Among the propositions condemned at Paris is 1277 we find "that God cannot produce the effect of a secondary cause without the secondary cause itself." Perhaps in response to this, one of the cornerstones of fourteenth century theology is the claim that God can do directly whatever can be done with the help of creatures. From this principle it follows immediately that the efficient causes of things are not essential to them and further that if God has any power over the past at all he can make one thing never to have been without altering anything else. The immutability of the divine power and Scotus' criterion combine to reduce necessity to something very like analyticity. In particular they combine to emphasise the transcendence of God and to eliminate time as an element in conceptions of modality.

This has consequences for the logic of modality (for instance, it makes it possible to accept the characteristic S5 thesis $Mp \supset LMp$, which is falsified in a modal tense logic in which sentences become necessarily true, and it has consequences too for our understanding of what modality is. The fourteenth century discussions of God's power in the period after Ockham emphasise the differences between the divine power and the powers of other things. In so doing they rob the notion of power of its explanatory centrality. We still think with the consequences.

2. Part II Modal Logic

By the early fourteenth century logicians were the possessors not only of an assertoric syllogistic that extended and considerably revised Aristotle's own, but also of a propositional logic which included a truth-functional account of conjunction and disjunction – though not of the conditional.[6] They turned to modal logic with these resources in hand.

For fourteenth century authors modals were, at their core, adverbial modifications of the copula of a sentence. Following a tradition going back to Aristotle they distinguished six 'basic' modals – necessary, contingent, possible, impossible, true and false. They also treated tenses

[6] At least not of the conditional if one does not identify what was termed 'ut nunc' consequence with the material conditional. I do not think this is the right way to understand ut nunc consequence but on this cf. Adams. M. 1973, pp. 5–37.

in a way largely parallel to one part of their treatment of modals and made forays into deontic logic. In what follows here I will concentrate on two of the four modes they themselves seem to have taken as central – necessary, possible. As is customary with us they recognised that impossibility and contingency could be defined in terms of these.

For Ockham the most basic modal sentence is one of the form

Q *A* (mode) is *B*

where Q is some determiner expression such as 'omnis' (all), 'aliquis' (some) or 'nullus' (none). *A* and *B* were taken to be names either proper or general which, in normal cases, 'supposited for' the objects they signified (i.e. of which they were names). Sentences without a determiner were taken to be singular if A was a proper name and 'particular'; that is equivalent to a sentence with an explicit 'aliquis' in front, if *A* was a general name. Within this framework in an assertoric sentence like

1) Some Humans are Greeks

'Humans' would supposit for the actual humans and 'Greeks' for the actual Greeks. The sentence would be true if there is overlap in the supposition of the two terms.

When we add a mode to such a sentence the sentence can, on Ockham's view, be understood in two ways. Thus

2) Some humans possibly are Greeks

could be understood 'in the divided sense' (*in sensu diviso*) or 'in the composite sense' (*in sensu composito*).

If we understand such a sentence in the divided sense, Ockham claims it has two readings and in each the supposition of the terms is affected in a way that depends on the particular modal involved. On both readings the supposition of the predicate is affected but on one reading, that of the subject is, and on the other it is not. Thus for 2) we get readings I might express this way:

2′ Some actual humans are possible Greeks.

And

2″ Some possible humans are possible Greeks.

2′ and 2″ are not sentences Ockham would have liked because they replace the adverbial modification of the copula with adjectival modifications of the terms and so are not, strictly speaking, even modal sentences by Ockham's lights. They do, however, as we will see, accurately represent the logical structure Ockham attributes to such sentences

If we understand 2 in the composite sense, it is, Ockham claims, equivalent to a sentence in which a modal adjective is predicated of a 'dictum'. A dictum is an accusative plus infinitive construction in Latin the closest natural analogue of which in English is a construction like "for humans to be Greeks" though we more commonly use "that humans are Greeks".[7] Thus, in Ockham's account, 2 taken in the composite sense is equivalent to

> 3. For some humans to be Greeks is possible
> (Aliquos homines esse Graecos est possibile)

or

> 3′. [It is] possible for some humans to be Greeks
> (Possibile est aliquos homines esse graecos)

which in the Latin differ only in whether the adjective 'possibile' is in subject or predicate position.

Strictly speaking, in these sentences a modal is the predicate (or subject) term and the other term, the dictum, functions very like a name for sentences.[8] Thus we have only what Quine would have called the first grade of modal involvement.

We are now in a position to represent Ockham's modal syllogistic. We can do so surprisingly simply: Take a complete set of rules for the sentential calculus and a complete assertoric syllogistic. Add two basic ideas:

[7] The difference between the apparently subjunctive 'possible for *A* to be *B*' and the indicative 'possible that *A* is *B*' may well be significant and bear directly on the issue whether possibility is to be understood in terms of powers; cf. Hacking 1967, pp. 143–168. I do not know how to settle the better English translation of the Latin accusative infinitive construction on grammatical grounds and will use both English forms indifferently.

[8] Indeed Buridan explicitly claims that dicta supposit for propositions.

a) that sentences in the divided sense should be understood so that the copula always ampliates the predicate but may or may not ampliate or restrict the subject. Treat the result as though it were an assertoric sentence with new terms. Apply the rules listed below and the standard assertoric syllogistic. Stir!
b) that sentences in the composite sense i.e. with a modal predicate or a modal term prefixed be treated inside syllogisms according to the rules of a propositional modal logic.

What slightly complicates the presentation of such a system is that Ockham, like all of his contemporaries as far as I have been able to determine, does not attempt to provide a complete and detailed account of his logic. For example, while he thinks that conjunction (*et*) and disjunction (*vel*) are commutative and associative follows from the truth conditions he gives for conjunctions and disjunctions he nowhere states that. Again while much of his modal syllogistic can be derived by applying a few rules to the assertoric syllogistic, he and his contemporaries prefer to divide the subject by figure and mood and treat each of these subdivision separately. In the Appendix I have attempted to provide a set of rules I think to be complete and to indicate where possible places in which Ockham states them or commits himself to them.

Ockham does not attempt himself to provide a minimal basis for his modal syllogistic so to determine the adequacy of the set of rules provided in the Appendix one would need, as far as I have been able to tell, to check them case by case against Ockham's resvults. I gladly leave that as an exercise for the reader!

3. Appendix

Ockham's Propositional logic

Let *A* and *B* be schemata for sentences

Reversible Rules (each of these can be read top to bottom or bottom to top)

Conjunction Commutation	Disjunction Commutation	Duplication	Double Negation Introduction	Double Negation Eliminaton
A et *B*	*A* vel *B*	*A*	*A*	non non *A*
B et *A*	*B* vel *A*	*A*	non non *A*	*A*

Conjunction

DeMorgan's Rules

Introduction

A
B
A et *B*

Elimination

A et *B*
A

A et *B*
Non (non *A* vel non *B*)

A vel *B*
Non (non *A* et non *B*)

non-reversible rules

A et *B*
A

A
A vel *B*

Si *A*, *B*
A
B

A
A

Meta-rules about consequence

If $A \vdash B$ and $B \vdash C$ then $A \vdash C$

Ockham's Modal Propositional Logic (OMPL)
Add to OPL the following non-reversible rules

Nec *A*
A

A
Pos *A*

Cont *A*
Pos *A* et Pos non *A*

Nec *A*
non pos non *A*

pos *A*
non nec non *A*

If $A \vdash B$ then Nec $A \vdash B$

(S.LII, c. 24.·If the antecedent is necessary, the consequent is necessary.)

Ockham's Assertoric Syllogistic

Simple Conversion

a)

No *a* is *b*
No *b* is *a*

b)

Some *a* is *b*
Some *b* is *a*

Subalternation

a)

Every *a* is *b*
Some *a* is *b*

b)

No *a* is *b*
Some *a* is not *b*

Let us, when convenient, abbreviate 'Every *a* is *b*' by A*ab*, 'Some *a* is *b*' by I*ab*
'No *a* is *b*' by E*ab* and 'Some *a* is not *b*' by O*ab*

Let –*X* be the contradictory of X then

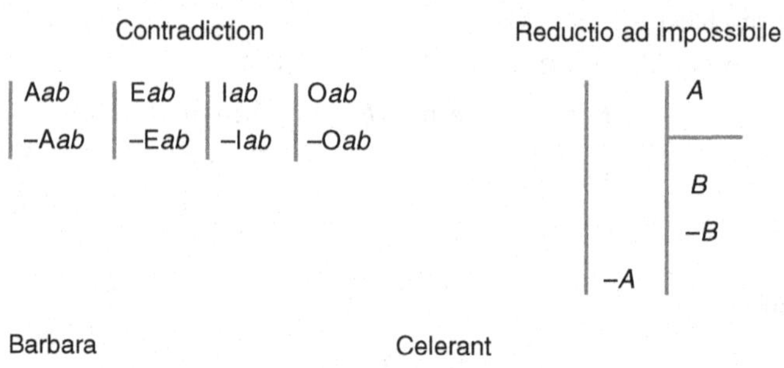

Barbara

| *Aab*
| *Aca*
|———
| *Acb*

Celerant

| *Eab*
| *Aca*
|———
| *Ecb*

Given this background we can extend the assertoric syllogistic to modal syllogistic in the divided sense by simply adding the following:

Let 'pos *X*' be the term that supposits for the things which can be *X*, 'nec *X*' the term which supposits for the things that are necessarily *X* and 'cont *X*' the term suppositing for the things which are contingently *X*.

Restriction Rules

| Every Pos *A* is *B*
| Every *A* is *B*

| Every *A* is *B*
| Every Nec *A* is *B*

| Every Pos *A* is *B*
| Every Cont *A* is *B*

| No Pos *A* is *B*
| No *A* is *B*

| No *A* is *B*
| No Nec *A* is *B*

| No Pos *A* is *B*
| No Cont *A* is *B*

Ampliating Rules

| Some Nec *A* is *B*
| Some A is *B*

| Some *A* is *B*
| Some Pos *A* is *B*

| Some Cont *A* is *B*
| Some *A* is *B*

Composite Sense Rules

| That *p* is necessary
|———
| | *p*
| |———
| | *q*
| That *q* is necessary

| That p is necessary
|———
| *p*

| That p is necessary
|———
| | *p*
| |———
| | *q*
| That q is necessary

The connection between the divided and the composite senses in the case of singulars.

That (This) *A* is *B* is Possible	(This) *A* can be *B*
(This) *A* can be *B*	That (this) *A* is *B* is Possible

8 *Theological and Scientific Applications of the Notion of Necessity in the Mediaeval and Early Modern Periods*

JACK MACINTOSH

In the mediaeval and early modern periods the notion of necessity was found important in both theological and scientific areas. For the mediaevals, following Aristotle, a scientific result had to be deduced from universal, necessary principles via a valid syllogism, to yield a universally necessary conclusion:

> It is clear [said Aristotle] that if the propositions from which a deduction proceeds are universal, then it is necessary for the conclusion of such a demonstration … to be eternal. There is therefore no demonstration of perishable things, nor any understanding of them simpliciter but only incidentally, because nothing holds of them universally but only at some time and in some way.[1]

In the late fifteenth century Petrus Garsia, bishop of Ussellus in Sardinia, noted that this left experimental results outside the scientific fold: "To assert that … experimental knowledge is science or a part of natural science is ridiculous … [Experimental knowledge] according to those of that opinion, is practical knowledge, whereas natural science in itself and in all its parts is purely speculative knowledge."[2] This view was still held in the early modern period, especially among the writers of logic texts, although others, following Francis Bacon, pointed to natural philosophy's need for a new notion of demonstration.

[1] *An. Post.*, Aristotle 1994, 75b 21–26.

[2] Garsia 1489, quoted Partington 1961, 2:10. Garsia was criticizing Pico della Mirandola.

This chapter began as a talk in the Logic Workshop organized by Max Cresswell, Adriane Rini, and Ed Mares. Adriane, Ed, and in particular Max have offered a variety of very helpful suggestions, as has Peter Anstey. Unfortunately, for reasons of space, not all of them have found their way into this version of the chapter.

Theologically, the derivations of truths about God and God's relation to humanity provided examples of necessity, but another, distinct, modal notion was also in play, that of necessary existence, which remained an important notion, theologically, throughout the early modern period. For the mediaeval views I shall concentrate on Aquinas, whose writings provide us with a variety of senses, sometimes overlapping, of necessity, and which also provide a clear account of the notion of necessary existence.

1. Necessity and Necessary Existence in Aquinas

There are a number of senses of *necessity* in Aquinas and other mediaeval authors. Here I consider four which remained important in the early modern period: necessity as the self-evident, tensed necessity, absolute necessity, and necessary existence.

1.1. *Self-Evidence*

Thomas notes that "self-evident" truths need not be *obviously* self-evident: necessary truths may be difficult to discover. Effectively dismissing the ontological argument when considering the question "Whether the existence of God is self-evident?"[3] Thomas allows a spectrum of possibilities: things which are obvious to all or almost all ("Humans are animals"), things which are 'obvious' only after a good deal of intellectual hard work[4] (revealed but deducible truths, scientific principles[5]), and things whose truth, let alone whose 'self-evidence', may be quite hidden from us:

> What is in itself the more certain may seem to us the less certain because of the weakness of our intellect.... Doubt about articles of faith is not due to the uncertain nature of the truths, but to the weakness of our human intelligence.[6]

[3] *Summa Theologiae* (hereafter *ST*) 1a 2.1. In the next century Ockham discussed this question in detail (Ockham 1970, I, D3, Q4, 432–442), but Thomas's succinct points are sufficient here.

[4] Cf. *De Veritate* 10.12c, where Thomas notes that similar points may be found in Boethius.

[5] *Summa Contra Gentiles* (hereafter *SCG*) 1.4; MacDonald 1993, 178–179.

[6] ST 1a 1.5 ad 1.

1.2. Tensed Necessity (Necessity per Accidens*)*

In ordinary language modal terms are typically tensed: It *was* possible for you not to be reading this sentence, but that is *no longer* possible. "It is impossible *per accidens* for Socrates not to have run, if he did run."[7] The sense of necessity involved, as Arthur Prior pointed out, following de Rivo,[8] is that of the "now-irrevocable," the "now-unpreventable." This *per accidens* sense of necessity is the sense in which all genuinely past or present tensed sentences are, if true, necessarily true, and if false, necessarily false.[9] Future events such as the sun rising tomorrow which are already knowable in their causes[10] also have *per accidens* necessity. As we shall see in authors such as Lowde and Cudworth, this notion continued to be philosophically, and particularly theologically, important in the early modern period. It follows at once that humans cannot know, nor indeed can the angels, any truths about "wholly future individuals."[11] Such knowledge belongs to God alone. As Prior pointed out, it follows that possibilities vanish (what was once possible is *no longer* possible), *and* come into existence: *until* item *x* exists, there are no possible truths about *it*.[12]

1.3. Absolute Necessity

Absolute necessity, or necessity *per se*, is typically contrasted with necessity *per accidens*, but it appears as a contrastive term in different contexts. It is, centrally, that whose negation is contradictory. Thus Thomas, considering the (absolute) possibility that the world could have been infinite in past time, writes: "By faith alone do we hold, and by no demonstration can it be proved, that the world did not always exist."[13]

[7] *De pot.* 1.3 ad 9.

[8] Prior 1967, 117.

[9] Genuinely, to rule out sentences such as 'Yesterday it was the case that two days from then, *p*', Ockham emphasizes the need to distinguish sentences which are *de futuro secundum rem*, from those which are so merely *secundum vocem*, and makes clear that this need not be a simple matter (Ockham 1978). See further MacIntosh 1998a.

[10] ST 1a 57.3c, ST 1a 86.4c, etc.

[11] ST 1ª 57.3c.

[12] Prior 1960, 688, reprinted in Prior 1968, 70. It also follows, though neither Prior nor Aquinas stresses this, that whatever properties an entity has initially it has with at least *per accidens* necessity.

[13] ST 1a 46.2 c, and cf. De Pot. 3.17 c. Despite this clear statement of his position modern discussions of the five ways often assume that Aquinas's argument

A world infinite in past time would be peopled with individuals not present in this world, and no branching from a finite world could lead to a world infinite in past time; conversely no branching from a world with an infinite past could lead to a world with a finite past,[14] a point which James Lowde in the seventeenth century was to note as theologically important.

Absolute necessity is contrasted with *suppositional necessity*:

There are two ways in which a thing is said to be necessary: absolutely, and by supposition.... absolutely necessary from the relation of the terms, as when the predicate forms part of the definition of the subject: thus it is absolutely necessary that humans are animals.... It is not necessary in this way that Socrates sits: ... though it may be so by supposition; for, granted that he is sitting, he must necessarily sit, as long as he is sitting.[15]

In *SCG* Aquinas remarks, concerning God and the necessity of supposition: "whoever wills something, necessarily wills whatever is necessarily required for it, unless there be a defect in him either because of ignorance or because he is led astray through passion from the right choice ... Thus [since these exceptives cannot apply to God] it is necessary that God will the rational soul to exist supposing He wills humans to exist."[16] Again, "Predestination most certainly and infallibly takes place, yet it does not impose any necessity." Anything which God wills "is necessary on the supposition that He so wills ... but is not necessary absolutely," and "the same must be said about predestination."[17] Aquinas points out that this allows us to see the scope fallacy involved in treating $\Box(p \rightarrow q)$ as if it were $(p \rightarrow \Box q)$:

Everything known by God must necessarily be ... may be taken *de re* or *de dicto*. If *de re* (Everything which God knows is necessary), it is divided, and

requires the impossibility of an infinite past sequence of causes. Whatever problems the five ways may have, this is not one of them, though such an infinite string would be a string of *accidental*, not of *essential* causes. See further Ibn Sīnā (Avicenna 2004), 6.2, 201ff.

14 But see Mackie 1974 for a provocative modern wonder about this point.

15 ST 1a 19.3c. Later Leibniz's usual case was necessity as that whose negation is contradictory, but suppositional necessity may be found in his distinction between *logical* and *hypothetical* necessity (Parkinson 1965, 109–110).

16 *SCG* 1:83.

17 *ST* 1a 23.6 c, ad 3. Cf. Aristotle's 'relative necessity.' (*An. Pr.* 30b39-40.)

false. If *de dicto* (This sentence, "that which is known by God is" is necessary) it is composite and true.[18]

Aquinas notes in passing that a conditional proposition may be (necessarily) true, though both antecedent and consequent are necessarily false (*If a man is a donkey, he has four feet)*.[19]

Finally we have the case of absolute necessity as opposed to something which is necessary because of a thing's *nature*. This notion overlaps with necessity of definition. "A human is a donkey" is impossible under both notions but, particularly where God is concerned, there are divergences. For example, because of God's nature, God cannot will evil.[20] That is distinct from God's inability to change the past, which is "impossible, not only in itself, but absolutely (*non solum per se, sed absolute*), since it implies a contradiction."[21] By contrast, God's inability to will evil is not impossible *absolutely*, but it is impossible in view of God's nature. Similarly, God cannot correctly be said to be sorrowful or in pain; God cannot hope or fear or repent.[22]

For mediaeval thinkers an explanation in terms of a thing's *nature* was found both clear and helpful, but one of the changes the seventeenth century's "new philosophy" brought about was a difference in the way in which *nature* was conceived. Briefly, for the mediaevals each kind of thing has a *nature* which explains many things about it: why is snow white? – because it is the *nature* of snow to be white. But if things are simply made up of tiny corpuscles, all composed of the same matter, such explanations no longer sufficed, or even had *any* explanatory value:

If you … Ask them, how white Bodies in Generall do rather Produce this effect of dazling the Eyes, then Green or Blew ones, instead of being told, that the former sort of Bodies reflect Outwards, and so to the Eye farre more of the Incident Light, then the Latter; You shall perchance be told, that 'tis their respective Natures so to act, by which way of dispatching difficulties, they make it very easy to solve All the Phænomena of Nature in

[18] *ST* 1a 14.13 ad 3.

[19] *ST* 1a 25.3 ad 2.

[20] *SCG* 1:95.

[21] *ST* 1a 25.4 ad 1.

[22] *SCG* 1:89. The case is different with God incarnate. Christ suffered pain and sorrow, knew fear and wonder, and, "without sin," was subject to anger, that "passion composed of sorrow and the desire of revenge (ST 3a 15.5–9)."

Generall, but make men think it impossible to explicate almost Any of them in Particular.[23]

Physical change was to be explained in terms of minute particles of matter, their motion, and the "texture" their relative positioning gave rise to, for only these could provide *intelligible* explanations: "if an Angel himself should work a real change in the nature of a Body [said Boyle], 'tis scarce conceivable to us Men, how he could do it without the assistance of Local Motion; since, if nothing were displac'd, or otherwise mov'd than before, (the like hapning also to all external bodies to which it related,) 'tis hardly conceivable, how it should be in it self other, than just what it was before."[24] Leibniz, though no corpuscularian, makes a similar point: "All things come about through certain intelligible causes ... And since we may perceive nothing accurately except magnitude, figure, motion, and perception itself, it follows that everything is to be explained through these four."[25]

1.4. Necessary and Contingent Being

Following Aristotle, Aquinas held that some things, of their nature, must cease to be in the course of time.[26] Others will not cease to be *naturally*, i.e., in the course of nature, though of course God could at any time cease sustaining them in existence.[27]

Something has necessary existence if it cannot cease to be naturally, if its *nature* does not guarantee its eventual corruption. For Aquinas the way for something of a given form to cease to be *naturally* is for the matter of which it is composed to shed that form and acquire another. Anything for which this is not a possibility has *necessary* existence. This means that there could be and, Aquinas believed, there were, two types of things that have necessary existence.

[23] Boyle, *Forms and Qualities*, *Works* 5:300–301.

[24] Boyle, *Excellency and Grounds of the Mechanical Hypothesis*, *Works*, 8:110.

[25] "On a Method of Arriving at a True Analysis of Bodies and the Causes of Natural Things", Leibniz 1961 7:265.

[26] *ST* 1a 2.3 c; cf. Aristotle, *De Generatione*, 337a4–5, 337a8–15, and *De Caelo* 281b25–282a25.

[27] *De Pot.* 5.3, *ST* 1a 10.5 ad 3, and elsewhere. Descartes, Boyle, and others in the early modern period retained the view that God sustained things in existence.

First, anything made of matter which cannot lose its form has necessary existence. No sublunar matter is like this, but mediaeval thinkers believed that the heavenly bodies were made of such matter.[28] As such they had necessary existence, as would Newtonian atoms on this account.[29] Secondly, anything that was not material at all, which was pure form, could not perish naturally, because it had no underlying matter that could lose one form and gain another. Thus the heavenly bodies, for one reason, and the angels, good and bad, as well as human souls and God, for another reason, were all necessary existents.[30] All of these but God had their necessity from another and all but God remain in existence only on God's sufferance.

There are other notions of necessity in Aquinas,[31] but these are the senses most important for early modern thinkers. It may be worth pointing out explicitly that contemporary philosophers who speak as if Thomas's *necessary existence* had something to do with "existence in every possible world" (when discussing the third way, for example) are wandering down an uninteresting anachronistic bypath. Nor is "existence in every possible world" the notion of necessary existence used by many seventeenth century writers since they clearly allowed possible worlds (that of the classical Epicureans, for example) in which the God of Christians did *not* exist.

2. The Early Modern Period: Theologians, Logicians, and Natural Philosophers

Early in the seventeenth century Descartes suggested that eternal truths, such as "mathematical truths ... have been established by God and depend on him entirely, just as all his other creatures do."[32]

[28] "The celestial bodies are said to be incorruptible because they do not have matter of the sort found in generable and corruptible things." *QD De Anima* art 14 sed contra. See also *ST* 1a 66.2c, and *ST* 1a 2ae 49.4c.

[29] Newtonian atoms never "wear or break in pieces; no ordinary Power being able to divide what God himself made one in the first Creation" (Newton 1730, *Qu* 31, 375–376).

[30] See, e.g., *ST* 1a 50.5c.

[31] See further MacIntosh 1998b.

[32] Descartes to Mersenne, April 15, 1630, *AT* 1:145. God can do everything that we understand to be possible, Descartes continued, but we must not think that our imagination reaches as far as his power (*AT* 1:146). There is a large literature on Descartes' claim. See, among others, Frankfurt 1977, Hacking 1980, Curley 1984, Normore 1993. Descartes had no idea *how* God might

This position, already rejected by the orthodox, following Aquinas, did not find general favour in the seventeenth century. It was in fact a very early doctrine: Fritz Zimmermann has noted that circa 700 AD there was an Islamic occasionalist treatise[33] suggesting that God could produce *any* combination of qualities. He could, for example, combine goodness with murder, so that murder would be good, not evil. It was precisely this moral cum theological facet of the doctrine, rather than the mathematical one, that most disturbed seventeenth century writers. Descartes' view, as applied to mathematical truths, was unacceptable, but as applied to moral truths, which they considered equally to be necessary truths, it was unacceptable in a much more important way.

Leibniz praised Bayle for combating "admirably the opinion of those who assert that goodness and justice depend solely upon the arbitrary choice of God."[34] As he remarked earlier, "We need not suppose with certain persons that the eternal truths, being dependent on God, are arbitrary and depend on His will, as Descartes, and afterward M. Poiret, seem to have held."[35] Poiret also came in for criticism in England: "Let us see," asked Lowde rhetorically, "how he answers that Objection, that according to his Principles, the hatred of God might have been good, and the love of him evil, if he had so appointed."[36]

2.1. Necessity in Proofs of God's Existence

As with the mediaevals, what many seventeenth century writers have to say concerning necessity is intertwined with their remarks on religion, and in particular with their attempts, endemic in the seventeenth century, to confound the atheist and demonstrate the existence of God. Here I consider three such attempts, those of Cudworth and Boyle, which seem to require an acceptance of Brouwer and S4, respectively, and of James Lowde, who notes that the relevant sense of 'necessity' in many of the discussions is that of necessity *per accidens*.

have made necessary truths either false or contingent; he simply was unwilling to say that God could not have done it.

[33] Sorabji 1980b, 298.

[34] *Theodicy*, §180, Leibniz 1961 6:221.

[35] *Monadology* §46, Leibniz 1961 6:614; cf *Theodicy* §§335, 351, Leibniz 1961 6:313–314, 322–323, and Leibniz 1981a, 1.1.5, 80.

[36] Lowde 1694, 103.

2.1.1 Cudworth

Descartes' arguments for God's existence met with varied responses in the seventeenth century. They were often recounted somewhat simplistically: an "*Axioma Necessarium*," said Henry More, is "an Axiome that is Necessary, or eternally true." He added, "Wherefore there being a *Necessary Connexion* betwixt *God* and *Existence*, this Axiome, *God does Exist*, is an Axiome Necessarily and Eternally true."[37] Others, seeing the full complexity of Descartes' argument, reacted differently. Cudworth's massive discussion in his *True Intellectual System* was intended "for the Confirmation of Weak, Staggering, and Sceptical Theists,"[38] and in consequence he was worried that Descartes' manoeuvres might be too complex for most people, who would "Distrust, the *Firmness* and *Solidity* of such *thin* and *Subtle Cobwebs*." Cudworth offered various alternative arguments, which might, he hoped, prove more "Convictive of the *Existence* of a God to the Generality."[39] First, a possibly unconvincing Cartesian argument: suppose that a putatively necessary being did *not* exist. That would show its *possible* non-existence. But then we would have the result that if the *necessarily* existing being *did* exist, it would exist not necessarily but contingently. But it "is a manifest *Contradiction*, that the same thing should Exist both *Contingently* and *Necessarily*."[40] However:

> We shall leave the Intelligent and Impartial Reader, to make his own Judgment concerning the forementioned Cartesian Argument for a Deity, drawn from its *Idea*, as including *Necessity of Existence* in it, that *therefore It Is*; ... it is not very Probable, that many *Atheists*, will be convinced thereby, ... they will rather be ready to say, that this is no *Probation* at all of a Deity, ... and a meer *Begging* of the *Question*; that therefore God *Is*, because he *Is*, or *Cannot But be*.[41]

Because this argument is less than clearly convincing,

> we shall endeavour, to make out [a] *Demonstration*, for the *Existence of a God*, from his *Idea*, as including *Necessary Existence* in it, some other ways.... Though it will not follow from hence, because we can *Frame an*

[37] More 1662, 22.
[38] Cudworth 1678, preface.
[39] Cudworth 1678, 725.
[40] Cudworth 1678, 723.
[41] Cudworth 1678, 724.

Idea of any thing in our minds, that therefore such a thing Really Existeth; yet nevertheless, whatsoever we can *Frame an Idea* of, *Implying* no manner of *Contradiction* in its Conception, we may certainly conclude ... that such a thing was not *Impossible to be*; there being nothing ... *Impossible*, but what is *Contradictious* and *Repugnant* to Conception. Now the Idea *of God* or a *Perfect Being*, can Imply no manner of *Contradiction* in it, because it is only the *Idea* of such a thing as hath all *Possible* and *Conceivable* Perfections in it; that is, all *Perfections* which are neither *Contradictious* in themselves, nor to one another.[42] ... Therefore *God* is at least *Possible*, or no way *Impossible* to have been. In the next place as this particular *Idea* of that which is *Possible*, includeth *Necessity of Existence* in it; from these *Two things* put together at least, the *Possibility* of such a Being, and its *Necessary Existence* (if not from the Latter alone) will it according to Reason follow, that *He Actually Is*.[43]

Cudworth takes himself to have shown that it is possibly necessary that God exists and then, implicitly assuming the Brouwer axiom in the form $\Diamond\Box p \rightarrow p$, he has his conclusion. Alternatively, given that *if* God exists, necessarily God exists, $g \rightarrow \Box g$, and given further the possibility of God's existing, $\Diamond g$, then assuming the equivalence of $g \rightarrow \Box g$ and $\Diamond g \rightarrow g$, the result follows immediately.[44]

Cudworth immediately moves to another argument, which depends on necessity *per accidens*: if God is possible then he exists, because if he did not exist, it would be absolutely impossible that he should come to exist. Cudworth draws the correct, disjunctive, conclusion: "Wherefore *God* is either *Impossible* to have been, or else *He Is*."[45] An equivalent conclusion is "God is either necessary to be, or else He does not exist," for with regard to God's existence, proving necessity, actuality, or possibility all amount to the same thing.[46] Cudworth has a variety of other arguments, of which perhaps the most

42 Cudworth's surprising restatement of "a thing as hath all *Possible* and *Conceivable* Perfections in it," as "all *Perfections* which are neither *Contradictious* in themselves, nor to one another," is one we shall meet again. See further Lovejoy 1906, 198–199.

43 Cudworth 1678, 724.

44 Since Cudworth is already, implicitly, invoking the Brouwer axiom, he may be allowed this equivalence, which follows, given that axiom. A similar, if less explicit, argument combining possibility and necessity to yield actuality may be found in Descartes' response to Caterus (*AT* 7:119, *CSMK* 2:85)

45 Cudworth 1678, 724.

46 See further MacIntosh 1991.

interesting is a sophisticated version of Zeno of Citium's "Nothing lacking consciousness and reason can produce out of itself beings with consciousness and reason,"[47] an uncompelling version of which was later offered by Locke.[48]

2.1.2 Lowde

Lowde, clearly invoking the notion of tensed necessity, remarks:

> Whoever grants the possibility of the Being of God, must either grant that he really is, or else will be forced to contradict himself.... [I]f God be not now, and actually was not existent from all Eternity; it is impossible for such a Being ... ever to begin to be.[49] So that an Atheist must prove not only that there is no God, but that it was from Eternity impossible, that there should be one; neither of which, he will ever be able to prove.[50]

Lowde also made use of tensed necessity to argue that it is incumbent on the believer to provide an *argument* to show that the world is *finite* in past time, for if the possibility of an infinite past world is allowed,

> it will be very hard ... by any other Argument to force 'em out of their Opinion: For I do not see, that it would be any absurdity to say, That supposing the World to be Eternal, there has been as many Years as Days; that is, an equal Infinite number of both, all Infinites being Equal; for Infinity can no more be exhausted by Years than Days; But the truth is, such is the nature of Infinite with respect to our Finite Capacity that the one is not a Competent Judge of the other: and when we enter into disputes of this nature, we are often entangled with unanswerable difficulties on both sides.[51]

Earlier Boyle had noted that it is a mistake to assume that "the common Rules whereby we judge of other things, [are] applicable to Infinites,"[52] and much earlier Ibn Ruṣd (1126–1198) pointed out, "Our adversaries believe that, when a proportion of more and less exists between parts, this proportion holds good also for the totalities, but this is only

[47] "Nihil quod animi quodque rationis est expers, id generare ex se potest animantem conpotemque rationis," quoted Cicero, *De nat. deor.* ii 22.

[48] *Essay*, 4.10.2–3. For discussion see MacIntosh 1997.

[49] The earliest clear statement of the point is Aristotle's "It is impossible for what once did not exist later to be eternal (*De Caelo* 283b6–12)." And note also "in the case of eternal things what may be is (*Physics*, 203b30)."

[50] Lowde 1694, 126.

[51] Lowde 1694, 131.

[52] *Things above Reason*, *Works*, 9:369.

binding when the totalities are finite."[53] However, Lowde's awareness that proper subsets and their supersets could have the same cardinality was uncommon in the early modern period. Indeed, the impossibility of such a result was regularly invoked to prove the world's finitude.

2.1.3. Boyle's S4 Argument from Miracles

1. There are "arguments from natural Reason, in favour of Religion." These may not amount to a *Demonstration*, but "if Physicks do but stand neuter, they may be sufficiently justify'd by History and Moral Arguments."[54] And, for any counter demonstration,

> it must be shewn that the Being said to be Impossible, has something inconsistent with the nature of a Being in general, or with the nature or Idæa that is essential to that particular Being. Upon these grounds I see not how it can be demonstrated, that the existence of God is impossible. For the best notion and the most Philosophical we know of the Deity, is, that 'tis a *Being Infinitly perfect*; or, to express the same notion a little more clearly, that 'tis *a Being as perfect as we are able to conceive*.[55]

So God's existence cannot be shown to be *im*possible, which Boyle treats as equivalent to God's existence *not* being impossible, i.e., using 'g' for 'God exists,' $\neg\Box\neg g$. We shall consider this claimed equivalence more closely shortly.

2. Now, let us consider the occurrence, or apparent occurrence, of miracles. Our testimony for miracles is as good as the testimony for any other notable historical occurrence, Boyle claims.[56] Indeed, his claim is milder than that of many of his contemporaries. "To deny and disbelieve all *Miracles*, is either to deny all Certainty of *Sense*, which would be indeed to make *Sensation* it self *Miraculous*; or else monstrously and unreasonably to derogate from *Humane Testimonies* and *History*," said Boyle's contemporary, Ralph Cudworth.[57]

[53] Ibn Ruṣd (Averroes 1954), 1:9.

[54] Among the "positive Reasons afforded by Philosophy to prove a Deity," Boyle offers "the Cartesian Idæa, the Originall of Motion, the use of Parts in Animalls, especially the Eye, the valves of the heart, the musculi perforantes & perforati, & the temporary [parts] of a foetus <& the Mother>" (*BP* 3:113, *BOA* 3.8.2, p. 310).

[55] *BP* 5:106, *BOA* 3.8.3, p. 311.

[56] Boyle offers a defence of the apostles' testimony to miracles at *BP* 3: 317–318, *BOA* 4.6.8, pp. 378–380.

[57] Cudworth 1678, 709.

3. There is only one real objection: "That the Miracles must be false because they suppose the power of God to worke them; and 'tis Impossible that there can be a God, or at least absurd to beleive there is one."[58] But this grants that *if* there is a God, miracles are possible: "this objection denyes not that the Historical proofe of the Miracles is *sufficient in its kind*, that is such as would engage us to believe the matters of Fact, if it were not impossible they should be true."[59] I.e., $g \rightarrow \Diamond m$. This will lead us, though Boyle does not make this explicit, to the result that if it is possible that God exists then it is possible that miracles are possible: $\Diamond g \rightarrow \Diamond\Diamond m$, and given S4, we have; if it is possible that God exists then miracles are possible: $\Diamond g \rightarrow \Diamond m$.[60] But there is a further point here. For since it was only their supposed impossibility that cast doubt on the historical testimony, if they are in fact *possible*, then the historical testimony may be allowed to ground them: $\Diamond m \rightarrow m$.
4. Let us grant the opponent's point that if God does not exist there are no miracles: $\neg g \rightarrow \neg m$.
5. Therefore, God exists.That is we have:

1. $\Diamond g$
2. $\Diamond g \rightarrow \Diamond m$
3. $\Diamond m \rightarrow m$
4. $\neg g \rightarrow \neg m$
5. g

Clearly lines 1 and 3 are dubious, and Max Cresswell has noted that line 3's contingency ensures the contingency of the conclusion. However, seventeenth century arguments for God's existence typically

[58] *BP* 4:60, *BOA* §3.8.1, p. 305.
[59] *BP* 4:70, *BOA* §3.8.1, pp 306–307.
[60] Explicitly:

1. $g \rightarrow \Diamond m$	Theological theorem
2. $\neg\Diamond m \rightarrow \neg g$	1, Sentential Logic
3. $\Box (\neg\Diamond m \rightarrow \neg g)$	2, Necessitation
4. $\Box\neg\Diamond m \rightarrow \Box\neg g$	3, Modal logic
5. $\neg\Box\neg g \rightarrow \neg\Box\neg\Diamond m$	4, Sentential Logic
6. $\Diamond g \rightarrow \Diamond\Diamond m$	5, df $\Diamond$
7. $\Diamond\Diamond m \rightarrow \Diamond m$	S4
8. $\Diamond g \rightarrow \Diamond m$	6,7, syll

aimed to show God's necessary existence, not the claim that necessarily, God exists. Boyle's argument for line 3 was, at the time, completely acceptable. Line 1, however, requires something more by way of defence. In general inability to *demonstrate* impossibility does not yield possibility. Further, Boyle's *"Being as perfect as we are able to conceive"* is not at all the same as "a *Being Infinitly perfect*," a transposition already to be found in Descartes and Cudworth. Descartes, responding to an objection of Caterus, simply invokes a being which has "all the perfections which can exist together."[61] And, as noted earlier, Cudworth suggests that "the Idea *of God* ... is only the *Idea* of such a thing as hath all *Possible* and *Conceivable* Perfections in it; that is, all *Perfections* which are neither *Contradictious* in themselves, nor to one another."[62] Leibniz saw clearly that a proof of the compatibility of God's perfections was needed, and he was not open to limiting God to all the compossible perfections. God has *all* the perfections, and we need to show that the notion of God does not contain a possibly unobvious contradiction as, for example, 'the greatest integer' does. At one time Leibniz thought that it would not be difficult to show this,[63] but it proved more difficult than he hoped.

Any apparently incompatible properties ascribed to God, Boyle suggests, may be due to the imperfections of commentators. And if the apparently incompatible properties really are incompatible and really are claimed of God? Drop one: "thô men shall forbear to ascribe to God any thing that it manifestly appears cannot belong to him, there will remain included in his *Idæa*, excellencies enough to make him the subject of our highest wonder & adoration."[64] In the case of apparent incompatibilities such as that highlighted by Hume in the next century between God's benevolence, omnipotence, omniscience, and the present state of his world, which should we drop? Boyle offers no suggestion, but perhaps his justification of line 1 above may now be allowed to stand, with the caveat that the (possibly diminished) deity whose possibility is invoked may not be a theologically acceptable deity.

61 *AT* 7:119; *CSMK* 2:84.

62 Cudworth 1678, 724.

63 "non est difficile ostendere, *omnes perfectiones esse compatibiles inter se*, sive in eodem esse posse subjecto." Leibniz 1961 7:261.

64 *BOA*, §3.8.3, 311.

3. Necessity in Demonstration

3.1. *The logicians*

By comparison with the work of logicians such as Buridan and Ockham, logic was in a severe decline in the early modern period. Textbook logic tended to be a mixture, albeit often an interesting mixture, of informal logic, epistemology, and psychology. Necessary propositions tended to be passed by in a sentence or two, for example:

> Of Enuntiations some are necessary, and some probable or contingent.... Necessarie Enuntiations are either when the Genus is attributed to the Species; as, *Man is a liuing creature*. Or the difference; as, *Man is reasonable*. Or the propertie; as, *All fire is hot*. The Contingents are when some accident is attributed to the Subject; as, *The horse is white. Man is a Physician.*[65]

Isaac Watts, after noting that there was a "Division of Propositions among the scholastick Writers ... into *pure* and *modal*," suggested that "there are many little Subtilties ... concerning ... these modal Propositions, suited to the *Latin* or *Greek* Tongues, rather than the *English*, and fit to pass away the idle Time of a Student, rather than to enrich his Understanding."[66] The central way in which the notion of necessity is found in the logicians lies in their treatment of *demonstration*.[67] A *scientific* demonstration required the valid deduction of a necessary truth from necessary principles, where the deduction provided an *explanation* or *cause* of the conclusion.

Aristotle distinguishes two types of demonstration: demonstration of "the fact" (*tou hoti; demonstratio quia*) and demonstration of the "reason why" or of the "reasoned fact" (*tou dioti; demonstratio propter quid*). Aristotle provides two examples, from the nearness of the

[65] Du Moulin 1624, 105–106.

[66] Watts 1726, 161. Nonetheless there are various examples of modal consequences among writers in the early modern period, for which see Ashworth 1974, esp. 170–186.

[67] See further *BOA*, §2.1, "The Difficulty of Demonstration," 70–81. For a discussion of *demonstration* as used by philosophers such as Bacon, Hobbes, and Locke see Jesseph 2013 and Jesseph 1999, ch. 5, as well as Peter R. Anstey's "Locke and the Problem of Necessity in Early Modern Philosophy" (Chapter 9, this volume).

planets and their lack of twinkling, and from that sphericity of the moon and its waxing.[68] Consider the two syllogisms:

I	II
The planets do not twinkle	The planets are near
What does not twinkle is near	What is near does not twinkle
The planets are near	The planets do not twinkle

Both are valid, acceptable syllogisms, but only the second provides a true *explanation*. It is not that the planets are near *because* they do not twinkle. However, their lack of twinkling *is* because of their nearness. The second syllogism offers a *causal* explanation of the phenomenon; it gives us the 'reason why', the demonstration *propter quid*.

This became the accepted view among early modern logicians. Galileo in 1590 looked with approval at the Aristotelian notion of demonstration.[69] Du Moulin, in 1624, remarked, "Science is a certaine knowledge of a thing certaine, whose proof is drawne from the cause." This is not sensory knowledge: "If a man know certainly a thing because he seeth it, or toucheth it, that is neither called Science, nor Faith, nor opinion, but *sense*, which knoweth only things singular: but Science is of things vniversall."[70] Gassendi concurred ("A syllogism whose premises are necessary ... is an apodeictic, demonstrative or scientific syllogism"[71]), as did Arnauld and Nicole: "A true demonstration requires two things: first, that the content include only what is certain and indubitable; the other, that there is nothing defective in the form of the argument."[72]

68 *An. Post.* 78a30–78b12. Aristotle offers a reason for the twinkling at *De Caelo* 290a 17–23: The visual ray being excessively prolonged becomes weak and wavering. [This] probably accounts for the apparent twinkling of the fixed stars and the absence of twinkling in the planets. The planets are near, ... but when [the light] comes from the fixed stars it is quivering because of the distance ... and its tremor produces an appearance of movement in the star. The incorruptibility of the heavenly bodies means that truths about them are eternally, or necessarily, true (*An. Post.*, 75b 21–26).

69 Galileo 1992; see further Wallace 1992, ch. 4, "Demonstration and Its Requirements in MS 27," 135–190.

70 Du Moulin 1624, 162, 163.

71 Gassendi 1658, 144.

72 Arnauld and Nicole 1996, part IV, ch. 8, 323–324. Pascal makes similar points in *L'Esprit Geometrique* (Pascal 1954).

In the next century, Isaac Watts noted two kinds of demonstration:

1. Demonstrations *a Priori*, which prove the Effect by its necessary Cause; as, I prove *the Scripture is infallibly true*, because it is *the Word of God, who cannot lye*.
2. Demonstrations *a Posteriori*, which infer the Cause from its necessary Effect; as ... I infer *there is a God, from the Works of his Wisdom in the visible World*. [This] is called *Demonstratio* του οτι, because it proves only the Existence of a Thing; the first is named *Demonstratio* του διοτι, because it shews also the Cause of its Existence.[73]

Throughout the early modern period, then, for logicians, a *scientific* demonstration not only had a necessary truth as its conclusion; it should give us the 'reason why': it should be a *causal* demonstration. This provided a difficulty about mathematics, to be noted shortly. It was, incidentally, this lack of a *causal* or *explanatory* connection that Aquinas found troubling in the notion of necessity held by Diodorus Cronos and by the Stoics:

Diodorus distinguished possibility and necessity according to the outcome: that which will never be the case is impossible; that which will always be the case is necessary; that which will sometimes be the case, and sometimes not, is possible. The Stoics ... distinguished between them according to external prevention. They said that which cannot be prevented ... is necessary; that which is always prevented ... is impossible; while that which can be prevented or not is possible.

Both these ways of making the distinction seem faulty. The former [because] it is not that something is necessary because it will always be the case; rather it will always be the case because it is necessary.... The latter [because] it is not that something is necessary because it is not prevented; rather, it cannot be prevented because it is necessary.[74]

However, there were problems with the logicians' account. It was, Bacon noted, not the kind of demonstration needed in natural philosophy. Consequently, Aristotle's point – "Our discussion will be adequate if it has as much clearness as the subject-matter admits"[75] was regularly invoked, and it was generally accepted that something other

[73] Watts 1726, 309.
[74] *In Perih.*, I, 14.8.
[75] *Nicomachean Ethics*, 1094b12–13, and cf. *Metaphysics*, 995a 15–16.

than the logicians' *demonstration* was appropriate in theology, in law, and in experimental philosophy.

Besides the Demonstrations wont to be treated of in vulgar Logick, there are among Philosophers three distinct ... *kinds* ... of Demonstration. For there is a *Metaphysical* Demonstration, ... such as *Nihil potest simul esse & non esse....* There are also *Physical* Demonstrations ... such as ... *ex nihilo nihil fit:* ... And ... *Moral* Demonstrations ... where the Conclusion is built, either upon some one *such* proof cogent in its kind; or some concurrence of Probabilities, that it cannot be but allowed.

And this *third* kind of Probation, though it come behind the two others in certainty, yet it is the surest guide, which the *Actions of Men*, though not their *Contemplations*, have regularly allow'd them to follow.[76]

3.2. The Mathematicians

Aristotelian *demonstration*, then, was at best incomplete. But the Aristotelian notion of demonstration led to a general question about mathematics: is mathematics a *science*? Aristotle considered mathematics to be a science (or a group of sciences), but, later writers wondered, is it truly scientific in the sense of providing the *reason why*, the *explanation* of its results? Moreover, some of its results apparently *cannot* provide such a (causal) explanation, for they are equivalence results. But if we have p if and only if q, then at most one of the conditionals can be such as to provide a demonstration of the "reasoned fact." Otherwise we would be faced with the claim that some result was its own explanation.[77]

Logicians also found a related problem with indirect mathematical proofs. In the *Port Royal Logic* Nicole and Arnauld picked out six "common mistakes in the geometers' method," including "Demonstrations by impossibility," for this method "can convince the mind but it cannot enlighten it,"[78] a point picked up by Watts in the next century:

Labour in all your Arguings to enlighten the Understanding, as well as to conquer and captivate the Judgment.... *Geometricians* sometimes break this Rule ... When they prove one Proposition only by shewing what Absurdities

[76] *Reason and Religion, Works* 8:281.
[77] See further Mancosu 1996, to which I am considerably indebted in this section.
[78] Arnauld and Nicole 1996, 255.

will follow if the contradictory Proposition be supposed ... as for Instance, When they prove all the *Radii of a Circle* to be equal, by supposing one *Radius* to be longer or shorter than another.... This ... forces the Assent, but it does not enlighten the Mind by shewing the true Reason and Cause why *all Radii are equal*, which is derived from the very Construction of a Circle....[79]

Mathematicians took the causality charge seriously, and Isaac Barrow, Newton's predecessor in the Lucasian Chair, produced an answer in the mid 1660s.[80] Barrow pointed out, anticipating Hume, that *no* case of presumed efficient causality could support a claim of necessity between cause and effect.[81] Only *formal* causality could yield the desired necessity, but *that* was precisely what mathematics provided:

Nor ... is [there] any other *Causality* in the Nature of Things, wherein a necessary Consequence can be founded. Logicians do indeed boast ... of Demonstrations, from *external Causes* either *efficient* or *final*; but without being able to shew one genuine Example ... For there can be no ... efficient Cause in the Nature of Things ... which is altogether necessary. For every Action of an *efficient Cause*, as well as its consequent *Effect*, depends upon the *Free-Will* and Power of *Almighty God*, who can hinder the ... Efficacy of any *Cause* at his Pleasure.... Hence it is possible that there may be such a *Cause* without a *subsequent Effect*.... There can therefore be no Argumentation from an efficient *Cause* to the *Effect* ... which is lawfully necessary.[82]

This, Barrow suggested, solves both the causal problem and the equivalence problem. The purported necessity of efficient causality is denounced as an impossible requirement; and the worry, if worry there be, about the lack of causality in mathematics is evaded by the shift to formal causality, with which mathematics abounds, and which allows equivalences as results.

Effectively, then, both natural philosophy and mathematics freed themselves from the 'causal' constraint the logicians found important. The notion of tensed, or *per accidens*, necessity continued to play an

[79] Watts 1726, 334–335.

[80] Published in 1683 as *Lectiones Mathematicae*, translated in Barrow 1734.

[81] Anders Kraal has pointed out to me that Dugald Stewart also noted Barrow's anticipation of Hume (Stewart 1805, 55–58).

[82] Barrow 1734, 88–89.

important role in theological contexts, but was little discussed by the logicians. The notion of *moral demonstration* (effectively, in natural philosophy, empirical demonstration) keep coming to the fore in legal, theological, and natural philosophy contexts provided a needed extension to the "Demonstrations ... in vulgar Logick."

9 *Locke and the Problem of Necessity in Early Modern Philosophy*

PETER R. ANSTEY

In the early modern period necessity was tied to each of a clutch of interlocking notions: causation, laws of nature, essence, demonstration, logical inference, mathematical proof, etc. So, for example, some argued that if existence is of the essence of God, then God exists necessarily. Again, some argued that if the laws of motion are not known a priori but are discovered by experience, then they are not necessary. There was, however, no attempt to develop a typology of necessities, such as our contemporary distinctions among logical necessity, epistemic necessity, metaphysical necessity and nomic necessity. Indeed, a central concern of current-day interpreters of early modern texts is to disentangle claims about necessity in one domain, say, causation, from those in another, say, logic. Nor was the problem of necessity a desideratum for early modern philosophers to solve. There is a sense then in which it is slightly anachronistic even to speak of the problem of necessity in early modern philosophy.

Nevertheless, the modal notion of necessity features prominently in early modern philosophy and it is important that we as modern-day interpreters of early modern texts find ways to understand and explain how it was understood. To that end, it will be helpful to find a general way of framing the problem of necessity in early modern philosophy. What I offer here is a two-stage analysis that aims to capture the broad contours of a cluster of problems that we would now call problems of necessity. As such, it provides a reference point against which different philosophers' accounts of modal notions can be compared and assessed. The chapter has two sections. Section 1 presents a general heuristic for studying the problem of necessity in the early modern period, including a crucial development in the early modern view of necessary facts about the world. Section 2 applies that heuristic, by way of example, to the philosophy of John Locke.

1. The General Problem of Necessity in Early Modern Philosophy

Most early modern philosophers admit that some propositions are true irrespective of the way the world is and would be true even if there were no world. Such truths include those of mathematics and self-evident principles, such as 'Whatever is, is'. Surprisingly, however, they were more concerned with the grounds of absolute certainty than with necessary truth per se. Debates about absolute certainty were more often framed around the notion of eternal truths (*aeternae veritates*) than around the notion of necessary truth. And many discussions related the absolute certainty of eternal truths to the will of God. Descartes, for example, notoriously claimed in a letter to Mersenne that

> [God] was free to make it not true that all the radii of the circle are equal — just as free as he was not to create the world (CSMK, Vol 3, 25)

Yet he claims in *The World* that

> knowledge of these [eternal] truths is so natural to our soul that we cannot but judge them infallible when we conceive them distinctly, nor doubt that if God had created many worlds, they would be as true in each of them as in this one. (Descartes 1985, vol. 1, 97)

This is no place to try to untangle Descartes' notion of necessity,[1] but what is evident, from the passage in *The World*, is that for Descartes necessary propositions provided a foundation from which other knowledge was derived. He tells us that one who has a sufficient grasp of the consequences of these truths 'will be able to have *a priori* demonstrations of everything that can be produced in this new world' (ibid.). It is Descartes' reference to 'this new world' that provides our entrée into the main forms of necessity that are the concern of this chapter. Descartes' imaginary world in *The World* is, to all intents and purposes, our world and it is necessary truths about our world that are our concern here.

Most early modern philosophers were committed to five theses about necessity as it pertains to the world: two metaphysical theses

[1] For a survey of the issues see Cunning 2014.

and three epistemological theses. First, they believed that there are necessary *facts* about the world.[2] Of course, we need to be careful not to load the term 'fact' with too much ontological baggage here. What we mean is that many early moderns believed that there are things about the world, broadly construed, that are necessary, or, again, that there are necessary features of the world, such as the co-occurrence of certain properties. I will, however, for better or for worse, continue to use the term 'facts'. Secondly, for many philosophers in the seventeenth and early eighteenth centuries, the most important of these necessary facts about the world are facts about the essential natures of things: that is, they are facts about essences. Turning to Descartes again, he claims in the Fifth Meditation:

> When, for example, I imagine a triangle, even if perhaps no such figure exists, or has ever existed, anywhere outside my thought, there is still a determinate nature, or essence, or form of the triangle which is immutable and eternal, and not invented by me or dependent on my mind. (Descartes 1985, vol. 2, 44–45)

Thirdly, most early modern philosophers believed that we can have epistemic access to some of these necessary facts. That is, we can and do know the essences of some things. This is important because, fourthly, many were also committed to the view that knowledge of necessary facts about the world is foundational. It is foundational in so far as this knowledge is not based on anything other than the facts themselves, unless, perhaps, it is based on the nature of God. Fifthly, as Descartes claims in the quote provided above, knowledge of necessary facts can generate new knowledge. The widespread view was that we are able to reason on the basis of these foundational necessary facts about the world in such a way as to generate systematic bodies of knowledge across whole domains, be it a science or *scientia* of nature, of morality or of politics.

If these five theses constitute a core of common belief about necessity, they also provide reference points for widespread disagreement about necessity. For example, while there was general agreement that there are essences in the world, the nature of these essences was

[2] On the emergence of the notion of fact as an epistemic category in the seventeenth century, see Serjeantson 2006, 158–164.

enormously controversial and hotly debated. And while there was agreement that we can know some of these essences, early modern philosophers disagreed over the manner in which we gain *epistemic access* to these modal facts. Again, while many held the view that knowledge of these necessary facts could lead to new knowledge about nature, they disagreed as to the mechanism by which this knowledge was generated.

Early in the seventeenth century the problem of how we gain new knowledge about nature on the basis of these necessary facts was commonly framed in a manner that derived from Aristotle. This was the view that systematic knowledge of nature, or *scientia*, could be derived by syllogistic reasoning from premises about necessary facts about the world. Indeed, when it came to the study of nature, many believed that an ideal demonstrative natural philosophy would be one that was both complete and based upon, inter alia, a finite set of propositions about the essences of natural substances and properties.[3] Thus, the question of epistemic access to necessary facts about the world, framed as knowledge of essences, was of the utmost importance: knowledge of necessary facts provides the basis for further knowledge. This bears reiterating: knowledge of modal facts about things in the world is a gift that keeps on giving; the certainty of these modal facts flows through to all other truths demonstrated from them.

Aristotle claimed that we could only obtain knowledge of essences through the use of *nous*, but there is sharp disagreement amongst commentators as to what *nous* is and the process by which the use of *nous* enabled the acquisition of knowledge of these particularly important modal facts.[4] If we substitute the early modern English term 'reason' for the Stagirite's *nous* we can set up the problem of epistemic access to necessary facts about the world as an inconsistent tetrad.[5]

1. We have knowledge of necessary facts about the world.
2. We can have knowledge of necessary facts about the world only if we acquire it by observation or reason.

[3] For Aristotle's theory of principles and knowledge acquisition in the *Posterior Analytics*, see McKirahan 1992.

[4] See, for example, the discussion in Perelmuter 2010.

[5] Some commentators use 'intuition' for *nous* and consider it a species of reason. Still others go further and identify intuition with induction. I choose to gloss over these distinctions here by sticking with the general term 'reason'.

3. We cannot acquire knowledge of necessary facts about the world from observations.
4. We cannot acquire knowledge of necessary facts about the world by reason.

This inconsistent tetrad provides a helpful heuristic for examining the views of leading philosophers of the early seventeenth century on the question of our epistemic access to essences. So, for example, Francis Bacon denied propositions 3 and 4 because he believed he had a new theory of induction that circumvented the problem of generalizing from particulars to necessary truths. Descartes denied proposition 4 because he argued that we can have innate knowledge of the essences of some things in nature.[6] Ralph Cudworth went even further than Descartes, claiming that there is a perceptive faculty in the soul that has immediate epistemic access to the essences themselves:

> The proper and immediate objects of science rightly so-called and intellection (τὰ νοητά), being the intelligible essences of things and their necessary verities, that exist nowhere but in the mind itself, the understanding by its active power is fully master of them, and comprehends 'not idols or images of them', but the very things themselves within itself.[7]

The Leibniz of the *New Essays* clearly affirmed proposition 3. He claims in the preface to that work:

> And we have reason to believe that these flashes reveal something divine and eternal: this appears especially in the case of necessary truths. That raises another question, namely whether all truths depend on experience, that is on induction and instances, or if some of them have some other foundation. For if some events can be foreseen before any test has been made of them, it is obvious that we contribute something from our side. Although the senses are necessary for all our actual knowledge, they are not sufficient to provide it all, since they never give us anything but instances, that is particular or singular truths. But however many instances confirm a general truth, they do not suffice to establish its universal necessity; for it does not follow that what has happened will always happen in the same way.[8]

[6] See, for example, Descartes, *Rules for the Direction of the Mind*, (*CSMK*, vol. 1: 22 and 44.)

[7] Cudworth 1996, 135.

[8] Leibniz 1996, 49.

This statement of inductive fallibilism is tantamount to an acceptance of the claim that we cannot reason from observation to knowledge of essences. And Leibniz uses this to claim that there must be another source of our knowledge of necessary truths over and above experience and that source is innate ideas.[9]

Needless to say, we could go on through the canon of leading philosophers from the period and analyse their views with respect to the tetrad. We could also explore the content of each proposition and find sites within important early modern texts that deal with each of them. Thus, we could analyse, say, book II of Bacon's *Novum organum* as a response to proposition 3. However, this would leave the picture incomplete, for I contend that there is an important second stage to the development of the problem of necessary facts about the world.

Anyone who is familiar with early modern logic texts and discussions of method from the period will know that a science, or *scientia*, in a particular discipline was founded not merely upon necessary propositions about the essences of its primary subject matter, but also upon a finite set of principles.[10] These principles included general principles applicable to all disciplines and proper principles that were *necessary* propositions that referred to the fundamental notions of a discipline. So, for example, in geometry there were proper principles about lines and points such as 'a line is a length without breadth'. These principles were subject to classification, the standard fault line being between speculative and practical principles following the division of the faculties of the soul as speculative and practical and the divisions of types of knowledge as speculative and practical. (There were other ways of dividing up principles, such as by the manner by which the mind grasped the principles in question.)

Once a philosopher had the general principles, the set of principles that pertained to his or her discipline, *and* propositions that specified the essences of the relevant species, that philosopher could then proceed to construct demonstrative syllogisms that through careful and systematic elaboration would eventually constitute a science. At least that was the theory. In practice, however, this method failed to deliver and, as is well known, one of the significant developments in

[9] Leibniz 1996, 51.

[10] See, for example, Blundeville 1617, 163–164.

early modern philosophy was the modification and constraining of this whole approach to knowledge acquisition.

So, in the early decades of the seventeenth century it was widely thought that in order to acquire knowledge in a discipline, one needed at least the following: a set of principles, a set of propositions about the essences of the relevant subject matter, and the syllogistic. Now, as we have seen, there was a deep problem about our epistemic access to essences, a problem that had beset Aristotle's *Posterior Analytics* and remained unsolved down to the time of Bacon. We arrive then at the key development that I am characterising as the second stage in the history of early modern necessity and that is the emergence of the concept of laws of nature. What I want to assert, and this is the central developmental claim of this chapter, is that *initially the laws of nature were grafted onto this neo-Aristotelian conception of the generation of knowledge, or* scientia, *because they were regarded as principles, but they were principles that expressed new modal facts*.

There is a sense in which the introduction of this notion, initially in the writings of Descartes, widened the range of modal facts that were to be discovered. Not only do natural philosophers need knowledge of the essence of, say, humans or heat, but they now needed to know necessary facts about the *behaviour* of natural bodies: bodies with gravity or bodies that can refract light, etc. Thus, the problem of epistemic access to modal facts seemed to broaden.

Five points are worth noting about this development. First, by the end of the century, it was pretty clear that these modal facts expressed as laws of nature were able to generate new knowledge. In fact, some regarded them as more effective than necessary propositions about essences at generating new knowledge of nature. Indeed, Katherine Brading has argued that these new modal facts actually provided a new way to discover the essences of things and therefore provided a partial solution to the problem of epistemic access to essences.[11] Second, these new modal facts that were understood as laws of nature were almost unanimously believed to be learned from experience and not innate and so they, with respect to the inconsistent tetrad, provided cases in which proposition 3 is false.[12] Third, as with knowledge

[11] See Brading 2011 and 2012.

[12] A heroic attempt to argue that the laws of mechanics are known a priori is that of Jean Le Rond d'Alembert in his *Traité de dynamique*, Paris, 1743.

of essences, however, the nature of the necessity involved in laws of nature was often inextricably tied to the will or mind of God. Thus, the theological complications that plagued the problem of knowledge of essences were not avoided in the case of laws. Descartes, for example, derived his laws of motion from the immutable nature of God.[13]

Fourth, most modern-day analyses of laws of nature involve relations, yet in the seventeenth century very few philosophers were realists about relations. On the whole when it came to relations nominalism reigned. This, it could be argued with hindsight, inhibited philosophical analysis of the nature of laws of nature, including their modal status. Fifth, even though the notion of laws of nature was new, philosophers had long discussed the notion of causation and 'cause' itself was a deeply theorized term in the seventeenth and eighteenth centuries and theories of causation invariably invoked modal notions. Thus, for example, Malebranche seems to have analyzed causal necessity in terms of logical necessity.[14]

In summary then, in the seventeenth century there was a general problem about our epistemic access to necessary facts in the world. Early in the century this problem was framed almost exclusively in terms of knowledge of essences of natural substances and properties. Later in the century, however, it was complicated by the introduction of a new type of necessary fact, laws of nature. These laws of nature were in some quarters regarded as additional or alternative principles that could complement a demonstrative natural philosophy. In other quarters, laws of nature came to be a means of generating new knowledge of nature in their own right, rendering the need for a demonstrative natural philosophy redundant.

Having established this two-stage heuristic for analysing the problem of epistemic access to necessary facts in early modern philosophy, I should now like to set out some guidelines by which to apply it on a case-by-case basis. First, we need to determine whether a philosopher believes that there are necessary facts and if so, to determine what that philosopher believed about the nature of essences. In examining a philosopher's views about essences it will be helpful to assess them in relation to the inconsistent tetrad. Second, we need to take account of a philosopher's views on the Aristotelian ideal of a

[13] See *Principles of Philosophy*, II, §§36–39, *CSMK*, vol. 1, 240–241.

[14] See the nuanced discussion in Fisher 2011.

demonstrative natural philosophy. This will include an examination of his or her views on the syllogistic, laws of nature, and principles. In addition we should address some or all of the following issues: his or her views on *innate principles* and ideas; his or her views on the nature of *causation*; his or her views on the nature of *sensory perception*; his or her theory of *demonstration* (and induction); his or her views of *God's relation* to the laws of nature; and his or her views of the nature of volition.

2. Locke on Necessity

In the remainder of this chapter I will focus on the views of the English philosopher John Locke. I should stress, however, that Locke is not an exemplar of early modern treatments of necessity, but merely one among a diverse constellation. Indeed, while some aspects of Locke's views on these matters were influential, there is no other philosopher whose views on the nature of necessity are quite like those of Locke. Moreover, Locke's views were not static, but developed over time and underwent significant change in the last decade of his life; nevertheless, they are particularly useful for illustrating the heuristic that I have presented here.

Locke accepted that *there are modal facts* about the world and that these are facts about the *essences* of things. He developed a sophisticated theory of essences. He defined essence as follows: '*Essence* may be taken for the very being of any thing, whereby it is what it is' (*Essay* III. iii. 15). He distinguished between real and nominal essence. Real essence is 'the real internal, but generally in Substances, unknown Constitution of Things, whereon their discoverable Qualities depend' (ibid.). Nominal essence is 'nothing else, but those abstract complex *Ideas*, to which we have annexed distinct general Names' (*Essay* III. iii. 17). That this is a genuine theory of essences is evident from the following modal claim:

> It is as impossible, that two Things, partaking exactly of the same real *Essence*, should have different Properties, as that two Figures partaking in the same real *Essence* of a Circle, should have different Properties. (Ibid.)

Locke accepted the ideal of a demonstrative natural philosophy though in a highly qualified sense. He believed that if we could penetrate to the inner natures of substances, their real essences, then we could deduce the qualities that they have:[15]

> These insensible Corpuscles, being the active parts of Matter, and the great Instruments of Nature, on which depend not only all their secondary Qualities, but also most of their natural Operations, our want of precise distinct *Ideas* of their primary Qualities, keeps us in an incurable Ignorance of what we desire to know about them. I doubt not but if we could discover the Figure, Size, Texture, and Motion of the minute Constituent parts of any two Bodies, we should know without Trial several of their Operations one upon another, as we do now the Properties of a Square, or a Triangle. Did we know the Mechanical affections of the Particles of *Rhubarb*, *Hemlock*, *Opium*, and a *Man*, as a Watchmaker does those of a Watch, whereby it performs its Operations, and of a File which by rubbing on them will alter the Figure of any of the Wheels, we should be able to tell before Hand, that *Rhubarb* will purge, *Hemlock* kill, and *Opium* make a Man sleep; ... But whilst we are destitute of Senses acute enough, to discover the minute Particles of Bodies, and to give us *Ideas* of their mechanical Affections, we must be content to be ignorant of their properties and ways of Operation; nor can we be assured about them any farther, than some few Trials we make, are able to reach. But whether they will succeed again another time, we cannot be certain. This hinders our certain Knowledge of universal Truths concerning natural Bodies: and our Reason carries us herein very little beyond particular matter of Fact. (*Essay* IV. iii. 25, underlining added)

How does one *demonstrate* that, say, rhubarb will purge and hemlock will kill? Locke had an idiosyncratic theory of demonstration. This was predicated on his theory of ideas. Space does not allow a full elaboration of this theory, but the bare bones are as follows.[16] Knowledge in its most basic form is the perception of the agreement or disagreement of ideas. This agreement or disagreement can be perceived immediately or mediately. Immediate perception of agreement is intuitive knowledge. Mediate perception is demonstrative knowledge.

[15] The claim that Locke was committed to a demonstrative natural philosophy in a rather strong sense is denied by some of Locke's interpreters. What worries them are Locke's claims in the *Essay* that God might superadd powers to bodies and that we cannot know what is the relation between primary and secondary qualities. See, for example, Stuart 1996 and Stuart 2013, 266–277.

[16] For a more detailed account of Lockean demonstration see Anstey 2011, chap. 7.

It is mediate in so far as a third idea, an intermediate idea that Locke calls a proof, is required for the perception to take place.

Demonstration, then, is the perception of the agreement or disagreement of two ideas with the assistance of a third idea. This primal form of knowledge is not propositional, but rather is a form of what Bertrand Russell called knowledge by acquaintance.[17] As such, demonstrative knowledge is not about the relation between propositions and does not involve the notion of validity. One important consequence of Locke's theory of demonstration is that he *rejected the syllogistic* as an efficacious means of acquiring knowledge of nature. Of course, Locke still needs to give an account of how the perception of the agreement or disagreement of the ideas is expressed propositionally, and he provides this in his account of the formation of mental and verbal propositions.[18] Nevertheless, Locke does admit to a severe limitation with regard to epistemic access to certain modal facts and it is to this that we now turn.

Instead of believing that demonstration concerns the relations between propositions, Locke claims that at the most basic level acts of demonstrative reasoning are related to perceiving the agreement or disagreement of ideas in one of four different ways:

1. *Identity*, or *Diversity*
2. *Relation*
3. *Co-existence*, or *necessary connexion*
4. *Real Existence* (*Essay* IV. i. 3)

It is clear that this list is related to modal facts. Let us take 'Co-existence, or necessary connextion'. This concerns the necessary connections between properties of substances such as the connection between the colour and malleability of gold. Since Locke gives particular attention to this form of agreement or disagreement and this concerns modal facts, we shall examine his view in more detail.

As we saw in the last extended quotation, Locke has a very pessimistic account of our prospects for determining the necessary connections between the properties of substances and between those properties and our ideas of them. He discusses this at length in *Essay* IV. iii. 9–17. He regards the determination of the connections between the

[17] See Russell 1976 and Anstey forthcoming.

[18] *Essay* IV. v.

properties of bodies to be 'the greatest and most material part of our knowledge concerning Substances' (*Essay* IV. iii. 9). By 'greatest' here he means important, for, our knowledge of these connections is 'very short'. This is because, in the first place, the simple ideas that make up the complex idea of a particular substance do not carry with them any 'visible necessary connexion, or inconsistency with any other simple *Ideas*' (§10). Locke seems to have in mind the manner in which some ideas 'carry' with them the additional idea that it is derived from memory or from sensation (*Essay* II. x. 2). There is no analogue of this when it comes to connections with other simple ideas.

Furthermore, most of the simple ideas that constitute our ideas of substances are ideas of secondary qualities, and the ontological grounds of these ideas, that is, the primary qualities of their insensible parts, are too removed from us for us to know which secondary qualities 'have a necessary union or inconsistency one with another' (§11). Nor do we have epistemic access to the connections between the underlying primary qualities and each individual secondary quality (§§12, 13). So we are ignorant of necessary connections between secondaries, and between secondaries and underlying primaries. We can know by intuition and demonstration that some primary qualities are necessarily connected, such as shape and extension (§14), and that the having of a determinate primary or secondary necessarily precludes the having of all others that fall under that determinable (§15). This modal knowledge derives from the ideas or concepts involved and not from observation. Furthermore, Locke believed that if we are going to make any headway on this problem for material substances it will probably be through the corpuscularian hypothesis (§16), but the situation is even worse when it comes to our ideas of the properties of spirits (§17).

Let us take stock so far. Locke believed in modal facts, he had a theory of essences, and he was committed to the ideal of a demonstrative natural philosophy. However, his view of demonstration bears little resemblance to the Aristotelian one. What then is Locke's view of the inconsistent tetrad?

1. We have knowledge of modal facts about the world.
2. We can have knowledge of modal facts about the world only if we acquire it by observation or reason.

3. We cannot acquire knowledge of modal facts about the world from observations.
4. We cannot acquire knowledge of modal facts about the world by reason.

If proposition 1 were restricted to knowledge of the real essences of *substances*, whether material or spiritual, clearly he would reject it, because he believes that we do not have epistemic access to the real essences of things.

There has been some debate as to whether Locke believes that we cannot have epistemic access to real essences in principle or as a matter of contingent fact. In my view this debate has been settled by Lisa Downing,[19] who has shown that, for Locke, it is a contingent matter that we cannot perceive real essences, but that with the development of more powerful microscopes we may one day have such knowledge. Nevertheless, Locke was supremely pessimistic on this point.

> And therefore I am apt to doubt that, how far so ever humane Industry may advance useful and *experimental* Philosophy *in physical Things*, *scientifical* will still be out of our reach: because we want perfect and adequate *Ideas* of those very Bodies, which are nearest to us, and most under our Command.... *Certainty* and *Demonstration* are Things we must not, in these Matters, pretend to. (*Essay* IV. iii. 26)

Yet Locke's view about our epistemic access to essences is not restricted to our ideas of substances, for, within his theory of ideas he allows for a broad class of ideas that he calls modes. Locke's use of the term 'mode' is not to be conflated with the ontological category of mode that was introduced by Suárez and popularized by Descartes. For Locke, modes are ideas that are modified in certain ways. Simple modes are those ideas that are modified by simple functions on that very idea, such as iteration. Thus, the numbers 2 and 9 are modifications of the idea of one or unity.[20] Complex modes, such as obligation or murder, are combinations of different simple ideas that are

> not looked upon to be the characteristical Marks of any real Beings that have a steady existence, but scattered and independent *Ideas*, put together by the Mind. (*Essay* II. xxii. 1)

[19] Downing 1992.
[20] *Essay* II. xvi. 2, 5.

In the case of our ideas of substances, we have no way of knowing whether our nominal essence, say, the complex idea of gold, corresponds to the real essence in the world. By contrast, in the case of our ideas of mixed modes, Locke claims that our nominal essences are the *same* as the real essences. If Locke means here that for any mixed mode the nominal essence is *identical* to the real essence, then the problem that besets our knowledge of real essences of substances is avoided: for mixed modes we have immediate epistemic access to real essences because the real essences just are those clusters of ideas that constitute the nominal essences. This promises to guarantee epistemic access to modal facts across the whole domain of mixed modes.

Unfortunately, however, this 'ideational' reading of Locke on mixed modes is not so straightforward. Part of the problem is that Locke explicitly claims that many of our ideas of mixed modes are derived 'By Experience and *Observation* of the things themselves' (*Essay* II. xxii. 2), and he often speaks of the transient nature of the real world correlates of our ideas of mixed modes, which, unlike substances, are 'not capable of a lasting Duration' (*Essay* III. vi. 42). Thus, the real essences of wrestling and murder are in the world, even if they never last very long. Moreover, when Locke says that for mixed modes the nominal essence is the *same* as the real essence, he could be equivocating between identity and equivalence. As for simple modes such as shapes and the geometrical propositions that are constructed from them, Locke appears to hold that the ideas are real and the propositions are certain in virtue of the nature of things in the world, and, in particular, the nature or essence of Euclidean space.[21] If this line of interpretation is the interpretation of best fit for the various passages in which Locke discusses ideas of simple and mixed modes, then it is clear that the certainty of propositions constructed from them depends upon the essences of things in the world.[22] Locke claims concerning someone who knows true propositions about triangles:

he is certain all his Knowledge concerning such *Ideas*, is real Knowledge: because intending Things no farther than they agree with those his *Ideas*, he is sure what he knows concerning those Figures, when they have barely *an*

[21] *Essay* IV. iv. 6.
[22] For a detailed discussion of these issues, see Anstey 2011, chap. 6.

Ideal Existence in his Mind, will hold true of them also, when they have a real existence in Matter. (*Essay* IV. iv. 6)

We as modern-day interpreters might want to gloss this by saying that Locke is grounding the *epistemic* necessity of, say, geometrical propositions, on the *metaphysical* necessity of the essential nature of actual (as he presumed it to be) Euclidean space.

What then is Locke's view of eternal truths? He is keen to stress that, in contrast to propositions that assert the existence of particular things, we are able to form universal and certain propositions from the ideas of the essences of things: 'Truths belonging to Essences of Things, (that is, to abstract *Ideas*) are eternal, and are to be found out by the contemplation only of those Essences' (*Essay* IV. iii. 31). Given that some of those essences of modes may be perceived by the senses – such as murder and wrestling – Locke cannot maintain that all eternal truths are known a priori. Nevertheless, eternal truths are propositions constructed as a result of our perceiving the agreement or disagreement of particular abstract ideas. The example he gives is 'Men ought to fear and obey God'. For Locke, this sort of proposition is an eternal truth because the certainty of the proposition is guaranteed by the relations between the abstract ideas:

because being once made, about abstract *Ideas*, so as to be true, they will whenever they can be supposed to be made again at any time past or to come, by a Mind having those *Ideas*, always actually be true. (*Essay* IV. xi. 14)

And yet, almost in the same breath, Locke can say that this proposition – 'Men ought to fear and obey God' – 'proves not to me the Existence of Men in the World, but will be true of all such Creatures, whenever they do exist'.[23] But this is equivalent to saying that 'the sum of the internal angles of all triangles is 180°' is certain and will be true of all triangles in Euclidean space whenever and wherever they do exist. Or again, that the proposition 'The Cartesian idea of space disagrees with the idea of a vacuum' is certain and will be true of all existing Cartesian worlds.[24] This is not to say, however, that all eternal truths are mere trifling propositions. The example of the derivation of

[23] *Essay* IV. xi. 13.
[24] See *Essay* IV. vii. 12–13.

geometrical truths shows that such knowledge can be instructive and not merely trifling.[25]

It might seem, particularly when we consider modal ideas, that Locke restricts intuitive and demonstrative knowledge to conceptual knowledge. But Locke's view is subtler than this. Take, for example, his distinction between certainty of truth and certainty of knowledge. The former, certainty of truth, occurs 'when Words are so put together in Propositions, as exactly to express the agreement or disagreement of the *Ideas* they stand for, as really it is'.[26] The latter, certainty of knowledge, is merely the perception of the agreement or disagreement of ideas. Only certainty of truth will give us real knowledge of modal facts about the world. In those cases when essences of modes are actually perceived by our senses we have the certainty of truth – we can know that wrestling involves motion – because the real essences are the same as the nominal essences. Importantly, however, we can also have certainty of truth in the case of verbal propositions that signify simple ideas, propositions such as that this snow is white. This is because, for Locke, as in modal ideas so 'in all simple *Ideas*' the real and nominal essence is the same.[27] Locke, therefore, believes that we can have knowledge of many contingent truths.

This takes us, finally, to Locke's view of principles and on this point I believe that Locke has often been badly misunderstood. Locke's view of principles changed over time. We fail to do Locke justice on this point if we presuppose that his position on principles was a static one. Furthermore, Locke's view of principles has been obscured or even overlooked because of the agendas and historiographical orientations of Locke's later interpreters. In my view, however, it is crucial to get the interpretation of Locke right on this point, not least because it enables us to see how Locke dealt with the issue of epistemic access to necessary facts about the world.

Let us start then with a very standard logic, Thomas Blundeville's *The Arte of Logick*, 1617. It is worth looking at Blundeville's discussion of certain knowledge before turning to Locke. Recall that on the traditional Aristotelian view what is required for *scientia* is necessary

[25] *Essay* IV. viii. 8. Locke regards propositions expressing identity as trifling; see *Essay* IV. viii. 1–3. For further problems in Locke's account of our epistemic access to modal facts, see Bennett 2000.

[26] *Essay* IV. vi. 2, underlining added.

[27] *Essay* IV. vi. 4.

propositions about essences, the syllogistic, and principles. The question, however, is how do we know what the principles are?

Blundeville begins by saying that certain knowledge resides in three things: universal experience, principles, and the judging of consequents, that is, deductive inference. We turn next to principles. Blundeville asks, 'What be Principles?'. His answer is

> Principles bee certaine generall conceptions and naturall knowledges grafted in mans minde of God to the intent that by the helpe thereof, he might invent such Arts as are necessarie in this life.... (Blundeville 1617, 163)

He then asks, 'How many Divisions doe the Schoole-men make Principles?' His answer, 'Divers'. And next he commands 'Rehearse those Divisions' Answer: 'The first is of Principles, some be called Speculative, and some Practive....'

These are precisely the doctrines that Locke attacks in book I of the *Essay*, which is not motivated by a rejection of innate ideas but with the rejection of innate speculative and practical principles.[28]

> There is nothing more commonly taken for granted, than that there are certain Principles both *Speculative* and *Practical* (for they speak of both) universally agreed upon by all Mankind; which therefore they argue, must needs be the constant Impressions, which the Souls of Men receive in their first Beings, and which they bring into the World with them, as necessarily and really as they do any of their inherent Faculties. (*Essay* I. ii. 2)

This is the rationale of book I of the *Essay*. Thus, to claim that it is an attack on innate ideas is to miss its polemical thrust entirely. But Locke is not just concerned to reject the view that principles are innate. He aims to show that, given his theory of demonstration, principles are superfluous in the acquisition of intuitive and demonstrative knowledge (see especially *Essay* IV. vii. 16–19). The *Essay* in its first edition included a forthright attack on any form of principle.

One particular form of principle that Locke rejected was maxims. The whole of *Essay* book IV chapter 7 is against maxims.[29] What are maxims? Let us turn back to Blundeville on this point. He tells us that maxims are a species of principle and that

[28] The term 'principle/s' occurs 156 times in book I of the *Essay*.

[29] See also *Essay*, IV. xxii. 1.

every Science has his proper principles: of which some bee called dignities or Maximes, and some Positions.

Wherefore are they called ... Maximes?

For that they are worthy to be credited for their selfe sake, for as soone as we heare them in such speech as we understand, wee naturally know them to bee true without any further proofe. (Blundeville 1617, 164)

This is exactly what Locke attacks. He opens the chapter saying:

There are a sort of propositions, which, under the name of maxims and axioms, have passed for principles of science: and because they are *self-evident*, have been supposed innate, without that anybody (that I know) ever went about to show the reason and foundation of their clearness or cogency. (*Essay* IV. vii. 1)[30]

One of the arguments Locke runs against maxims is that any instance of intuitive knowledge will be a maxim. Not only did Locke reject a role for special maxims, he also had little use for laws of nature. There is only a handful of references to laws or rules of nature in the *Essay* and they play no significant explanatory role. It was only after he came to understand something of the achievement of Newton's *Principia* that Locke realised that, in fact, laws of nature can generate new knowledge and that they function in the way that principles do in the Aristotelian approach to *scientia*.

The first inkling that Locke was altering his views with regard to the role of principles in natural philosophy is in 1693 in a new section written for *Some Thoughts Concerning Education*, a section that does not derive from the earlier correspondence with Edward Clarke, which formed the backbone of the book. Locke tells us:

the incomparable Mr. *Newton*, has shewn, how far Mathematicks, applied to some Parts of Nature, may, upon Principles that Matter of Fact justifie, carry us in the knowledge of some, as I may so call them, particular Provinces of the Incomprehensible Universe. (Locke 1989, §194, 248, underlining added (=§182 in the first edition))

[30] Compare Leibniz, who says in the *New Essays*, summarizing Locke's position: 'Whatever use one may make of maxims in verbal propositions, they cannot yield us the slightest knowledge of substances which exist outside us'. He replies, 'I am of an entirely different opinion'; Leibniz 1996, 423.

It is the 'Principles that Matter of Fact justifie' that are the basis of Newton's success. What might these principles be? In the *Conduct of the Understanding*, begun around 1697, we get another insight, for there Locke tell us

> there are fundamental truths that lie at the bottom, the basis upon which a great many others rest, and in which they have their consistency. These are teeming truths, rich in store, with which they furnish the mind, and, like the lights of heaven, are not only beautiful and entertaining in themselves, but give light and evidence to other things, that without them could not be seen or known. Such is that admirable discovery of Mr. Newton, that all bodies gravitate to one another, which may be counted as the basis of natural philosophy; which, of what use it is to the understanding of the great frame of our solar system, he has to the astonishment of the learned world shown; and how much farther it would guide us in other things, if rightly pursued, is not yet known. (Locke 1823, vol. 4, 282, underlining added)

One cannot help but regard the fundamental truth that 'all bodies gravitate to one another' as one of the principles that 'Matter of Fact justifie'. This may be reinforced by Locke's *Elements of Natural Philosophy*, purportedly composed at this time. There Locke claims:

> Two bodys, at a distance; will put one another into motion by the force of attraction: which is unexplicable by us, tho' made evident to us by experience, and so to be taken as a Principle in Natural Philosophy. (Locke 1823, vol. 3, 304–305, underlining added)

However, the authorship of the *Elements* has recently been called into question, and it may be some time before we can be confident that Locke was the author of this comment.[31] Nevertheless, I believe that this was a decisive shift in Locke's thinking about how we can gain demonstrative knowledge of nature, but it all came too late and was never grafted in to the *Essay* and has largely been ignored in Locke scholarship. It is, however, the final piece in the picture that constitutes Locke's treatment of natural necessity.

3. Conclusion

This chapter has claimed that the problem of gaining epistemic access to necessary facts about nature and the processes of deriving further

[31] See Milton 2012.

truths from them is central to the early modern conception of necessity. The heuristic that I have set out, including the inconsistent tetrad, is a useful means for determining just what view of epistemic access to necessary facts any particular philosopher from the period maintained. This is clearly illustrated in the philosophy of John Locke. Furthermore, changes to Locke's conception of the role of principles within natural philosophy nicely illustrate the manner in which the new modal facts pertaining to laws of nature forced early modern philosophers to revise the traditional manner in which necessity was handled in the period.

Of course, this heuristic does not provide an exhaustive treatment of the notion of necessity in early modern philosophy – it does not, for example, give us much purchase on the problem of free will in the period – but it does provide a starting point. The general features of the early moderns' conception of the nature of necessary facts about the world, the processes through which we come to have knowledge of them and the knowledge derived from them by demonstration, were to have a long-term legacy. One can even construe some of the salient features of the syntactic conception of scientific theories, so popular in the twentieth century, as descending from this view.

10 *Leibniz's Theories of Necessity*

BRANDON C. LOOK

1. Introduction

It is not uncommon to say that rationalism, interpreted in a strict, consistent, and thoroughgoing manner, entails necessitarianism – the view that everything that happens necessarily and that there are no unactualized possibles. For, according to the principle of sufficient reason (PSR), there is a reason why everything is so and not otherwise, or, put differently, nothing exists for which there is not a reason why it is so and not otherwise. In the realm of the apparently contingent happenings of the world, this means that every event has a reason for its coming to be, and every event is itself the reason for the occurrence of some other event. Every event is, it seems, necessitated by some previous event. More worrisome for some philosophers, if every event or action is necessitated by some particular event or action, then it is not the case that an agent *could have done otherwise*. He or she is not free; he or she is not morally responsible. Moreover, if God is good and the creator of the universe, then God must bring into existence the best world; it is impossible for God to have actualized another world; hence, strictly speaking, there are no unactualized possibles. That, at least, is the short version of the story.

And it is not so far from the truth to say that something like this story – or rather the worry that it manifests – was running through the mind of Gottfried Wilhelm Leibniz (1646–1716) as he contemplated the implications of his own budding version of rationalism. Writing in 1689, Leibniz remarks that his reflections on the principle of sufficient reason led him "very close to the view of those who think that everything is absolutely necessary" (A VI iv 1653/AG 94).[1] What pulled him

[1] In this chapter, the following abbreviations will be used: "A" followed by series, volume, and page number = Leibniz 1923-; "AG" = Leibniz 1989; "AT" followed by volume and page number = Descartes 1996; "C" = Leibniz 1903; "CP" = Leibniz 2006; "CSMK" followed by volume and page

back from this "precipice" was the consideration of possible things that neither are nor will be nor have been. In other words, let the simplistic argument be that rationalism implies necessitarianism. Leibniz simply rejected the consequence, and by *modus tollens* concluded that the antecedent needed refinement. But how? Leibniz, more than any other philosopher, placed great weight on the PSR; he really did believe that there was a reason for everything. Indeed, given Leibniz's other views on the nature of individual finite substances and divine omniscience, omnipotence and benevolence, it is fair to say that one of the central tensions in his philosophical system relates to determinism and necessitarianism: can one give an account of necessity and determinism that does justice to the rational order of the universe while it preserves the possibility of human freedom? Moreover, concerning the nature of necessity, it is also important to point out that the problem for Leibniz was not simply to give an account of the nature of mathematical truths and their like – the "eternal truths" for those working in the seventeenth century – but also, as has been suggested, to give an account of those truths – seemingly "contingent" from our contemporary point of view – that are, for Leibniz, necessary *ex hypothesi*.

To resolve this tension and to preserve both the commitments of rationalism and the possibility of human and divine freedom, Leibniz employed an argumentative strategy that carefully distinguished metaphysical or absolute necessity from hypothetical necessity.[2] But there was much more to Leibniz's account. It is generally recognized that Leibniz developed two theories of our modal notions of necessity, possibility, and contingency. According to his first account, propositions can be considered necessary or contingent per se; likewise, individuals, or better their essences, can be considered per se possible or impossible. According to some scholars, Leibniz abandoned this per se theory of modality account in favor of a second, according to which

number = Descartes 1985; "DSR" = Leibniz 1992; "G" followed by volume and page number = Leibniz 1875–1890; "Grua" = Leibniz 1948; "H" = Leibniz 1985; "L" = Leibniz 1969; "LLP" = Leibniz 1960; "MP" = Leibniz 1995; "RB" = Leibniz 1996; "T" = *Theodicy*, followed by section number. Whenever possible the original language text will be cited along with the most common English language text. Passages lacking a reference to an English language text are given in the author's translation; an asterisk indicates that the cited translation has been altered.

[2] Leibniz also adds moral necessity – the obligation to choose the best. But this will not play a role in the story I wish to tell here.

a proposition is necessary if it can be resolved into a statement of identity in a finite number of steps, while a proposition is contingent if its analysis can go on to infinity without arriving at a statement of identity.[3] Yet, neither account has proven to be very satisfying to interpreters of Leibniz's thought. What does it mean to say that something is per se possible or per se necessary? And how helpful is the infinite analysis account in practice? Some scholars, in fact, have suggested that Leibniz should have simply gone over the precipice and accepted that the principle of sufficient reason implies strict necessitarianism. Recently, however, a number of scholars have suggested that Leibniz's first account of per se possibility and necessity is in fact superior to the infinite analysis account and can be used to block the inference from the principle of sufficient reason to necessitarianism.[4] In this chapter, I shall argue that the infinite analysis account and the per se account of modality complement each other. Finally, it should be rather obvious that Leibniz's invocation of possible worlds in his philosophy is echoed in contemporary accounts of modality that appeal to possible-worlds semantics. While Leibniz does not *explicitly* define necessity and contingency in terms of possible worlds, I shall show that his conception of possible worlds underlies his accounts of our modal notions and that his standard account of necessity and contingency can easily be translated into possible-worlds semantics.

2. Leibniz's Theory of Per Se Possibles

It is common to point to the fact that Leibniz spent much of his career situating his philosophy in opposition to the thought of Spinoza. This is a common story because it is true. And even in the simple story sketched earlier linking rationalism and necessitarianism, it is easy to see Spinoza as the main opponent.[5] When we try to tell a historically

[3] This view is expressed best by Sleigh (1990). However, Blumenfeld (1988) and Adams (1994) argue that the per se account of modality is preserved throughout Leibniz's career, a view I share, as will become clear. Newlands (2010) makes a similar case – though his focus differs from the general view presented here. The reader interested in this subject should also consult the following important works: Ishiguro (1990), Mates (1986), Parkinson (1965), and Russell (1937). The most recent major contribution to this discussion can be found in Griffin (2013).

[4] See, for example, Lin (2012).

[5] While not classed among the "rationalists" and not arriving at the position in the same way, Hobbes espoused a version of necessitarianism that was just as

faithful story of Leibniz's philosophy, however, we need to be a little more careful. For well before Leibniz had even seen a copy of Spinoza's *Ethics*, he was very much concerned with giving an account of our modal notions that would respect both his rationalism and his other philosophical commitments.

Indeed, in 1672–1673, Leibniz, a twenty-something, budding philosopher just arrived in the intellectual capital of Paris, began to compose a long dialogue, *Confessio Philosophi (The Philosopher's Confession)*, that addressed the problem of evil and (for him) the related issue of the nature of necessity and contingency. His first attempt to explicate modal notions runs like this:

> Those things are *contingent* that are not necessary; those are *possible* whose nonexistence is not necessary. Those are *impossible* that are not possible, or more briefly: the *possible* is what can be conceived, that is, (in order that the word *can* not occur in the definition of possible), what is conceived clearly by an attentive mind; the *impossible* what is not possible; the *necessary* that whose opposite is impossible; the *contingent* that whose opposite is possible.[6] (A VI iii 127/CP 55)

There are several points to note here. First, the sense of Leibniz's definition of necessity is essentially identical to what one will find through much of the rest of his career, though it is usually expressed in this way: "Something is necessary if its contrary implies a contradiction." In fact, Leibniz will immediately describe necessity in these terms on the next page of the dialogue, as we shall see. Second and more interesting, Leibniz clearly links possibility with conceivability. But he also

disturbing to the young Leibniz. Spinoza became the preferred philosophical sparring partner because Spinoza's philosophy was vilified throughout Europe after his death and because Leibniz recognized that his own system was in many ways very close to that of Spinoza. The literature on this subject is extensive; for more a good starting point is, again, Newlands (2010) as well as my own Look (2006).

6 A brief remark on the text: Sleigh's translation is perfectly fine, but the reader should be aware that Leibniz does not explicitly say "those things are *contingent* that are not necessary," but rather "*contingents* are what are not necessary." I mention this because Leibniz talks about the modal properties of *beings*, essences, and actions (e.g. the necessary being, God, or as we shall see in a moment, the necessity or contingency of *sins*) and the modal properties of propositions or truths. Naturally, these two ways of speaking are related, but Leibniz is not always as careful as one would wish him to be.

recognizes that conceivability itself is a kind of modal notion; that is, "conceivability" implies that the thinker *can* conceive something. Hence, Leibniz removes the modal component and says that the possible is what is conceived clearly by the attentive mind. Sadly, in much of the rest of the *Confessio*, Leibniz forgets this point and falls back to talk about the link between possibility and conceivability.

More important for the purposes of this essay, however, is the point that immediately follows this set of definitions.[7] The theologian puts the following argument to the philosopher.

> What is your response going to be to the argument proposed previously: the existence of God is necessary; the sins included in the series of things follow from this; whatever follows from something necessary is itself necessary. Therefore sins are necessary.

The conclusion, of course, is heretical and, in fact, can be generalized to be a statement of necessitarianism. For it would be not just sins but *every* state of affairs that follows necessarily from the necessary existence of God. The most natural response for an orthodox theist would, of course, be to reject the second premise of this argument – the sins included in the series of things follow from the necessary existence of God. But Leibniz's philosopher does not take this route; instead, in the final draft of the dialogue, he says:

> I reply that it is false that whatever follows from something necessary *per se* is itself necessary *per se*. Certainly it is evident that nothing follows from truths except what is true; nevertheless, since a particular proposition can follow from purely universal propositions, as in [the syllogistic forms] Darapti and Felapton, why not something contingent or necessary on the hypothesis of another from something necessary *per se*?

Now, in Leibniz's first draft of this passage, the philosopher simply claims that "it is false that whatever follows from something necessary is itself necessary." In other words, Leibniz wishes to reject the third premise – that whatever follows from something necessary is itself necessary. It might seem that Leibniz is hereby denying a fairly elementary theorem of modal logic, namely, the axiom of the weakest system

[7] For a very nice discussion of these issues, see Sleigh's introduction to CP (pp. xxiv–xxv).

of modal logic K: $\Box(p \rightarrow q) \rightarrow (\Box p \rightarrow \Box q)$. But his invocation of the "*per se* modalities" is meant to be a way around this issue. The philosopher thus continues:

I will establish this from the very notion of necessity. Now I have defined the *necessary* as something whose contrary cannot be conceived; therefore, the necessity and impossibility of things are to be sought in the ideas of those very things themselves, not outside those things. It is to be sought by examining whether they can be conceived or whether instead they imply a contradiction. For in this place we call *necessary* only what is necessary *per se*, namely, that which has the reason for its existence and truth in itself. The truths of geometry are of this sort. But among existing things, only God is of this sort; all the rest, which follow from the series of things presupposed – i.e., from the harmony of things or the existence of God – are *contingent per se* and only hypothetically necessary, even if nothing is fortuitous, since everything proceeds by destiny, i.e., from some established reason of providence. Therefore if the essence of a thing can be conceived, provided that it is conceived clearly and distinctly..., then it must already be held to be possible, and its contrary will not be necessary, even if its existence may be contrary to the harmony of things and the existence of God, and consequently it never will actually exist, but it will remain *per accidens* impossible. Hence all those who call impossible (absolutely, i.e. *per se*) whatever neither was nor is nor will be are mistaken. (A VI iii 127–128/CP 55–57*)

The fundamental idea behind Leibniz's account of per se possibility appears quite simple: an essence (or a state of affairs) is possible just in case it is conceivable, just in case there is no inherent contradiction in that essence (or state of affairs) considered in and through itself. Per se necessity, on the other hand, can be characterized thus: a truth or being is said to be necessary just in case the *ground* for the truth or being is contained within itself. Therefore, Leibniz believes himself to have avoided the heretical conclusion of the theologian's argument since the sins or states of affairs that follow from the necessity of God's being (and creation)[8] are not necessary per se but rather merely hypothetically necessary. Moreover, since the necessitarianism

[8] I avoid the question of whether and in what sense, for Leibniz, God's choice of worlds and act of creation was necessitated. As mentioned in fn. 2, God is in some sense constrained by "moral necessity" to choose the best; but this is not the same as being constrained by absolute necessity.

of the theologian's argument also goes hand in hand with a denial of unactualized possibles, Leibniz takes himself to have dispatched with that as well in this passage.

While it is easy to get a general sense of what Leibniz was trying to say with his per se modalities, further reflection cannot but lead to the conclusion that this is all fairly vague. There arise a number of obvious questions: how do we know that we have clearly and distinctly conceived of an essence, so that we know that there are no contradictions among its properties? What does it mean for a being or a truth to contain the ground of its being within itself?

As mentioned earlier, it has been argued that Leibniz largely abandoned this per se account of modality. While it never again constituted Leibniz's *main* account of modality, the idea remained with Leibniz throughout his career. Consider, for example, the following from Leibniz's *Theodicy* (1710):

> When one speaks of the *possibility* of a thing it is not a question of the causes that can bring about or prevent its actual existence; otherwise one would change the nature of the terms, and render useless the distinction between the possible and the actual.... That is why, when one asks if a thing is possible or necessary, and brings in the consideration of what God wills or chooses, one alters the issue. For God chooses among the possibles, and for that very reason he chooses freely, and is not compelled; there would be neither choice nor freedom if there were but one course of possible. (T §235: G VI 257–258/H 278–279)

In short, possibility is a property intrinsic to the nature or essence of a thing; it is prior to and independent of existence itself. Moreover, Leibniz's per se account of modality is connected with two philosophical commitments in the metaphysics of modality that remained remarkably stable for the rest of his career: the distinction between hypothetical and absolute necessity, and the view that the modal properties of beings and propositions are real, tied to their essences (in the case of beings), and not mere conventions.[9] Indeed, the distinction between what is hypothetically necessary and absolutely necessary became an essential part of Leibniz's account of human freedom. Several years after beginning the composition of the *Confessio*, he showed a

[9] Neither of these points is unique or novel to Leibniz, of course. Where Leibniz is genuinely insightful and unique will be shown in a moment.

version to Nicholas Steno, and in the notes on his conversation with Steno, Leibniz writes the following:

Absolute necessity is when a thing cannot even be understood to be otherwise, but it implies a contradiction in terms, e.g., three times three is ten. *Hypothetical necessity* is when a thing can be understood to be *otherwise* in itself, but *per accidens*, because of other things already presupposed outside itself it is necessarily such and such.... *The series of things is not necessary by an absolute necessity*. (A VI iv 1377/CP 119)

The point should now be familiar: something is per se necessary if and only if its negation entails a contradiction in terms; something is hypothetically necessary if and only if it follows from something else but could be conceived differently. For example, it is conceivable (in some sense) that Judas did not betray Christ, though it is necessary that he did so given the particular world-series of which he was a member. Regarding the role of the consideration of essences in Leibniz's theory, consider the following passage from a letter to Simon Foucher, written in 1675:

This possibility, impossibility, or necessity (for the necessity of something is the impossibility of its contrary) is not a chimera we create, since we do nothing more than recognize it, in spite of ourselves and in a consistent manner. Thus of all things that there actually are, the very possibility or impossibility of being is the first. Now, this possibility or this necessity forms or composes what we call the essences or natures and the truths we commonly call eternal – and we are right to call them so, for there is nothing so eternal as that which is necessary. Thus the nature of the circle with its properties is something existent and eternal. That is, there is a constant cause outside us which makes everyone who thinks carefully about the circle discover the same thing. (A II i 387–388/AG 1–2)

Here the view is that possibility and necessity are prior to and independent of human thought. The essence of a being is simply the set of properties that either are or are not per se or through themselves internally consistent. Thus, when there is no contradiction of properties within the essence of a being, it can be said to be a possible existent; when there is a contradiction, the thing in question is impossible. An unactualized possible is simply a being that is possible per se, but whose essence is impossible given the other essences in this world, the world that has the most perfection and reality.[10] Now, the last sentence in this passage might appear somewhat strange given what we have seen thus far, and it is deserving of two brief comments. First, what Leibniz says here is really nothing other than what he says often throughout his entire oeuvre: that the eternal truths are grounded in the infinite intellect of God. Second, Leibniz's point here foreshadows his more sophisticated account of necessity and possibility, the infinite analysis account. There are uniformity and unanimity in our thoughts about mathematical objects, or mathematical or eternal truths, precisely because we can all arrive at an understanding of the relevant concepts in a finite number of steps. And it is precisely this doctrine that we will turn to now.

3. Infinite Analysis

In a short piece, *On Freedom*, written in 1689, Leibniz gives the autobiographical sketch mentioned in the Introduction to this essay. He claims that after considering the implications of the principle of sufficient reason, he "was very close to the view of those who think that everything is absolutely necessary" but that "consideration of possibles ... brought [him] back from this precipice" (A VI iv 1653/ AG 94). There is much more to the story of Leibniz's rejections of necessitarianism than just this, however, for he goes on in this piece to link the solution to the problem of freedom and contingency to the nature of truth:

> Recognizing the contingency of things, I further considered what a clear notion of truth might be, for I hoped, and not absurdly, for some light from that direction on how necessary and contingent truths could be distinguished. Now, I saw that it is common to every true affirmative proposition, universal and particular, necessary or contingent, that the predicate is in the subject, that is, that the notion of the predicate is involved somehow in the notion of the subject. And this is the source [*principium*] of infallibility in every sort of truth for that being who knows everything *a priori*. But this seemed only to increase the difficulty, for if the notion of

[10] I will expand upon this point later, but just to give away part of the story now: necessary truths are grounded in the principle of contradiction; contingent truths are grounded in the principle of the best or the principle of sufficient reason; the PSR will require that there be a ground for a truth or a being outside itself.

the predicate is in the notion of the subject at a given time, then how could the subject lack the predicate without contradiction and impossibility, and without changing that notion? At last a certain new and unexpected light shined from where I least expected it, namely, from mathematical considerations on the nature of infinity. (A VI iv 1654/AG 95)

Leibniz believes himself simply to be following Aristotle in holding that a true (affirmative) proposition is one in which the notion of the predicate is contained in the notion of the subject. Indeed, as he will say elsewhere, "Every true proposition can be proved" (A VI iv 776/LLP 77). The problem that he immediately admits here is that in many (indeed, most) cases subjects – i.e. *substances* that exist in a world – will have contradictory predicates: the young Leibniz had a full head of hair; the old Leibniz was bald. But the solution is not simply to time-index the properties of subjects; the solution is connected with Leibniz's doctrine of the complete individual concept of a substance, the classical statement of which is to be found in §8 of the *Discourse on Metaphysics* (1686): "The nature of an individual substance or of a complete being is to have a notion so complete that it is sufficient to contain and to allow us to deduce from it all the predicates of the subject to which this notion is attributed" (A VI iv 1540/AG 41).

This view of the nature of substance and its properties leads Leibniz to reflections on the infinite. For, he believes, "no substance [is] so imperfect that it does not contain the entire universe, and whatever it is, was, or will be, in its complete notion (as it exists in the divine mind), nor is there any truth of fact or any truth concerning individual things that does not depend upon the infinite series of reasons" (A VI iv 1655/AG 95). It is at this point, Leibniz tells us, that he realized "a most profound distinction between necessary and contingent truths":

Namely, every truth is either primitive [*originaria*] or derivative. Primitive truths are those for which we cannot give a reason; identities or immediate truths, which affirm the same thing of itself or deny the contradictory of its contradictory, are of this sort. Derivative truths are, in turn, of two sorts, for some can be resolved into primitive truths, and others, in their resolution, give rise to a series of steps that go to infinity. The former are necessary, the latter contingent. (A VI iv 1655/AG 95–96*)

In short, the fundamental distinction between necessary truths and contingent truths is that the former can be reduced to identities in a finite number of steps; the latter cannot be reduced to identities and thus cannot be demonstrated.

In contingent truths, even though the predicate is in the subject, this can never be demonstrated, nor can a proposition ever be reduced [*revocari*] to an equality or to an identity, but the resolution proceeds to infinity, God alone seeing, not the end of the resolution, of course, which does not exist, but the connection of the terms or the containment of the predicate in the subject, since he sees whatever is in the series. (A VI iv 1656/AG 96)[11]

Leibniz's account of necessity as a property of propositions amenable to *finite* analysis seems relatively straightforward. A proposition is necessary (or expresses a necessary truth) if and only if, in analyzing the subject and predicate terms, one arrives at a statement of identities. A simple example can be found in arithmetic: "2 + 1 = 3." Since "2 = 1 + 1" and "3 = 1 + 1 + 1," we can easily show that the original statement can be reduced to "1 + 1 + 1 = 1 + 1 + 1." Here we have an obvious case of an identity and one where we can say that the concept of the predicate – in this case, "4" – is contained in the concept of the subject, "2 + 1" (or "(1 + 1) + 1").[12] It should also be clear that Leibniz's definition of what is *necessary* will be adopted by Kant, *but* Kant will call this *analytic*. Indeed, Leibniz's account of necessity as essentially involving conceptual analysis ("reduction") will sound familiar to many contemporary philosophers.

Leibniz's account of contingency via infinite analysis, on the other hand, might seem less satisfactory. Of course, what Leibniz intends is clear enough: a proposition expresses a contingent truth if and only if an identity between subject and predicate terms cannot be demonstrated. To clarify his position, Leibniz often compares the distinction between necessary and contingent truths to rational and irrational numbers.

In proportions, while the analysis sometimes comes to an end, and arrives at a common measure, namely, one that measures out each term of the proportion through exact repetitions of itself, in other cases the analysis can be continued to infinity, as happens in the comparison between a rational number and an irrational number, such as the comparison of the side and

[11] Leibniz makes this point in many writings, but a particularly nice formulation can also be found at A VI iv 1650/AG 28.

[12] Leibniz famously gives this as an example of how the truths of mathematics and logic work in his *New Essays*, bk. IV, ch. 2 (A VI vi 366–367/RB 366–367).

the diagonal of a square. So, similarly, truths are sometimes provable, that is, necessary, and sometimes they are free or contingent, and so cannot be reduced by any analysis to an identity, to a common measure, as it were. And this is an essential distinction, both for proportions and for truths. (A VI iv 1657/AG 97)

The *reason* or *ground* for any contingently true proposition is to be found in another contingently true proposition, and this chain of reasons goes back ultimately to the creation of the world (or rather to God's free choice of *this* particular world). We, with our finite minds, are not able to grasp this chain of reasons, but God can grasp the entire chain of reasons as well as the comparative perfection of *this* world and that of all other possible worlds.

The "infinite analysis" account of modality has not met with many supporters. Indeed, one of the most acute scholars of Leibniz's thought has remarked, "I have to confess that I can find no intuitive plausibility whatsoever in this 'solution.' It is hard to see what the length of the reduction of a proposition would have to do with whether the proposition is false of some possible world."[13] But is it really so bad? The issue is not the *length* of the reduction, but the act of reduction. Necessary propositions are demonstrable in a finite number of steps, and the actual number of steps is usually not great. Contingent true propositions, on the other hand, simply do not admit of a reduction to identical terms. Now, I believe that part of the resistance to Leibniz's account is to be found in a knee-jerk suspicion of rationalist metaphysics. For what Leibniz actually goes on to say is not very radical or strange at all. After claiming that God alone sees the totality of the world-series and thereby the connection (not identity!) of terms or the containment of the predicate in the subject, Leibniz writes that

two ways remain for us to know contingent truths, one through experience, and the other through reason – by experience when we perceive a thing sufficiently distinctly through the senses, and by reason when something is known from the general principle that nothing is without a reason, or that there is always some reason why the predicate is in the subject. (A VI iv 1656/AG 96)

[13] Mates (1986: 108). For now we will bracket Mates's invocation of possible worlds in the analysis of modality. The take away is simply: "This is silly; why should it matter how long the analysis goes on?"

In other words, we know necessary truths ("truths of reason") through analysis and demonstration, and we know contingent truths ("truths of fact") through experience and the principle of sufficient reason. Many contemporary philosophers seem to be skeptical of the principle of sufficient reason,[14] but, again, Leibniz's view of contingency should not alarm anyone: he is simply saying that a proposition is contingently true just in case the demand for a ground can keep being pushed back.

The distinction between the necessary and the contingent is connected to Leibniz's "two great principles of all our reasoning": the principle of contradiction and the principle of sufficient reason. We see this throughout his writings, though the connection becomes more explicit and more closely connected to the infinite analysis account later in his career. For example, in a relatively early piece, *On Freedom and Necessity* (1680–1684?), Leibniz says the following:

> There are two primary propositions: one, the principle of necessary things, *that whatever implies a contradiction is false*, and the other, the principle of contingent things, *that whatever is more perfect or has more reason is true*. All truths of metaphysics, or all truths that are absolutely necessary, such as those of logic, arithmetic, geometry and the like, rest on the former principle, for someone who denies them can always be shown that the contrary implies a contradiction.... And so all truths that concern possibles or essences and the impossibility of a thing or its necessity (that is, the impossibility of its contrary) rest on the principle of contradiction; all truths concerning contingent things or the existence of things, rest on the principle of perfection. (A VI iv 1445/AG 19)

Strictly speaking, the principle of the best is not exactly the principle of sufficient reason. But what Leibniz is saying is clear enough: the reason or ground for the existence of contingent things is found outside the thing itself and in God's free choice of the world.[15] And in a passage from the *Monadology* (1714) that all students of philosophy are familiar with, Leibniz writes:

[14] Real philosophers can deal with brute facts!

[15] One might think that Leibniz has moved on from his per se account of possibility here. But Leibniz is talking about the principle of sufficient reason in connection with *contingents*; the nature of unactualized possibles must still be considered per se.

There are also two kinds of *truths*, those of *reasoning* and those of *fact*. The truths of reasoning are necessary and their opposite is impossible; the truths of fact are contingent, and their opposite is possible. When a truth is necessary, its reason can be found by analysis, resolving it into simpler ideas and simpler truths until we reach the primitives. This is how the speculative *theorems* and practical *canons* of mathematicians are reduced by analysis to *definitions*, *axioms* and *postulates*. And there are, finally, *simple ideas*, whose definition cannot be given. There are also axioms and postulates, in brief, *primitive principles*, which cannot be proved and which need no proof. And these are *identical propositions*, whose opposite contains an explicit contradiction. (§§33–35: G VI 612/AG 217)

Conceptual analysis and demonstration aided by the principle of contradiction can thus reveal truths to be necessary; while those propositions that cannot be revealed to be reducible to statements of identity are contingent. Their *truth* can only be grounded in an appeal to the principle of sufficient reason, which ultimately will be a question of why God chose to create one particular over others.

4. Leibniz's Possible Worlds and the Nature of Necessity

Of all of Leibniz's philosophical theses, the most famous or (thanks to Voltaire) infamous is that we live in the best of all possible worlds. More exactly, according to Leibniz, there are an infinite number of possible worlds, and God chose to bring into existence the world that is most perfect.[16] Because of this commitment to the infinity of possible worlds Leibniz has been seen as a kind of forefather to contemporary modal theories, in particular possible-worlds semantics. Indeed, we can ask, did Leibniz himself endorse a kind of possible-worlds semantics for our modal concepts? The short answer is no – at least not explicitly. The longer and more interesting answer is that he had all the conceptual resources to give a possible-worlds semantics of necessity and contingency, and his standard accounts of modality can easily be translated into the terms of possible-worlds semantics.

Before we examine Leibniz's view, let us very briefly lay out possible-worlds semantics. The claim here is that our basic modal

[16] Even more specifically, the best of all possible worlds is the world that is simplest in hypotheses and richest in phenomena (*Discourse on Metaphysics* §6: A VI iv 1538/AG 39).

concepts – necessity, contingency, possibility, and impossibility – can be defined simply in terms of truth in a world or truth in worlds.

(a) *Possibility:* A proposition is possible if and only if it is true in some possible world. A being is possible if and only if it exists in some possible world.
(b) *Contingency:* A proposition is contingently true if and only if it is true in this world and false in another world.
(c) *Necessity:* A proposition is necessarily true if and only if it is true in every possible world.
(d) *Impossibility:* A proposition is impossible if and only if it is not true in any possible world.

The crucial insight of possible-world semantics is that it explains modal concepts without making reference to modality at all.[17] Now, the precise nature of possible worlds has, of course, been a subject of great debate in contemporary modal metaphysics, and fortunately we can avoid this debate altogether.

For Leibniz, however, a *world* is a set of individual substances, or as he puts it in the beginning of *On the Ultimate Origination of Things*, a world is a "collection of finite things"[18] (G VII 302/AG 149). But it is false to think that a world is just any old collection of things. According to Leibniz, "The universe is only a certain kind of collection of compossibles; and the actual universe is the collection of all possible existents, that is, of those things that form the richest composite" (Letter to Bourguet: G III 573/L 662*). The point is general, however: a world can contain only those individuals that are jointly possible; that is, a world cannot contain individuals that have contradictory properties.[19] Further, according to Leibniz, "there are as many possible worlds as there are series of things that can be conceived that do not imply a

[17] In a way, this is like Leibniz's recognition in his early work that there is something wrong with describing what is possible as that which is conceivable, for conceiv-*ability* is implicitly modal.

[18] Leibniz sometimes speaks as if a world is the collection of substances *along with* the laws of nature. However, insofar as each substance is to have a complete individual concept or insofar as each substances has a law of unfolding of its own perceptions and appetitions, it is just as fair to say that the physical laws are derivative of the concepts, essences, or perceptual states.

[19] The issue of compossibility in Leibniz is interesting and has generated nice work among Leibniz scholars; a wonderful place to begin with this issue is Brown (1987).

contradiction" (*Discussion with Gabriel Wagner* (1698), Grua 390). And, Leibniz believes, there are an infinite number of possible worlds that God did not choose to bring into existence. "There are, in fact, an infinite number of series of possible things. Moreover, one series certainly cannot be contained within another, since each and every one of them is complete" (*On Contingency* (1686): A VI iv 1651/AG 29). In saying that a world is a set of compossible things and that each world is complete in itself, however, Leibniz is saying that a world is a kind of collection of things that God *could* have brought into existence. But not even God can bring into existence a world in which there is some contradiction among its members or their properties.[20]

In contemporary discussions of modality, it is not uncommon to talk about possible worlds in terms of "world-books."[21] It will perhaps be of some interest to readers of this volume to learn that Leibniz employs exactly this kind of language in the fascinating conclusion to the *Theodicy*. He tells the story of the high priest Theodorus, who witnesses Sextus Tarquinius, complaining to Jupiter about his fate. Moved by Sextus's story, Theodorus seeks counsel in the temple of Pallas Athena and is there shown the "palace of the fates." Athena says, "Here are representations not only of that which happens but also of all that which is possible. Jupiter, having surveyed them before the beginning of the existing world, classified the possibilities into worlds, and chose the best of all" (*Theodicy* §414: G VI 363/H 370). She continues,

> I have only to speak, and we shall see a whole world that my father might have produced, wherein will be represented anything that can be asked of him; and in this way one may know also what would happen if any particular possibility should attain unto existence. And whenever the conditions are not determinate enough, there will be as many such worlds differing from one another as one shall wish, which will answer differently the same question, in as many ways as possible.... But if you put a case that differs from the actual world only in one single definite thing and in its results, a certain one of those determinate worlds will answer you. These worlds are all here, that is, in ideas. I will show you some, wherein shall be found, not absolutely the same Sextus as you have seen (that is not possible, he carries

[20] There is a sense in which Leibniz's account of per se possibles underlies this account of compossibility.

[21] See, for example, Plantinga (1978).

with him always that which he shall be) but several Sextuses resembling him, possessing all that is already in him imperceptibly, nor in consequence all that shall yet happen to him. (*Theodicy* §414: G VI 363/H 370–371)

Theodorus is led into one of the halls of the palace and, observing a great volume of writings in the hall, asks what it is. "It is," Athena tells him, "the history of this world which we are now visiting…; it is the book of fates" (*Theodicy* §415: G VI 363/H 371–372).

It was claimed previously that Leibniz did not explicitly adopt the kind of possible-worlds semantics endorsed by Lewis and others in his own definitions and explanations of our basic modal concepts. But we are now in a position to see that Leibniz's account can easily be translated into these terms. For what is possibility but being included in one of the representations of what God might have created? Put differently, a being is possible if and only if it is included in a world-book. Further, let us return to the distinction between the absolutely necessary and the hypothetically necessary. In §13 of the *Discourse on Metaphysics:*

the one whose contrary implies a contradiction is absolutely necessary; this deduction occurs in the eternal truths, for example, the truths of geometry. The other is necessary only *ex hypothesi* and, so to speak, accidentally, but it is contingent in itself, since its contrary does not imply a contradiction. And this connection is based not purely on ideas and God's simple understanding, but on his free decrees and on the sequence of the universe. (A VI iv 1547/AG 45)

Leibniz then moves on to discuss Julius Caesar's crossing of the Rubicon (and predicate "crosses the Rubicon" in the concept of Caesar):

For it will be found that the demonstration of this predicate of Caesar is not as absolute as those of numbers or of geometry, but that it supposes the sequence of things that God has freely chosen, a sequence based on God's first free decree always to do what is most perfect and on God's decree with respect to human nature, following out of the first decree, that man will always do (although freely) that which appears to be best. But every truth based on these kinds of decrees is contingent, even though it is certain; for these decrees do not change the possibility of things, and, as I have already said, even though it is certain that God always chooses the best, this does not prevent something less perfect from being and remaining possible in itself, even though it will not happen, since it is not its impossibility but its imperfection which causes it to be rejected. And nothing is necessary whose contrary is possible. (A VI iv 1548/AG 46)

The point is simple: a proposition is contingently true in this world if and only if it is dependent upon the first free decree of God; in other words, a proposition is contingently true when its ground is in the actualization of this particular world. In the case of Julius Caesar, his action of crossing the Rubicon was free, that is, not necessitated, because the contrary does not imply a contradiction.[22] In other words, there is a possible world in which Caesar did not cross the Rubicon; and, therefore, it is not metaphysically necessary that he cross the Rubicon.[23] Further, as we have already seen, Leibniz defines a possible proposition as a proposition that will never result in a contradiction in its analysis; but given Leibniz's account of possible worlds, this simply means that the proposition is true in some world. Finally, and most important for the purposes of this contribution, since necessary truths are not dependent upon the first free decree of God but are instead revealed by analysis to be identities, they are in fact propositions that are true in *all* possible worlds. And so, while the "official" Leibnizian account of necessity, contingency, and possibility is given in terms of per se possibles and infinite analysis, it is a not a great jump to possible-worlds semantics.

5. God and Necessity

Leibniz's notion of necessity is tied up with two of his arguments for the existence of God in ways that might interest readers of this volume. The first case concerns the ontological argument, of which Leibniz offered an important critique and revision.[24] Let the standard ontological argument be expressed in the following manner:

[22] More precisely, for Leibniz, an action is free if it is (a) contingent, (b) spontaneous, and (c) the result of intelligence (i.e. choice). See T §288: G VI 287/H 303.

[23] In my view, Leibniz denies transworld identity. Therefore, it is better to say that there is a possible world in which *a* Caesar did not cross the Rubicon.

[24] Leibniz focuses on Descartes's version of the ontological argument, not on Anselm's. There are some minor differences that I discuss in my Look (forthcoming), but they need not slow us down here.

(1) God is the *ens perfectissimum* (i.e. the most perfect being, or the being containing all perfections).
(2) Existence is a perfection.
(3) Therefore, God exists.

Leibniz's criticism of the argument so construed is that it is incomplete or, rather, question-begging, because it assumes that an *ens perfectissimum* is possible. There are expressions, however, that seem to correspond to something, but that in fact contain contradictions. For example, Leibniz says, we can think of and talk about a "greatest number"; but, in fact, such a number does not exist, and closer inspection reveals a contradiction within the concept.

As Leibniz writes in one of his first publications, the *Meditations on Knowledge, Truth and Ideas* (1684), the Cartesian ontological argument only licenses the conclusion "if God is possible, then it follows that he exists" (A VI iv. 588/AG 25). Leibniz therefore sets himself the task of proving the possibility of an *ens perfectissimum*; that is, he wishes to prove the possibility of a necessary being. To accomplish this task and fill the lacuna in the ontological argument, it must be shown that it is possible for some object to possess all perfections. In a number of texts from 1676, Leibniz gives us arguments that demonstrate precisely this possibility. Now, according to Leibniz, "perfections, or simple forms, or absolute positive qualities, are indefinable or unanalyzable" (A VI iii 575/DSR 97); similarly, a perfection is a "simple quality which is positive and absolute, or, which expresses without any limits whatever it does express" (A VI iii 578/DSR 101). Leibniz now has what he needs for a *reductio ad absurdum*. Suppose there are two perfections, A and B, that cannot possibly belong to the same subject; that is, suppose the two perfections are necessarily incompatible. If the proposition "A and B are incompatible" is necessarily true, then, according to Leibniz, it is either self-evident or demonstrable. But it is impossible for us to know by intuition that A and B are incompatible. Nor can we demonstrate this since demonstration requires the analysis of concepts and A and B are simple qualities, that is, unanalyzable. Therefore, we cannot demonstrate that they are necessarily incompatible. But since it cannot be shown that A and B cannot be in the same subject, it must be possible for them to be in the same subject, and since the same reasoning applies to any two or more simple, positive qualities, all perfections are compatible.[25], It is now possible to rewrite the ontological argument in the following way:

(1) God is the *ens perfectissimum*. (Definition)
(2) Existence is a perfection.
(3) If the *ens perfectissimum* is possible, it exists.
(4) The *ens perfectissimum* is possible.
(5) Therefore, God exists.

Leibniz was particularly proud of this argument, writing, "I am perhaps the first to have shown that a being to which all affirmative attributes belong is possible" (A VI iii 395–396/DSR 47–49). But does this argument actually prove God's *necessary* existence? Might it not simply be a contingent fact that all perfections are instantiated in this case? In short, is there a difference between an *ens perfectissimum* and an *ens necessarium*?

Leibniz was aware of these kinds of questions, and he sought to clarify the issue in other texts from this period. He writes, for example, that "a necessary being is the same as a being from whose essence existence follows. For a necessary being is one which necessarily exists, such that for it not to exist would imply a contradiction, and so would conflict with the concept or essence of this being" (A VI iii 583/DSR 107). Now, obviously, if it is possible for a being to have all perfections and if existence is a perfection, then existence is one of the perfections of the *ens perfectissimum*. Further, if existence belongs to the essence of the *ens perfectissimum*, then, according to Leibniz's new definition, the *ens perfectissimum* is a necessary being. It is now possible, however, to reformulate the ontological argument without reference to an *ens perfectissimum*, but rather in terms of an *ens necessarium*. Indeed, according to Leibniz, if a necessary being is a being whose essence entails existence, "we have a splendid theorem, which is the pinnacle [*fastigium*] of modal theory and by which one moves in a wonderful way from potentiality to act: If a necessary being is possible, it follows that it exists actually, or, that such a being is actually found in the universe" (A VI iii 583/DSR 107).[26] What

[25] See A VI iii 578–579/DSR 101–103 (= L 167); A VI iii 575–576/DSR 97–99.

[26] Leibniz repeats this idea and terminology throughout his career, writing in 1688 (?) again of the "pinnacle of modal theory [*fastigium doctrinae modalis*]" (A VI iv 1617/MP 76) and after 1700 of "a modal proposition which would be

exactly is this "pinnacle of modal theory"? It is none other than the fundamental axiom of the strongest of our modal logics, S5: "if possibly necessarily *p*, then necessarily *p*" ($\Diamond\Box p \rightarrow \Box p$).[27] In the case of the argument for the existence of God, we would seem to have this axiom working as a premise: "if it is possible that God necessarily exists, then God necessarily exists."[28] Leibniz is clear, however, that this is a "privilege" of the necessary being, for only it has such a nature that, given its possibility, it exists.[29] But this version of the argument takes us beyond the simple ontological argument sketched at the beginning of this section; it is a version of what has come to be called the "modal ontological argument."[30]

Leibniz offers a second argument for the existence of God that has relevance to issues in the nature of necessity – namely, his argument from the eternal truths. On Leibniz's view, the existence of God is demonstrable because only an infinite intellect can ground the eternal truths, i.e. the truths of logic and mathematics. "If there were no eternal substance, there would be no eternal truths; so this too affords a proof of God, who is the root [*radix*] of possibility, for his mind is the very region of ideas or truths" (A VI iv 1618/MP 77). According to Leibniz, there must be a reason why a true proposition is true – in contemporary parlance, every truth requires a "truthmaker" – and there must be something that explains not only their truth but also their *reality*. As he puts it in the *Monadology* §44, "if there is reality in essences or possibles, or indeed, in eternal truths, this reality must be grounded [*fondée*] in something existent and actual" (G VI 614/AG 218). We know that, for Leibniz, the nature of truth is that the concept of the predicate is contained in the concept of the subject and that in the case of necessary truths an identity can be demonstrated. And we could, therefore, claim that mathematical and logical truths are simply

one of the greatest fruits of all of Logic [*une proposition modale qui seroit un des meilleurs fruits de toute la Logique*]" (G IV 406).

27 The S5 axiom, as presented in most textbooks, is $\Diamond p \rightarrow \Box\Diamond p$, but this is logically equivalent to the formula shown earlier.

28 Admittedly, this formulation allows for a kind of *de re/de dicto* confusion. But Leibniz's intention ought to be clear enough.

29 See, for example, A VI iv 542/MP 13; A VI vi 438/RB 438; G IV 359; G IV 402; G VI 614/AG 218 (*Monadology* §45).

30 For more on this issue, see Malcom (1960), Oppy (1995), Gödel (1995), as well as my own (2006) and (forthcoming).

true by definition; they are "analytic truths" in the Kantian sense and might simply be grounded in human language and convention. Leibniz rejects this kind of nominalist position that he had seen in, among others, Hobbes.[31] The obvious alternative position is that of Plato, for whom the abstract objects of mathematics exist or subsist independently of anything else, and propositions are true or false independent of subjective features. But this position is also rejected by Leibniz. For, on his view, the objects of logic and mathematics are simply not the kinds of things that can subsist by themselves; only substances can truly be said to exist, and essences, possibilities, or truths require a mind. Therefore, Leibniz believes, it is "the divine understanding which gives reality to the eternal truths" (T §184: G VI 226/ H 243).

It is important to realize that Leibniz's view that the eternal truths are grounded in the divine understanding is explicitly opposed to the Cartesian position that the eternal truths owe their existence to the divine will.[32] "It is highly erroneous," he writes,

> to suppose that eternal truths and the goodness of things depend on the divine will; for every act of will presupposes a judgement of the intellect about goodness – unless by a change of names one transfers all judgements from the intellect to the will; and even then it cannot be said that the will is a cause of truths, since the judgement is not such a cause. The reason for truths lies in the ideas of things, which are involved in the divine essence itself. (A VI iv 1618/MP 77)

For Leibniz, then, the ideas of things, whether essences, mere *possibilia*, or the eternal truths of logic and mathematics, are very much dependent upon God. Indeed, they are only possible because there exists an infinite intellect. Leibniz makes this point even more clearly in his comments on Wachter's *Elucidarius cabalisticus*: "The very reality of the essences, indeed, that by which they flow into existence, is from God. The essences of things are coeternal with God like numbers. And the very essence of God comprehends all other essences, to the extent

[31] In 1677, Leibniz addressed the Hobbesian position in a short and fascinating piece. See A VI iv 20–25/AG 268–272.

[32] This is one of the most notorious and striking positions of Descartes, whose voluntarism rendered God capable of freely decreeing that, for example, 2 + 2 = 5. (See, for example, his letter to Mersenne from 15 April 1630: AT I 145/ CSMK III 23).

that God cannot be perfectly conceived without them."[33] Moreover, because essences and the eternal truths are contained within the essence of God, when we come to know the eternal truths and indeed even when we clearly and distinctly perceive things in this world or reflect on possible essences, we establish a connection with God.

God is the only immediate external object of souls, since there is nothing except him outside of the soul which acts immediately upon it. Our thoughts with all that is in us, insofar as it includes some perfection, are produced without interruption by his continuous operation.... And it is thus that our mind is affected immediately by the eternal ideas which are in God, since our mind has thoughts which are in correspondence with them and participate in them. It is in this sense that we can say that our mind sees all things in God. (G VI 593–594/L 627)

Of course, this kind of view is foreign to most contemporary philosophers. Nevertheless, it is interesting and important to see how one of the great mathematicians and logicians of all time sought to explain the nature of mathematical truth, the nature of necessity, navigating between conventionalist nominalism and Platonist realism.

6. Conclusion

I began this chapter by mentioning the connection between rationalism and necessitarianism. The driving question could be stated this way: how should a rationalist conceive of necessity, contingency, and possibility? In seeking to preserve human freedom in the face of the apparent necessitarian consequences of rationalism and determinism, Leibniz asked one of the most fundamental of all philosophical questions: What is truth? And he came to realize that there is a fundamental difference between true necessary propositions and true contingent propositions, ultimately relating to the nature of the infinite. His compatibilist response to the threat of necessitarianism is one that preserves the determinism and grounding of events within this world, while still allowing for human freedom. Further, he claims that the necessary and the contingent rest on the two great principles of all our reasoning: the principle of contradiction and the principle of sufficient

[33] Beeley 2002, 5/AG 273*.

reason. With regard to the nature of necessity, we learned that propositions are necessary just in case their negation implies a contradiction. At the same time, however, we saw that necessary propositions can also be explained by Leibniz in very contemporary terms as being true in all possible worlds, precisely because they do not require explication in terms of the principle of sufficient reason and God's first free decree to bring a particular world into existence. Leibniz also seems both traditional and contemporary insofar as his preferred version of the ontological argument depends upon a strong axiom of modal logic: if possibly necessarily p, then necessarily p. But Leibniz also seems to think that the necessary truths of logic and mathematics are only possible if there is the infinite intellect of God, for truths and essences depend upon a mind for their reality. While I suspect that the reader will not be convinced of all (or any) of Leibniz's views as presented here, I do hope to have shown that Leibniz is a fascinating example of a philosopher whose thought, especially on matters modal, is at times traditional and orthodox and at times modern and radical, one who has an important place in the history of necessity.

11 *Leibniz and the Lucky Proof*

JONATHAN WESTPHAL

There are a number of different strands to Leibniz's thinking about necessity. From as early as the compatibilist determinism of the Letter to Magnus Wedderkopf in 1671,[1] which is the setting for all his subsequent thinking about contingency and necessity, to the *Theodicy* of 1710, and beyond, the strands are twisted together in various different ways. Leibniz begins as a determinist, and then has to consider how contingency and freedom fit in. Among the elements in his thinking about the analysis of modality are the following:

(i) The impossible is or is on analysis the *contradictory* and the possible is what is or is on analysis the *consistent*.

(ii) The necessary is what is *essential* to a thing, or possible consistently with its own nature, if it exists at all, as in a definition, and the possible and contingent is what actually *exists*, if it is created, except God, whose essence is existence.

(iii) The necessary is what is real, though incompletely or not determinately described, such as an abstract plane rather than a real tabletop.

(iv) The necessary is what is *internal* to a thing, derived from its cause, when it comes into existence, and the contingent is what is *external* to it.

(v) The necessary is what obtains in all possible worlds, and the merely possible is what obtains only in some.

(vi) A necessary truth can be finitely analyzed, whereas a contingent truth is a truth whose analysis is infinite, in the sense that it

[1] Gottfried Wilhelm Leibniz, "Letter to Magnus Wedderkopf," May 1671, Leibniz 1969, p. 146.

The material in this chapter is taken from a paper read at the Northern New England Philosophical Association annual meeting, University of Massachusetts, Dartmouth, October 17–19, 2014. I am grateful to the participants for comments and criticisms.

"contains" an infinitely long list of predicate terms. This list converges asymptotically on the complete individual concept associated with the subject term, the truth known by God in a single apprehension. We beings with finite minds can know necessary truths by analysis, since the list of their predicate terms is finite in number, but we can only know or "prove" contingent truths a posteriori. The contingent truths do allow an a priori proof, but it is not one that is available to us.

(vii) The necessary truth is the eternal truth in the mind of God.

I am very uneasy about thinking that these elements or groups of elements form different, complete, and incompatible theories of contingency and necessity in Leibniz's mind, as Robert Adams has suggested.[2] For Adams there are three distinct theories. I shall call them (A) Neo-Spinozism, (B) the Finite Analysis Solution, and (C) the Possible Worlds Analysis. Adams observes that it is hard to find (C) in Leibniz's texts,[3] that it is a reading of the twentieth-century semantics of modal logic back into Leibniz, and that it may not count as a complete theory in Leibniz, if it is there at all. (A), neo-Spinozism, is the view that necessity is what is in the nature of a thing. By 1677 the necessitarian position [the view that 'flatly and without qualification,' 'everything that ever happens is necessary'][4] was replaced by a theory that Leibniz frequently repeated, publicly and privately, to the end of his career, and which must be regarded as his principal (and most confident) solution to the problem of contingency."[5] In a comment on Spinoza (1678) Leibniz writes, "Certainly nothing but these things could have been produced but according to the very nature of things considered in itself [per se] things could have been produced otherwise."[6] Neo-Spinozism tells us what necessity *is* (the sense of "necessity," if you like), but also *which* events are necessary – *all* of them. When Leibniz was "pulled back from this precipice,"[7] it was not by abandoning his determinism, but by finding a form of compatibilism more sophisticated than "The free is the necessary and uncoerced," and one integrated with a theory of contingency.

[2] Adams 1994, pp. 9–35.
[3] Ibid., p. 46.
[4] Ibid., p. 12.
[5] Ibid., p. 12.
[6] Ibid.
[7] Leibniz, "On Freedom," Leibniz 1969, p. 263.

(B), the Finite Analysis Solution, is only a theory about what necessity is. It claims very roughly that necessity is what can be finitely analyzed. The "solution to the version of the problem of contingency that has so fascinated his twentieth-century readers"[8] is that necessary truths are those that can be analyzed in a finite number of steps, whereas for contingent truths the analysis "proceeds to infinity," and only God can see its reason in a single stroke.[9] In (C), the Possible Worlds Analysis, necessity is truth through all possible worlds. Contingent truths are truths only "at" particular possible worlds or groups of possible worlds. The following passage from "Necessary and Contingent Truths," dated *c.* 1686 by Parkinson,[10] contains this conception.

Propositions of essence are those which can be demonstrated, by the resolution of terms; these are necessary, or virtually identical, and so their opposite is impossible, or virtually contradictory. The truth of these is eternal; not only will they hold whilst the world remains, but they would have held even if God had created the world in another way. Existential or contingent propositions differ entirely from these. Their truth is understood *a priori* by the infinite mind alone, and cannot be demonstrated by any resolution. These propositions are such as to be true at a certain time; they express, not only what pertains to the possibility of things, but also what actually exists, or would exist contingently if certain things were granted – for example that I am now alive, or that the sun is shining.[11]

(A), according to Adams, "must be regarded as Leibniz's principal (and most confident) solution to the problem of contingency."[12] (C) is to be found earlier in the passage, and so this text is one of the few in Leibniz where (C) is to be found easily, according to Adams; but then (B) can also be detected in the same passage. (A) is most in line with the metaphysics of the creation, since with "the nature of a thing" we are looking both at God's definition or blueprint for creation, and at the power that keeps everything in its being as what it is, but (B) certainly adds a snap and a crackle that are of the essence of Leibniz's thinking on the topic.

What, therefore, is the ultimate reason for the divine will? The divine intellect ... what then is the reason for the divine intellect? The harmony of

[8] Adams, p. 26
[9] Leibniz, "On Freedom," Leibniz 1989, p. 99.
[10] Adams, pp. 44–45, esp. note 70.
[11] Leibniz, "Necessary and Contingent Truths" (*c.* 1686), Parkinson 1973, p. 98.
[12] Adams, p. 12.

things. What is the reason for the harmony of things? Nothing. For example no reason can be given for the ratio of 2 to 4 being the same as that of 4 to 8, not even in the divine will.[13]

If (A), (B), and (C) were separate theories, then one might well expect to find difficulties with one or another of them that do not apply to the others. This would in turn provide a kind of evidence that the theories *are* distinct. I do not myself feel any tension among (i)–(vii), or any difficulty in reconciling their different elements, and I can see only a single many-sided theory. One should also add to Adams' list (i), as a separate element, the suggestion that the necessary truth is one whose denial is a contradiction. Why is this not a separate theory, if the (A)-(C) are? Where are the contradictions among (A), (B), (C) and (i)? And how would they contradict the proposition (viii) that necessary truths are those which exist eternally in the mind of God. The views are different in emphasis, certainly, but that is all. There is as far as I can see no explicit rejection of any of them, anywhere in Leibniz's work, even after he had discovered the others, and I cannot see an obvious route from any one to a negation of any of the others. Contra Adams I also believe that we would do well to note Leibniz's delight[14] in having discovered (B), as well as the twentieth century's recognition of its technical excellence and amazing cleverness, and its wonderful fit with the metaphysics of the infinite complexity of the contingent universe.

Still, Adams has described a criticism of (B), which William Irvine has christened "the Lucky Proof," that would drive wedges among the three theories. Adams offers no solution to the criticism. If the criticism is a success, then (A), (B), and (C), as well as (i) and (vii), may not be consistent, in which case they are distinct theories.

In this chapter I propose to defend Leibniz against the criticism without entering into the wider project of demonstrating the consistency of (A), (B), (C), (D) and (E). The criticism, according to Adams, is a "difficult problem" that Leibniz "ignores." I shall be arguing that the problem is not difficult, that it has an obvious solution, and

[13] Leibniz, "Letter to Wedderkopf," Leibniz 1969, p. 146.

[14] In "The Source of Contingent Truths," Leibniz 1989, p. 98 ff., one can see that Leibniz is so thrilled with his discovery that he lays out the description of necessary and contingent truth, in the manner of a mathematician describing a dual construction or laying out dual proofs, in a column parallel to one describing commensurable and incommensurable proportions. It is a very interesting question why he should have taken *truth* and *proportion* to be duals.

that given the solution, it would not have appeared as a problem to Leibniz. In short, it is a misunderstanding. I shall also argue briefly that what would or should have been Leibniz's solution to this problem itself creates a second problem, one which Adams regards as "deep." It is related to the question how we can *refer* to anything with an infinite analysis in the first place. But first to the Problem of the Lucky Proof.

> Even if infinitely many properties and events are contained in the complete concept of Peter, at least one of them will be proved in the first step of any analysis. Why couldn't it be Peter's denial? Why couldn't we begin to analyze Peter's concept by saying "Peter is a denier of Jesus and..."? Presumably such a Lucky Proof must be ruled out by what counts as a step in an analysis of an individual concept, but so far as I know, Leibniz does not explain how this is to be done.[15]

It is clear enough what Adams and Leibniz mean by "analyzing Peter's concept"; it consists of listing the things that are true of Peter: that he was a denier of Jesus, that he denied Jesus three times, that he was the first of the twelve to be called, and so on. So it is not easy to see why Adams supposes that Leibniz's best defense is in the understanding of what a step in an infinite proof is. For we *do* have a concept of just such a step in the straightforward understanding of the *finite* proof. The only difference is that the infinite proof has infinitely many of them – there is always another – so that we cannot complete it, but the nature of the step remains the same. Nor can God complete the proof, as Leibniz points out,[16] but God can apprehend the entire a priori proof of the contingent truth in a single act of knowledge,[17] just as we can apprehend, with sufficient understanding, the a priori proof of the necessary truth in one go.

However, we might find, perhaps to our surprise, that some concept we are analyzing turns out to be contradictory. There is Leibniz's

[15] Adams, p. 34.

[16] "But in contingent truths, even though the predicate is in the subject, this can never be demonstrated, nor can a proposition ever be reduced [*revocari*] to an equality or to an identity, but the resolution proceeds to infinity, God alone seeing, not the end of the resolution, of course, which does not exist, but the connection of the terms or the containment of the predicate in the subject." Leibniz, "On Freedom," Leibniz 1989, p. 96.

[17] G. W. Leibniz, "The Source of Contingent Truths," ibid., p. 99.

example of the concept of the greatest possible speed,[18] an example that seems reasonable enough, especially given relativity theory. Or one can imagine the concept of some supposedly historical character that turns out, in the small print, to be contradictory. Suppose we were to begin the analysis of "Peter is one who denied Jesus." We begin the analysis, as the objection of the Lucky Proof imagines, with our first predicate: *a denier of Jesus*.

Peter is a denier of Jesus
one who remembered the words of the Lord when the rooster crowed.
.
.
.
X
–*X*
.
.
.

During the course of the a priori proof or analysis we arrive at a step that gives us *X*, followed by a step that gives us -*X*, as would happen in the attempted proof of the proposition that there is some wheel that turns with the greatest speed. We cannot know that there will *not* be such a contradiction at some future point in the analysis. Since there are always more steps to be examined, there might always turn out to be a contradiction tucked away among the infinitely many predicates "contained" within the subject. In an elaboration of a related difficulty that Leibniz faces, Hawthorne and Cover write,

> Suppose, as Leibniz seems to assume, that a determination of *s* as possible requires an examination of all of *s*'s predicates. Only an infinite analysis will suffice to determine *s* as possible. The point can be made easily enough. A subject-predicate statement of the form "*s* is F" (reading 'is' essentially, not existentially) is true, according to Leibniz, if *s* the F is a possible existent. As we have been at pains to emphasize, the possible existence of *s* the

[18] Leibniz, "Meditations on Knowledge, Truth and Ideas," Leibniz 1989 p. 25, Leibniz 1961 IV, p. 424.

F requires not only that F be contained in the concept expressed by "*s*" but also that this concept denotes a possible existent. Should it emerge that "*s*" expresses an incoherent concept, then it will not be a possible individual concept in the divine mind, and thus will not correspond to one of the creative potentialities contained in God's nature. Granting the existence of a finite proof that the predicate F is indeed contained in the subject concept expressed by "*s*," it remains to be proven that this is a possible concept. Suppose, as Leibniz seems to assume, that a determination of *s* as possible requires an examination of all of *s*'s predicates. Only an infinite analysis will suffice to determine *s* as possible. But then *no matter what F is*, an infinite proof sequence is required to determine that *s* the F is a possible existent. "Caesar is a rational animal" – indeed even "Caesar is Caesar" – turns out to be no less contingent than "Caesar crossed the Rubicon."[19]

Yet in Hawthorne and Cover's discussion of Leibniz's well-known argument that the ontological argument requires a proof of the consistency of the perfections and thereby of the concept of the most perfect being, they notice neither the content nor the form that Leibniz's own proof takes. He does not consider the infinite analysis of "perfect" or "the most perfect"; nor does he need to, for two reasons.

(a) As to content, there is the definition of the perfection. "I call every simple quality that is positive and absolute a *perfection*."[20] Analyses ending in simple concepts are finite, provided that they have a finite number of terms.

(b) As to form, Leibniz's proof proceeds not by an infinitely prolonged inspection but by *reductio ad absurdum*, with the *reductio* assumption "*A* and *B* are compatibles," where *A* and *B* are perfections. When this assumption produces a contradiction, we can see that its negation is true, and we can do it in a *finite* number of steps.[21] Now (a) does not matter, since the proofs of

[19] Hawthorne and Cover 2000, pp. 155–156.

[20] Leibniz 1961, VII, p. 261. There is perhaps the basis here for a Leibnizian distinction between necessary truths and conceptual truths. The necessary truths are the finite ones containing only simple concepts, whereas the conceptual ones are the finite ones containing concepts that themselves have infinite analyses. What the Lucky Proof would then show is that "Adam is the first human" is a conceptual truth.

[21] Ibid. There are ways other than the infinite survey. Hidé Ishiguro notes that God's knowledge involves an infinite analysis of the concepts, which God alone can carry out. "Leibniz believes that God can do this, not because He

the possibility and then the existence of the most perfect being are a priori proofs of necessary truths, but there can be *reductio* proofs of contingent truths as well. Galileo proves the *contingent* truth that heavier bodies do not fall faster than light ones by *reductio*. Suppose that heavier bodies do fall faster than light ones. Consider two bodies, one heavier than the other. Let them be connected by a physical line, and drop the system into a free fall. Since the first body is heavier than the second, the second will function as a drag on the first, so the heavier body will fall more slowly than it would have done alone. But since heavier bodies fall faster than lighter ones, the system of the two bodies as a whole will fall faster together than the heavier body alone, and so the heavier body will fall faster than the heavier body alone.[22]

(c) There is a further difficulty with the claim of Hawthorne and Cover (that "Only an infinite analysis will suffice to determine *s* as possible"), in the existence of the possibility of non–a priori knowledge offered by Leibniz in the "Meditations on Knowledge, Truth and Ideas" of 1684.[23] Leibniz states there that the possibility of a thing can be known a priori or a posteriori. "The possibility of a thing is known *a posteriori* when we know through experience that a thing actually exists, for what actually exists or existed is at very least possible."[24] So there can be knowledge of contingent truths even without their *a priori* proofs, as from the fact that an infinite list of predicates cannot be produced it does not follow that we do not know that the concept is consistent.

Still, if we were to allow the premise that "only an infinite analysis will determine the possibility of '*s* is F', or, for Leibniz, '*s* the F is a possible existent', then, Hawthorne and Cover write, '… *no matter what F is*, an infinite proof sequence is required to determine that *s* the F is a possible existent.' "

I have suggested that the premise (only an infinite analysis will determine whether a truth is necessary) of the argument that we need

sees the end of the analysis, there being no such end, but because He knows the principle which combines the concepts to be analyzed (i.e. the principle of perfection). It is like our using mathematical induction to give a finite proof of a truth concerning an infinite totality" (Ishiguro 1990, p. 195).

22 Galileo 1989, pp. 107–109.

23 Leibniz 1989, p. 26, Leibniz 1961, IV, p. 425.

24 Ibid.

infinite proof sequences for every truth is false, for Leibniz, but if it were true, how would it follow that identities such as "Caesar is a rational animal" are not necessary truths? Hawthorne and Cover's argument is that for Leibniz "*s* is F" is "*s* is a possible F." The possibility of F requires an analysis. But the analysis will be infinite. I have also suggested that in some cases either the analysis of possibility terminates short of infinity, or it is unnecessary: the two cases are the *reductio* proof and the object known empirically to exist.

(d) There also remains the case of objects, mathematical and logical, that only enter into restricted or necessary relations. Hawthorne and Cover have not included the vital role of the definition in their discussion. One might think that one could by their argument show that "A triangle is a three-angled closed plane figure made by joining three non-collinear points by straight line segments" is contingent. No doubt some senses of "figure" have an infinite analysis, but in geometry "figure" has a finite definition: any set of points in a plane. So *triangle* and then *figure* as well as the other four concepts in the concept of a triangle as defined above have finite analyses, as do all mathematical and logical terms. They are like "single-criterion" concepts;[25] or they might be called "finite-criterion" concepts. The mathematical or logical term, as Leibniz writes, "se peut distinctement expliquer." This is incidentally for Leibniz why terms for secondary qualities can be made intelligible, and why, when they are, we can call them primary qualities.[26]

Certainly Hawthorne and Cover are right to suggest that the infinite analysis doctrine of contingent truths must accomplish two things, or two things at least. "It will (i) explain why the proof of any contingent claim of the form '*s* is F' must work through the whole infinite complex that is contained in the concept of *s*, and it will do so (ii) without making it the case that every statement of the form '*s* is F' is contingent."[27]

[25] The idea of the "one-criterion concept" is Hilary Putnam's. See e.g. "Is Semantics Possible," Putnam 1975, p. 141, and "The Analytic and the Synthetic," ibid., p. 68.

[26] Leibniz 1981, book II, ch. 8, p. 130, Leibniz 1961, V, p. 116.

[27] Hawthorne and Cover, 2000, p. 156.

Nevertheless, it seems odd, obviously wrong even, that the infinity of the contingent is linked merely to the need for consistency, and not directly to infinite complexity as such. The infinity of the contingent has a much deeper and wider role in Leibniz's conception of the relation between the contingent and the necessary. The finite and the mathematical can be grasped by a finite mind in such a way that it is completely clear, so that there is nothing in it that is not clear. The contingent, on the other hand, contains unresolved elements that do not become clear short of infinity. *That* seems to be the essential point of the conception of the infinity of the contingent, however matters stand with the Lucky Proof.

Hawthorne and Cover give their own very interesting account of *why* the analysis of the complete individual concept should be finite in the case of mathematical subjects but infinite in the case of contingent real entities. They frame their analysis with Leibniz's concept of *a balance*, on which the infinitely many "micro-inclinations" to which the monad is subject are weighed. Short of infinity one will be unable to determine which way the balance will tip, and whether the predicate F*s* or ~F*s* will turn out to be true. So we have here an account of why an analysis must proceed forever.[28] "Leibniz's best answer would have been ... that 'sinned' decomposes to a disjunction of the infinite sets of micro-inclinations that would each suffice for sinning."[29]

An advantage of Hawthorne and Cover's account is that it links the logical side of Leibniz's doctrine to several important elements in the metaphysical side of infinite divisibility and composition in a helpful way that seems fully Leibnizian. Yet they write, "Now in a sense this is clearly a speculative reconstruction of Leibniz's view."[30] Besides, there is a shorter and simpler logical argument, equally Leibnizian, for the conclusion that the Lucky Proof is not valid. I cannot see anything in it that is inconsistent with Hawthorne and Cover's argument or with Leibniz's metaphysics of the infinity of the monad. Furthermore, this argument is already Leibniz's own, as we can already know from his metaphysics of identification and individuation, and so there is no need for a 'speculative reconstruction'.

[28] Ibid., section 3 ff.
[29] Ibid., p. 162.
[30] Ibid., p. 162.

When we perform an analysis on *s*, Adam, for example, the reason we cannot reach an end of the predicates in a finite time, or even an infinite one (as there is no end) is that there are an infinite number of predicates. The constructor of the Lucky Proof seeks just one predicate ("the first human," for example) and finds it, by chance, at the top of the list of Adam's predicates. What the Lucky Proof objection then asks is a question about the list of predicates. Why do we need to continue to consider them all, of course in the end failing, unless we have the mind of God, since we already have the one we are after? One possible answer, as we have seen, is that there might be a hidden contradiction, and that this will not be discovered short of infinity. Another is that a proper proof that the subject indeed has the predicate will need to deduce the presence of the predicate from further facts about Adam, *plus petites*, so to speak, and that the task of sorting through these and setting them on the balance that determines whether arbitrary F*s* or ~F*s* cannot be completed.

As I see it the difficulty lies not in a restriction on or some feature of the *predicates* of Adam, but in what it takes to understand and refer to Adam, the *subject*, in the first place. It is not as though we take the subject, clearly and distinctly conceived in Leibniz's sense,[31] having it within our grasp, and then start riffling through the pack of predicates looking for matching items. *We do not clearly and distinctly conceive the subject*; nor can we refer to it specifically through all the possible worlds containing things like it, since we lack the infinity of the marks which is the "complete being."

Consider the very important passage in the Letter to Arnauld of May 1686 "about My Proposition that the Individual Notion of Each Person Includes Once and for All Everything That Will Ever Happen to Him."[32]

I have said that all human events can be deduced not simply by assuming the creation of a vague Adam, but by assuming the creation of an Adam

[31] "Knowledge is *clear* when I have the means for recognizing the thing represented. Clear knowledge, again, is either confused or distinct. It is confused when I cannot enumerate one by one marks [*nota*] sufficient for differentiating a thing from others, even though the thing does indeed have such marks and requisites into which its notion can be resolved." "Meditations on Knowledge, Truth and Ideas" of 1684, Leibniz 1989, p. 24, Leibniz 1961, IV, p. 423.

[32] Leibniz 1989, p. 41–42, Leibniz 1961, II, p. 37.

determined with respect to all those circumstances, chosen from among an infinity of possible Adams. This has given Arnauld the occasion to object, not without reason, that it is as difficult to conceive of several Adams, taking Adam as a particular nature, as it is to conceive of several mes. I agree, but when speaking of several Adams, I was not taking Adam as a determinate individual. I must therefore explain myself. This is what I meant. When one considers in Adam a part of his predicates, for example that he is the first human, set in a garden of pleasure, from whose side God fashioned a woman, and similar things conceived *sub ratione generalitatis*, in a general way (that is to say, without naming Eve, Paradise and other circumstances that fix individuality), and when one calls Adam the person to whom these predicates are attributed, all this is not sufficient to determine the individual, for there can be an infinity of Adams, that is, an infinity of possible persons, different from one another, whom this fits ... Therefore we must not conceive of a vague Adam, that is a person to whom certain attributes of Adam belong, when we are concerned with determining whether all human events follow from positing his existence; rather we must attribute to him a notion so complete that everything that can be attributed to him can be deduced from it.[33]

If we were to apprehend Adam through the senses, or through the concept that includes what we know about him, we would not know which of the infinity of Adams we were apprehending, except that it is *this* one.[34] But if we are to give an a priori proof of one of his predicates, such as *being the first human*, we must know *which* Adam we are talking about. There is for Leibniz to speak strictly only one Adam.

[33] Leibniz 1989, pp. 72–73.

[34] This is connected to the importance of the sample in Hilary Putnam's theory of reference. "The key point is that the relation same_L is a *theoretical* relation: whether something is or is not the same liquid as *this* may take an indeterminate amount of scientific investigation to determine," i.e. in indeterminate or infinite amount of analysis through all the possible worlds. See "The Meaning of 'Meaning,'" Putnam 1975, p. 225. At the end of the day, of course, for Leibniz there is only *one* world containing *this*, something bearing the relation same to the liquid in front of us. Connecting this as well to Leibniz's epistemology, we can see that what goes for the senses also goes for the restricted or vague *Sinn* given by a finite number of predicates. "We use the external senses as a blind man uses his stick, following the comparison used by an ancient writer, and they allow us to know their particular objects, which are colors, sounds, odors, flavors and the tactile qualities. But they do not all us to know what these sensible qualities nor what they consist in." Leibniz, "On What Is Independent of Sense and Matter" (1702), Leibniz 1989, p. 186; Leibniz 1961 VI, p. 499.

Without knowing which of the possible Adams this one is, though we can refer to him "as a blind man uses a stick," we do not know in metaphysical strictness to whom we are referring, and we have not demonstrated that it is indeed *Adam* of whom we are predicating *being the first human*, or at any rate *this* Adam, the actual Adam, or the one in our world.

What the Lucky Prover overlooks is that the Lucky Predicate, which turns up first on his search list, can be predicated of someone, but for Leibniz there is no telling *who* that someone is without the complete individual concept. The Lucky Proof does not demonstrate the proposition that *Adam* or indeed anyone in particular, in metaphysical strictness, is the first human, but only that someone or other is, someone incompletely specified, who resembles our Adam or the actual Adam, whoever *he* may be, in certain specified respects, though he fails to resemble him in others, infinite in number. Since the list of Adam's predicates is infinite, so the number of possible Adams and the number of possible worlds containing someone like Adam, the first human, is also infinite. We cannot distinguish *this* Adam, who is the first human and was buried in the Cave of Machpelah, as *per* a traditional Jewish belief, and *that* Adam, who was not.

> The most important point in this is that individuality involves infinity, and only someone who is capable of grasping the infinite could know the principle of individuation of a given thing.[35]

When we initiate a Lucky Proof, we can be asked whom we mean by "Adam." The answer will be that he is the first human, to be found in Paradise, from whose side God fashioned Eve.... Our subject term contains three predicates, and the three little dots. We begin to analyze Adam's concept, and the first thing we hit on is '... is the first human.' We conclude that 'Adam is the first human' is necessary. If "Adam" were a synonym for "the first human" plus the other two predicates alone, then following Adams, then 'Adam is the first human' would not be contingent. But "Adam," unlike "triangle," has for Leibniz an inescapable and complete metaphysical failure of reference, in our mouths, though not in God's. It is only synonymous with the description of the *complete* individual concept. In the Lucky Proof, we can if we wish

[35] Leibniz 1981, pp. 289–290; Leibniz 1961, V, p. 268.

suppose that “Adam” is just the first human, found in Paradise, from whose side God fashioned Eve. Yet we are still in no position, short of infinite analysis, to say that *this* Adam we are talking about is the actual Adam, our Adam, the Adam in this world. So we cannot say that the Lucky Proof has shown that the proposition “*Adam* is the first human” (which one?) is not contingent. The word “Adam” is vague. Some of the Adams, though not all, are the first human. This fact is, I believe, what blocks Adams’s Lucky Proof, and it is in perfect accord with Leibniz’s metaphysics and epistemology.

12 Divine Necessity and Kant's Modal Categories

NICHOLAS F. STANG

1. Introduction

The Pölitz transcripts of Kant's lectures from the 1780s[1] on rational theology contain a fascinating discussion of modality:

> On this point rests the only possible ground of proof for my demonstration of God's existence, which was discussed in detail in an essay I published some years ago. Here it was shown that of all possible proofs, the one that affords us the most satisfaction is the argument that if we remove an original being, we at the same time remove the substratum of the possibility of all things. – But even this proof is not apodictically certain; for it cannot establish the objective necessity of an original being, but establishes only the subjective necessity of assuming [*annehmen*] such a being. But this proof can in no way be refuted, because it has its ground in the nature of human reason. For my reason makes it absolutely necessary for me to assume a being which is the ground of everything possible, because otherwise I would be unable to know what in general the possibility of something consists in [*worin etwas möglich sey*]. (Pölitz RT, Ak. 28:1034)[2]

The "essay published some years ago" to which Kant here refers is his 1763 work *The Only Possible Ground of Proof in Support of a Demonstration of the Existence of God* (henceforth, *OPG*). While the exact interpretation of that work is controversial,[3] it is clear that he there defends an interconnected set of claims about possibilities and their grounds:

1 Kant lectured on rational theology during the winter semesters 1783–1784 and 1785–1786, and it is unclear which set of lectures is the basis of the Pölitz transcripts. See Kant (1996), 337–338 for more.

2 In the rational theology lectures, see also Volckmann RT, Ak. 28:1176; and Danziger RT, Ak. 28:1259. The translation of the Pölitz text is from Kant (1996), with minor modifications.

3 Cf. Adams (1994); Watkins and Fisher (1998); Chignell (2009), (2012), and (2014); Stang (2010); Yong (2014); and Abaci (2014).

(1) It is logically possible that p iff p does not entail a contradiction. This is the definition of logical possibility. (Ak. 2:77)
(2) Not all logical possibilities are *real* possibilities (Ak. 2:77, 85). Kant does not explicitly offer a developed theory of what real possibility is, but it is clear that real possibility is more "metaphysically robust" than mere logical possibility. He also lays down certain conditions on it.[4] For instance,
(3) If it is really possible that p then some existing being grounds the fact that it is really possible that p (Ak. 2:78, 79). As with (2), Kant does not explicitly offer a developed theory of the relation between real possibilities and their grounds.
(4) All real possibilities must be grounded in a "first" or "highest" ground of real possibility (Ak. 2:85–87). This first ground of real possibility is said to exist with *absolute* real necessity, which Kant defines to mean, if this being did not exist, nothing would be really possible (Ak. 2:83).
(5) Since there are real possibilities, there is a first ground of real possibility (Ak. 2:157). Kant argues that this being has the traditional divine attributes (simplicity, omniscience, omnibenevolence, etc.) and identifies it with God.

Of course, Kant's actual theory is much more complex than that, but for now this brief outline will suffice.[5]

In this passage from the 1780s Kant appears to be saying that the 1763 argument does not constitute a proof of the existence of God or of a first ground of real possibility. Whereas in 1763 Kant argued that the fact that something is really possible *entails* that there exists a ground of all real possibility, he now appears to think that this fact rationally requires us to seek an explanation of it, and the only rationally satisfying explanation is that there is a ground of all real possibilities, so we are rationally required to assume – I want to remain neutral on exactly what epistemic attitude we are supposed to adopt so I am going to use the nontechnical term 'assume'[6] – that there is such a being.

[4] One of the most important elements of Kant's modal theory in *OPG*, the principle that real possibility requires a "material element," is not as important for understanding the Pölitz passage, so I omit it here. Cf. Ak. 2:77, 79–81.

[5] I examine all of these claims in more detail in Stang (2010) and in Stang (2016).

[6] I will also sometimes speak of "hypothesizing" this being, or the hypothesis that there is such a being. I mean this to be synonymous with assuming that there is such a being.

This raises a number of questions, including, what is the nature of this rational necessity, the necessity that we assume an original being? And what epistemic status does this resulting assumption have, if it is not known to be true? However, I want to focus on the necessity of the being that is assumed, which Kant goes on to characterize:

> For in addition to the logical concept of the necessity of a thing (where something is said to be absolutely necessary if its nonexistence would be a contradiction, and consequently impossible), we have yet another rational concept of real necessity. This is where a thing is *eo ipso* necessary if its nonexistence would cancel all possibility. Of course in the logical sense possibility always precedes actuality, and here I can think the possibility of a thing without actuality. Yet we have no concept of real possibility except through existence [*Existenz*], and in the case of every possibility that we think *realiter* we always presuppose [*setzen … voraus*] some existence [*Daseyn*]; if not the actuality of the thing itself, then at least an actuality in general which contains the data for everything possible. Hence all possibilities presuppose [*voraussetzen*] something actually given, since if everything were merely possible, then the possible itself would have no ground; so this ground of possibility must itself be given not merely as possible but also as actual. But it must be noted that only the subjective necessity of such a being is thereby established, i.e. that our speculative reason sees itself necessitated to presuppose [*vorauszusetzen*] this being if it wants to have insight into *why* something is possible, but the objective necessity of such a thing can by no means be demonstrated in this matter. (Pölitz RT, Ak. 28:1036)[7]

Kant here distinguishes between a "logical" and a "real concept" of the necessity of a thing. I will focus on the former: a being exists absolutely necessary just in case that being exists, and were it not to exist, nothing would be really possible. This is the very same definition of absolutely necessary existence Kant gave in 1763 in *OPG* (see (4) earlier).[8] He also introduces a concept of "real" possibility and a principle governing real possibility – if it is really possible that p then the fact that it is really possible that p is grounded in something that actually exists – that are familiar from that earlier work (see (3), as well as Ak. 2:77–78).

[7] This crucial text is also cited and discussed in Adams (1994), 182, note 9 on that page; and Chignell (2007b), 349.

[8] Ak. 2:81–83. See Stang (2010) for a discussion of this interpretation of absolutely necessary existence.

Readers of *CPR* know that Kant frequently distinguishes between logical and real possibility and claims that we cannot always infer from logical possibility to real possibility. However, it is often unclear what the Critical Kant means by 'real' possibility and whether, in fact, he means one thing, or whether he has a number of distinct notions of possibility that he simply groups under the heading 'real' to distinguish them from logical possibility.

In this chapter I am going to argue that in the Critical period Kant does have a unified theory of real possibility and that this theory has the same general structure as his pre-Critical theory of real possibility. In Section 2 I consider what it could mean, within Kant's Critical system, to apply modal concepts such as <*necessary*>[9] to a hypothesized being that cannot be an object of experience; in Section 3 I argue that the concept of real possibility of which Kant speaks in the Pölitz lectures is the modal category of possibility and that this is a generalized version of the pre-Critical concept of real possibility. It is this concept that unifies the different kinds of real possibility Kant accepts in the Critical system. In Section 4 I argue that for each kind of Critical real possibility there is a corresponding kind of absolute necessity. I conclude by identifying which kind of absolute necessity is being discussed in the Pölitz lectures.

2. Schematized and Unschematized Categories

One important feature of the Pölitz passages quoted earlier is that these modal concepts – absolutely necessary existence and real possibility – are being applied beyond the range of experience. We are rationally necessitated to hypothesize an absolutely necessary being, but no object of experience is absolutely necessary.[10] Without opening the Pandora's box of competing interpretations of Kant's transcendental idealism, I would like to say a few things about what kind of object we are hypothesizing. The object we are hypothesizing is what Kant calls a "noumenenon in the negative sense," which means "a thing insofar as

[9] I follow Anderson (2004) and use italicized expressions flanked by angle brackets to denote the corresponding concept. E.g. "<*necessity*>" denotes the concept corresponding to the English (German) expression "necessity" (*Nothwendigkeit*).

[10] A559/B587, A561/B589.

it is not an object of our sensible intuition" (B307).[11] I will simply refer to such objects as "noumena." Noumena are "supersensible" [*übersinnliche*] because they are not the right kind of thing for us to sense or experience; because they are not in space and time, we could not sense them, even if our sense organs were vastly finer, and we cannot infer their existence from any empirical causal laws.[12] In hypothesizing the existence of an absolutely necessary being that grounds all real possibility, we are using modal concepts to think about a noumenon. Although Kant denies that we 'cognize' [*erkennen*] noumena through the categories, he maintains that we not only *may* think of them using the categories but that we *must* do so. In fact, we cannot think about any objects without thinking of them under the categories.[13]

My question in this section is, what are we doing when we apply a modal category to a noumenon? This is an instance of a larger question in Kant's philosophy: what are we doing when we apply any category whatsoever to noumena? In other words: what is the content of a category when it is applied to noumena?

Although we can (and must) think of all objects under the categories, there is an important difference in the epistemic status of these thoughts for objects we can experience (phenomena) and those we cannot experience (noumena).[14] Kant claims that the categories have 'objective validity' when applied to phenomena, but lack it when applied to noumena.[15] Categories *as applied to empirical objects* have objective validity in part because they can be *schematized*, i.e. provided with a priori rules for application to objects given in experience. I will henceforth refer to them as "schematized categories."

[11] Some readers will wonder whether any supersensible being could be an object for us. By "object" here I mean only "object of thought." I am not claiming that such beings could be objects in a more substantive sense, e.g. objects of cognition or of experience. See A105, A108, B137.

[12] See A226/B273, and Ak. 8:205. In this chapter, I remain neutral on the relation between "noumena in the negative sense" and what Kant calls "things in themselves" [*Dinge an sich (selbst)*] and, indeed, on whether "things in themselves" functions referentially as a name for a *kind* of object, or adverbially for a way of considering objects (as they are in themselves), as Prauss (1974) famously argued.

[13] A253, Ak. 5:54, 103.

[14] On the "one object" interpretation, empirical and nonempirical conceptions of objects.

[15] A93/B126, A111, A128, A239/B298.

Categories applied to empirical objects can be schematized, the categories in general cannot, and because of this, categories have objective validity for empirical objects, while categories lack objective validity for objects *überhaupt*; consequently, in applying categories to empirical objects we cognize those objects, but in applying categories to supersensible objects (noumena) we do not cognize anything about those objects. Sometimes, rather than categories as concepts of objects in general, I will just talk about "unschematized" categories. The question from earlier can be rephrased: what are we doing when we think of an object under the unschematized categories *<necessary>* or *<possible>*? What is the content of such thoughts?

Before going any further in trying to answer *that* question, I want to investigate further what is lacking or missing or defective about unschematized categories. The section of the *CPR* in which Kant goes into the most detail about the distinction between categories as concepts of empirical objects (phenomena) and as concepts of objects in general is 'On the ground of the distinction of all objects in general into phenomena and noumena.' It contains the following interesting discussion of the difference between schematized and unschematized categories:

> For every concept there is requisite, first, the logical form of a concept (of thinking) in general, and then, second, the possibility of giving it an object to which it is to be related. Without this latter it has no sense, and is entirely empty of content. (A239/B298)

This is the familiar Kantian point that in order to cognize objects using some concept we must first be able to *prove* the *real* possibility that there are objects falling under that concept.[16] The *logical* possibility of a concept's being instantiated consists in that concept's being logically consistent (not containing any marks that are mutually self-contradictory), but logical possibility does not suffice for *real* possibility (whatever that is, exactly). Kant here is claiming that the reason unschematized categories do not have objective validity for noumena is that we cannot prove the real possibility of noumena falling under them; we can only prove the real possibility of phenomena falling under them, so only empirical (schematized) categories have objective validity.

[16] Cf. the footnote to Bxxvi.

However, Kant goes on in the next paragraph to suggest that unschematized categories – the categories as applied to *all* objects in general, empirical as well as nonempirical objects – are defective for cognitive use by us in a more radical sense:

> That this is also the case with all categories [*that we cannot use them to cognize anything about objects in general*] however, and the principles spun out from them, is also obvious from this: That we cannot even give a real definition of a single one of them, i.e. make intelligible the concept of their object, without immediately descending to conditions of sensibility, thus to the form of the appearances, to which, as their sole objects, they must consequently be limited, since, if one removes this condition, all significance, i.e., relation to the object, disappears, and one cannot grasp through an example what sort of thing was really intended by concepts of that sort. (A241/B300 – underlined material added in B edition)

Kant here identifies the lack of objective validity of the categories in general with our inability to give real definitions of them for all objects in general (phenomena and noumena). He identifies a real definition of a category as making "intelligible" [*verständlich*] to ourselves how it is really possible for objects to fall under them; this is most naturally read as meaning that a real definition of a concept reveals not merely *that* it is really possible for objects to fall under it, but reveals or makes comprehensible to us *why* it is really possible. In his own copy of the A edition, Kant wrote in the margin of this passage: "we cannot explain their [the categories'] possibility." Giving a real definition of a concept is proving that an object of that concept is really possible by eliciting the *grounds* of that real possibility and thereby explaining that real possibility.

This connection between real definitions and knowledge of the grounds of possibility is further substantiated in Kant's lectures on logic, in which he distinguishes between nominal and real definitions.[17] A nominal definition of a concept reveals what we "think" in the concept, the marks we have assembled together to form that concept, what

[17] The principal discussions of real and nominal definition, and the associated notions of real and logical essence, are in the Jäsche Logic (Ak. 9:61, 140–145) and the transcripts of Kant's logic lectures (Ak. 24): Blomberg Logic (113–118, 268–273), Vienna Logic (838–840, 913–925), Dohna-Wundlacken Logic (727-729,756-760), Philippi Logic (408–409, 456–459), Busolt Logic (634–635, 656–660), and Pölitz Logic (535, 573–575).

Kant sometimes calls the logical essence of the concept. Kant thinks that nominal definitions are relatively trivial for empirical scientific purposes. It may be important to make sure we all mean the same thing by our words – that we are all using "water" to denote the clear, drinkable liquid in rivers and lakes – but this does not by itself tell us anything substantive about *water*, the stuff to which our concept refers.[18] *Water* itself, rather than the concept <*water*>, is the object of a real definition. This means that when Kant says that real definitions are definitions of *things* [*Sacherklärungen*][19] he does not mean to restrict real definitions to *individual* things, which Kant thinks are *never* susceptible to real definition;[20] real definitions, if they are possible at all, would be of the generic kinds of individual objects that fall under our concepts. A real definition reveals the essential inner marks of the object that make it possible and that explain its other properties. For instance, a real definition of *water* (if we could give one, which Kant denies) would reveal the inner properties of water that are essential to it and that explain its manifest properties (clarity, drinkability, etc.). In doing so, a real definition would reveal not only *that* water is really possible, but *why*. It would reveal the grounds of the *real possibility* of water.

However, Kant is quite skeptical about our ability to give real definitions of 'given' concepts (as opposed to concepts we *make*), either a posteriori given concepts of natural kinds such as water, or a priori concepts given by the nature of our understanding itself, the categories. Kant allows that for "given" concepts, both empirical and a priori, we may be able to analyze the concept to determine a few marks contained in it, although in general we will not be able to analyze the concept fully and thus give a complete nominal definition.

Kant admits that we can give definitions *of a sort* of the categories for all objects *überhaupt*, both phenomena and noumena, and gives a series of examples of such pseudodefinitions:

> If I leave out persistence (which is existence at all times), then nothing is left in my concept of substance except the logical representation of the subject, which I try to realize by representing to myself something that can occur solely as subject (without being a predicate of anything). (A242/B300)

[18] Ak. 24:116, 271, 757, 839, 918. Cf. A728/B756.
[19] Jäsche Logic §106 (Ak. 9:143); cf. Ak. 24:839, 757.
[20] Ak. 24:118, 268.

> From the concept of a cause as a pure category (if I leave out the time in which something follows something else in accordance with a rule), I will not find out anything more than that it is something that allows an inference to the existence of something else. (A243/B301)
>
> No one has ever been able to define possibility, existence, and necessity except through obvious tautologies if he wanted to draw their definition solely from the pure understanding. For the deception of substituting the logical possibility of a **concept** (since it does not contradict itself) for the transcendental possibility of **things** (where an object corresponds to the concept) can deceive and satisfy only the inexperienced. (A244/B302)[21]

The 'obvious tautologies' to which Kant alludes in the third passage are principles such as that the necessary is that whose nonbeing is not possible.[22] The third passage also tells us what is incomplete in these putative definitions – it is not that they are false, but that they do not show *that* it is really possible for an object to fall under the concept, much less *why*. At most they show that the concept is logically consistent. Kant does not deny, for instance, that a cause *in general* is something from which the existence of something else follows, but this principle does not show us what makes it really possible for there to be a cause. Similarly, it is surely the case that a necessary being is one whose nonexistence is not possible. But this does not tell us what would make such a being necessary. Given that Kant has just invoked the notion of a real definition – and thus implicitly contrasted the real definition of the categories with mere nominal definitions – it is natural to think of these as partial nominal definitions. They unpack conceptual connections between categories (e.g. the interdefinability of the necessary, the possible, and the contingent) but they make no progress in making intelligible to us the real possibility that objects in general fall under these concepts.

3. Real Possibility and the Categories

What does "real possibility" mean here? In the case of mathematical concepts and schematized categories the relevant notions of real possibility are constructability in pure intuition and compatibility with

[21] Cf. A459/B487 and, especially, A593/B621.

[22] This standard interdefinition of possibility and necessity is repeatedly asserted by Kant in his metaphysics lectures. See Ak. 28: 418, 498, 556, 557, and 633.

the forms of experience (the former may be an instance of the latter), respectively (A239–240/B298–299). But those notions of real possibility clearly do not apply to all objects in general. When Kant claims that we cannot give a real-definition of a single one of the categories as concepts of objects in general, which I have interpreted to mean that we cannot identify the grounds of the real possibility of an object in general falling under the categories, he, presumably, does not have these notions of real possibility in mind.

Before continuing, I want to point out that the modal categories should be treated slightly differently than the other categories. The problem of the real definition of <*cause-effect*>, for instance, is to determine the ground of the real possibility of objects in general standing in cause-effect relations, i.e. one object being such as to posit the existence of an object distinct from it. But applying the same scheme to the problem of the real definition of <*possibility*> we get that the problem is to determine the ground of the real possibility of objects in general being possible. In other words, we get an iterated modality: the ground of the real possibility of possibility. This is very indirect and artificial. It is much more natural to think that what we want from a real definition of the category <*possibility*> is to determine the grounds of real possibility of objects in general; no modal iteration is necessary. After all, possibility is already included in the idea of a real definition of a concept, so it makes sense that when we arrive at the question of the real definition of <*possibility*> itself we do not ask about the grounds of the real possibility of possibility: we ask about the grounds of real possibility *period*. To take a related example, Kant's complaint about what he calls the 'nominal definition' of necessary existence (A593/B621)[23] – a being exists necessarily just in case its nonexistence is impossible – is not that this fails to tell us what makes a necessary being possible (iterated modality) but that it fails to tell us, if there is a necessary being, what grounds the necessary existence of that being, i.e. *why* it is not possible for that being not to exist. Likewise, what is missing in our grasp of the category <*possible*> is an account of what grounds the possibility of all objects in general – not an account of what grounds the real possibility of the possibility of all objects in general.

But this means that the question being raised about the nonmodal categories – e.g. what is the ground of the real possibility of objects in

[23] Kant makes the same point in *OPG* at Ak. 2:81.

general being causes? – is the same kind of question being raised about the modal category – what is the ground of the *possibility* of objects in general? – only if the notion of 'real possibility' involved in the first question is the *same* as the notion of possibility involved in the *second* question, which is itself the modal category *<possibility>*. This means that the sense of 'real possibility' involved in a demand for a real definition of the nonmodal categories is the same notion of possibility involved in a demand for a real definition of the modal category *<possibility>*. In other words, the unschematized category *<possibility>*, applied to all objects in general, is the sense of real possibility at issue when Kant denies we have insight into the grounds of real possibility of objects in general falling under the categories.

The categories, including the modal categories, are concepts we apply when we think about objects in general. So if the unschematized category *<possibility>* (henceforth, *<possibility>*$_{UC}$) is a concept of real possibility, then it must be the highest or most general concept of real possibility. If there were a more general concept of real possibility than *<possibility>*$_{UC}$ then in thinking about that kind of real possibility we would not be using the unschematized category, and this is impossible; any thought about objects requires thinking about them under the categories.

What does it mean that the unschematized category is the most general concept of real possibility? First of all, it means that *<possibility>*$_{UC}$ is a concept of real possibility, rather than of logical possibility. This means that *<possibility>*$_{UC}$ is a different concept from *<logical possibility>*. Consequently, it is logically possible that they pick out distinct extensions[24] in any given domain; it cannot be determined through conceptual analysis alone whether, for instance, all and only the propositions[25] about nonempirical objects that are logically possible are

[24] Some readers might balk at the idea of modal concepts having "extensions," especially given Kant's claim that "the categories of modality … do not augment the concept to which they are ascribed in the least, but rather express only the relation to the faculty of cognition" (A219/B266). I take this to mean that the extensions of modal concepts (the "things" to which they can be correctly applied) are not objects (*<possibility>* does not pick out a class of objects, the possible objects, as opposed to the impossible objects).

[25] By "proposition" I mean merely the *content* of a judgment rather than the act of judging itself. See previous note.

also really possible in the most general sense.[26] This, I take it, is Kant's point when he repeatedly distinguishes real possibility from logical possibility, and denies that we can always infer from logical possibility to real possibility,[27] even in the case of nonempirical objects.[28]

The other thing I mean by claiming that the unschematized category of possibility is a concept of real possibility is that it is part of the unschematized category of (real) possibility that a (real) possibility is grounded in some aspect of actuality. This is what Kant means in the Pölitz lectures when he writes: "Yet we have no concept of real possibility except through existence, and in the case of every possibility which we think *realiter* we always presuppose some existence" (Ak. 28: 1036).[29] I interpret this as a claim about the unschematized category of (real) possibility in general, <*possible*>$_{\mathrm{UC}}$. I take this to be a partial nominal definition of the category similar to the principle: the necessary is that whose nonbeing is not possible (A244/B302). This is *why* our concept of "real possibility" carries with it the rational necessity to inquire into the ground(s) of real possibility: in thinking about noumena we have to think about them using the unschematized category of (real) possibility and this carries with it the thought that (real) possibilities for noumena – as for all objects in general – have grounds in actuality.[30]

To anticipate slightly, I will argue that Kant implicitly distinguishes several different *kinds* of real possibility. What makes them all kinds of *real* possibility is that they all instantiate the nominal definition of the highest concept of real possibility, <*possibility*>$_{\mathrm{UC}}$; they are all conceptually distinct from logical possibility and they are grounded in actuality. More precisely,

[26] It is, I take it, a conceptual truth that everything that is really possible is logically possible.

[27] E.g. Bxxvi, A244/B302 (and the footnote added to that page in the B edition), and Ak. 20:325.

[28] E.g. A596/B624. In context, Kant's point is that we cannot infer that God is really possible just because the concept <*God*> is logically consistent. This only makes sense if real possibility also applies to nonempirical objects, for if it did, we could conclude that God is not really possible because he is not the object of any experience.

[29] See also Ak. 28:310, as well as Refl. 3931 and 5034.

[30] This is only the *beginning* of such an explanation. My complete story of this 'rational necessity' is told in Stang (2016).

(*Real Possibility*) For any kind of possibility $\Diamond_x p$ and its associated kind of necessity $\Box_x p$ (where $\Diamond_x p$ iff $\sim\Box_x\sim p$), $\Diamond_x$ is a kind of real possibility (and $\Box_x p$ is a kind of real necessity) only if

(i) *Nonlogicality*. It is not a conceptual truth that $\Diamond_L p \supset \Diamond_x p$ (and it is not a conceptual truth that $\Box_x p \supset \Box_L p$) and,

(ii) *Groundedness*. If $\Diamond_x p$ then the fact that $\Diamond_x p$ is grounded in some feature of actuality.

The subscript "*x*" indicates the different kinds of possibility; e.g. "$\Diamond_L$" refers to logical possibility.

This also shows us how to relate Kant's Critical theory of real possibility in general (his theory of <*possibility*>$_{UC}$) to his pre-Critical theory of real possibility. As I explained in the Introduction, in *OPG* Kant argues for two principles about real possibility: (i) not all logical possibilities are real possibilities, and (ii) all real possibilities must be grounded in an actually existing being. In particular, Kant argues that all real possibility must be grounded in a substance, God. Kant's Critical theory of real possibility in general (his theory of <*possibility*>$_{UC}$) is a generalization of the pre-Critical theory that maintains the same overall structure. With respect to clause (ii), it generalizes from the specific pre-Critical doctrine that real possibilities must be grounded in an *existing* (causally efficacious) substance the requirement that real possibilities be grounded in some feature of how things *actually* are (whether these are features of substances, or mere forms of experience, for instance).

We can now see that Kant's claim that we can give real definitions of the categories only for empirical objects means, in the case of the modal categories, that we only know the grounds of real possibility for empirical objects. If *p* is some proposition, and *p* is really possible, then we can know the grounds of the real possibility that *p* only if *p* is a proposition concerning empirical objects only. But empirical objects have two kinds of grounds: "immanent ones," i.e. the forms of experience of cognizing subjects and causal grounds in other empirical objects, and "transcendent" ones, i.e. noumenal objects that ground empirical objects by causally affecting subjects and producing the sensory matter of experience.[31] We only know the "immanent" grounds

[31] Kant talks of phenomena having noumenal grounds at A380, Ak. 4:314–315, 453, 8:203, 205. Elsewhere, he claims that noumena causally affect us, thereby

of real possibility for empirical objects, so the only kinds of real possibility for which we can give real definitions are those with immanent grounds.

To illustrate this point, consider two kinds of real (nonlogical) possibility Kant invokes in *CPR* without explicitly distinguishing them:

(*Formal*) It is formally really possible that p iff p it is compatible with the forms of experience that we experience objects for which p is true.[32]

(*Empirical*) It is empirically really possible that p iff p is compatible with the natural laws and the past up to time t, where p is the proposition that alteration E occurs at t.[33]

For instance, it is formally possible both that I go to the movies tonight, and that I do not, since either state of affairs is compatible with the a priori forms of experience. However, if the natural laws are deterministic, as Kant thinks they are,[34] they determine a unique empirically possible future; so the proposition that I go to the movies tonight is either empirically *necessary* or empirically *impossible*. These are both kinds of real possibility because both of them are distinct

producing the matter of experience: A190/B235, A387, A494/B522, Ak. 4:289, 4:314, 4:318, 4:451, and 8:215. On my view, noumena are grounds of phenomena in that they causally affect our minds, producing sensations, which are synthesized into experience of phenomena. For a defense of the claim that noumena causally affect us, see the classic Adickes (1924), as well as the introduction to Ameriks (2003).

32 This is the kind of real possibility Kant defines at A218/B266. It is the kind of real possibility Kant typically has in mind in *CPR* when he talks about "possibility" in general or "real possibility" in particular, e.g. A111 and A127.

33 This is the kind of possibility that corresponds to the kind of necessity Kant defines at A218/B266 and discusses further at A226–227/B279–280. Those are discussions of empirical necessity, which is related to what I am calling "empirical possibility" in the standard way: it is empirically possible that p iff it is not empirically necessary that $\sim p$. See also Kant's lectures on metaphysics (Ak. 28:417, 29:814) and Refl. 4298 and 5177. I am departing from Kant's terminology here. Kant typically defines possibility in terms of the possibility of an *object* – see his metaphysics lectures (28:410, 543 and 29:811, 960). For ease of exposition, I have defined possibility in terms of the possibility of a proposition. These are interconvertible definitions: an object a is possible just in case the proposition that a exists is possible.

34 A108, A113, A114, A127–128, B165, A159/B198, A216/B263, and *Prolegomena* § 36.

from logical possibility; not all logical possibilities are compatible with the forms of experience (e.g. the propositions of non-Euclidean geometry) and many logically possible propositions about the future are not compatible with the laws and the past up to this moment (e.g. any logically consistent but *false* proposition about the future). Both kinds of real possibility are grounded in actuality: our *actual* forms of experience (space, time, and the schematized categories) and the *actual* natural laws and past events.

In both cases, we know what grounds real possibility. For instance, we can construct a triangle in pure intuition and thereby know *why* it is formally possible: it is compatible with our spatial form of intuition. Likewise, by coming to know the natural laws we come to know not merely that various actual events are empirically possible but *why*; we come to know why they are continuations of the past that are compatible with those laws. But we do not know the noumenal (nonimmanent) grounds of the real possibility of empirical objects, because we in general do not know the noumenal (nonimmanent) grounds of the real possibility of objects. So we only know one "half" or "side" of the real possibility of empirical objects. And we cannot give a real definitions of *<possibility>*$_{\text{UC}}$ for all objects *überhaupt* because we do not know in general the grounds of the real possibility of objects. We only know the immanent grounds of real possibility for empirical objects.

However, we can *think* of the noumenal grounds of the real possibility of objects, even for these kinds of real possibility (formal, empirical) that apply only to empirical objects. For instance, a series of events that is different from the series of actual events is compatible with our forms of experience, if it is a series of law-governed alterations in permanently existing and reciprocally interacting substances.[35] Consequently, it is formally possible. Likewise, such a series of nonactual events is empirically impossible because it is not a continuation of the past that is compatible with actual laws (since these laws are deterministic). But there is another kind of real possibility we might ask about: does this nonactual empirical series have a noumenal ground? To make this more concrete, could we be noumenally affected so that

[35] I will leave unanswered here whether it is compatible with the forms of experience that these events be governed by natural laws different from the actual laws. The question of the modal status of natural laws for Kant is difficult; I address it in Stang (2016).

we would experience this nonactual empirical series? It is compatible with my forms of experience that I receive a completely different set of sensory matter, and thus have an experience with the same *form* as my actual experience but a different *matter*, but that is not the question. What is at issue is a question about the noumenal ground of the real possibility of my experiencing a different empirical series. I take it that is what Kant has in mind in the "Postulates of empirical thinking in general" when he writes "whether other perceptions than those which in general belong to our entire possible experience and therefore an entirely different field of matter can obtain cannot be decided by the understanding, which has to do only with the synthesis of that which is given" (A231/B283). It is not something we can know; we do not know the noumenal grounds of real possibility, even of the real possibility of propositions exclusively about empirical objects. This is the notion of real possibility that I have called "noumenal$_1$" in Figure 12.1. I also think this is the notion of real possibility involved in Kant's theory of transcendental freedom and his reconciliation of our freedom in action with the empirical necessity of all of our actions, but I do not have the space to defend that claim here.[36]

Finally, for propositions involving noumena (as well as empirical objects) we might ask whether they are really possible, using the unschematized category of (real) possibility, *<possibility>*$_{UC}$.[37] In doing so, we would be thinking about a kind of real possibility whose grounds must be noumenal, because noumena ground empirical objects, and not vice versa (so empirical objects cannot be among the grounds of the real possibility of noumena or any state of affairs involving noumena). We cannot in principle know what the grounds of this kind of real possibility would be. This is the notion of real possibility that I have called "noumenal$_2$" in Figure 12.1. Some readers might be skeptical that Kant accepts any such notion of real noumenal possibility. However, note the following claims Kant makes in the Pölitz lectures:

[36] I argue for this in Stang (2011), 459–460 and in greater detail in Stang (forthcoming-*a*).

[37] Kant seems to be applying modal concepts directly to noumena – rather than merely considering noumena as putative grounds of the possibility of empirical objects – at Refl. 5184, 5723, and 5177. I also think that this kind of possibility is involved in Kant's resolution of the Fourth Antinomy, but I do not have space to defend that claim here.

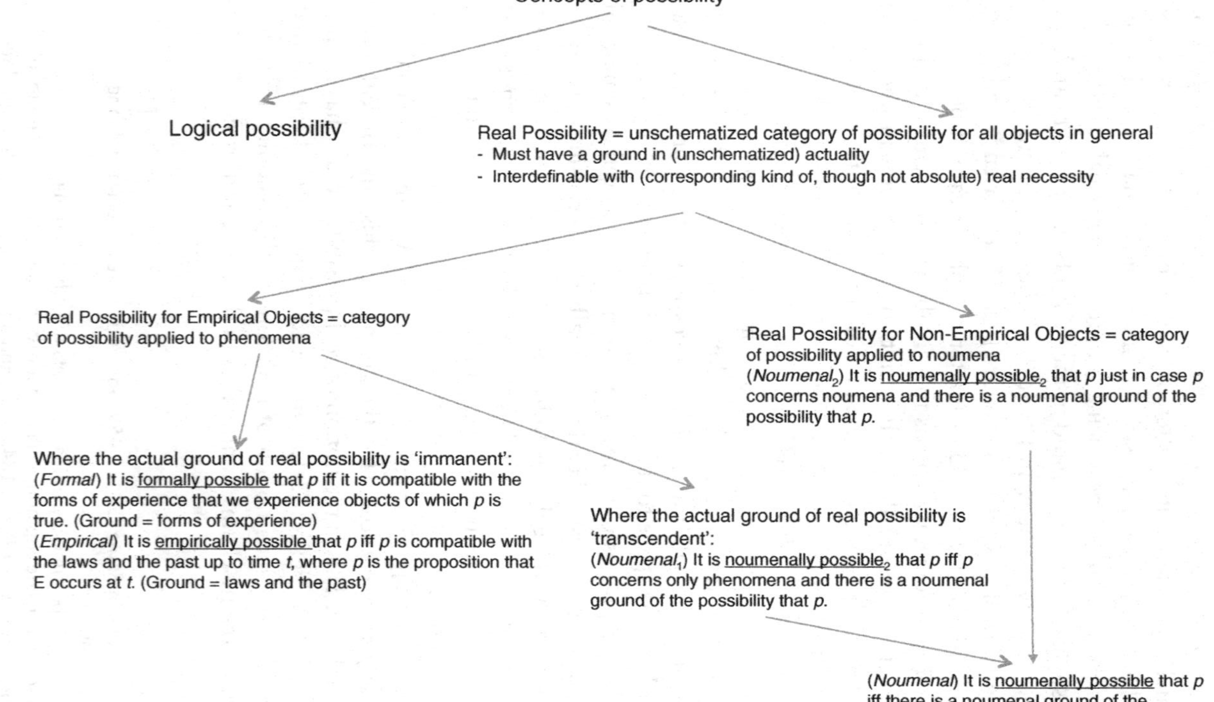

Figure 12.1 Kant's concepts of possibility.

If we remove an original being, we at the same time remove the substratum of the possibility of all things.... My reason makes it absolutely necessary for me to assume a being which is the ground of everything possible ... otherwise I would be unable to know what in general the possibility of something consists in.... A thing is *eo ipso* necessary if its nonexistence would cancel all possibility ... in the case of every possibility which we think *realiter* we always presuppose some existence ... hence all possibilities presuppose something actually given. (Ak. 28:1034–1036–underlining by NS)

Kant's repeated emphasis that the absolutely necessary being must be hypothesized to ground the real possibility of "all" things, or real possibility "in general," is naturally read as meaning that the absolutely necessary being grounds the real possibility of both phenomena *and noumena*. This only makes sense if there is at least one concept of real possibility that applies to noumena, what I have called noumenal$_2$ real possibility.[38]

Finally, noumenal$_1$ and noumenal$_2$ real possibility are species of a single conception of real possibility, which I have relabeled noumenal real possibility *simpliciter* in Figure 12.1: the concept of any kind of real possibility grounded in noumena. We cannot give a real definition of this kind of possibility, for we do not know the noumenal grounds of real possibility.

4. Absolutely Necessary Existence

To return to the original passage from the Pölitz theology lectures, I am now in a position to offer my interpretation of what Kant means by "real possibility" there: noumenal real possibility, real possibilities for all things insofar as those real possibilities have noumenal grounds, which we think through the unschematized category of (real) possibility, <*possibility*>$_{UC}$. The reason "real possibility" there means "noumenal real possibility" is that Kant claims we must hypothesize an absolutely necessary being to ground all such real possibilities and, as I pointed out at the beginning of Section 2, an absolutely necessary being must be a noumenon (no phenomena are absolutely necessary). So the real possibility at issue in the Pölitz text is grounded in noumena; it is noumenal real possibility, the genus of which noumenal$_1$

[38] The Pölitz lectures are not the only place where Kant discusses what I am calling "noumenal$_2$ possibility"; see A255/B310.

and noumenal$_2$ real possibility are species. But Kant's primary concern in that passage, and the jumping-off point for our whole discussion, was his invocation of the notion of "absolutely necessary existence" and his Critical reinterpretation of the 1763 argument for the existence of an absolutely necessary ground of real possibility, or, as we can now identify it, of noumenal real possibility.

The first thing we need to understand is what "absolute necessity" means in the Pölitz text. Fortunately, it means the same thing it does in *OPG*. Consider these two passages, the first of which is from Pölitz, the second of which is from *OPG*:

In addition to the logical concept of the necessity of a thing (where something is said to be absolutely necessary if its nonexistence would be a contradiction, and consequently impossible), we have yet another rational concept of real necessity. This is where a thing is *eo ipso* necessary if its nonexistence would cancel all possibility. (Pölitz RT; Ak. 28:1036)

Something may be absolutely necessary either when the formal element of all that can be thought is cancelled by means of its opposite [*the 'logical concept of the necessity of a thing' – NS*], that is to say, when it is self-contradictory; or alternatively, when its non-existence eliminates the material element and all that data of all that can be thought. (*OPG*; Ak. 2:82)

The "data of all that can be thought," in the context of *OPG*, is a reference to real possibility. Consequently, I interpret "absolute necessity" in both texts as follows:

(1) It is absolute necessary that $p =_{def}$ were it not the case that p, nothing would be really possible.

The right-hand side of this definition is a counterpossible conditional: it says that *per impossibile* if p were false, nothing would be really possible. This is how I interpret Kant's talk of the "cancellation" of possibility in Pölitz (quoted previously) and the "elimination" of possibility in *OPG*. Earlier in the same paragraph in *OPG* Kant writes, "When I cancel all existence whatever and the ultimate real ground of all that can be thought therewith disappears, all possibility likewise vanishes" (Ak. 2:82). I take this to be clear support for the counterpossible interpretation of absolute necessity in (1).[39]

[39] I further defend this interpretation in Stang (2010).

One natural question is whether absolute necessity, so defined, can be simply identified with real necessity, i.e.

(2) It is absolutely necessary that *p* $=_{def}$ it is really necessary that *p*.

It cannot; if it were, by the interdefinition of real necessity and real possibility, this would entail that:

(3) It is absolutely necessary that *p* iff it is not really possible that ~*p*.

Given the definition of absolute necessity from earlier, (1), and the principle that if it is really possible that ~*p* then there is a ground of the real possibility that ~*p*, this would entail that:

(4) If there is no ground of the real possibility that ~*p* then (if it were the case that ~*p*, nothing would be really possible).

But this cannot be right, because if ~*p* simply lacks a ground of real possibility, it does not follow that its nonobtaining would cancel all real possibility. For instance, if I do not have a transcendentally free will, then, in the relevant sense of possibility, the possibility of omitting an act I actually commit lacks a ground of real possibility; it does not follow that were I to have done otherwise, nothing would be really possible! My actions do not have that exalted status; only God does.

This is a short argument to the effect that the concept of absolute necessity invoked in the original Pölitz passage is not interdefinable with the concept of real possibility (which I have identified with the unschematized category of possibility). But recall how Kant gets to that concept of absolute necessity: it is part of the concept of real possibility that real possibilities have grounds in actuality. But reason requires me to think of things as unified, to impose in thought a unity that I do not know to obtain. This gives me reason to think that there is a single unified ground of real possibility. While I am not attempting to reconstruct Kant's reasons in the 1780s for claiming that we are rationally necessitated to hypothesize such a first ground of real possibility, this short sketch does suffice to make an important point. Whereas Kant had argued in 1763 in *OPG* that real possibility could not be collectively grounded by a plurality of distinct grounds, he now claims less ambitiously but more plausibly that it is rationally unsatisfactory if there are a plurality of distinct grounds of real possibility, none of which has priority over any others. The most rationally satisfactory conception of modal reality is one on which there is a unique ground of real possibility – or at least a "first" ground, one that grounds the grounding of real possibility in "lower level" grounds, such that the nonexistence of that first ground would cancel all real possibility. So rationally we have to hypothesize such a being, because it is part of the most rationally satisfactory explanation of why anything is really possible.

If my interpretation is correct up to this point, we should expect that – because quite generally we are required to hypothesize an absolutely necessary first ground of real possibility of all objects in general – for each lower level kind of real possibility, we should hypothesize an absolutely necessary (in some sense) first ground of that kind of real possibility for all objects of the relevant sort. Thus, for each lower-level kind of possibility, we should expect to find some absolutely necessary principle *p* such that

(1*) It is absolutely necessary$_x$ (absolute-$\Box_x$) that $p =_{def}$ were it not the case that *p*, nothing would be really possible$_x$ ($\Diamond_x$).

The subscript *x* refers to the different kinds of possibility, from earlier. For instance, we should expect that there is some *p* such that the right-hand of this definition is true:

(1_F) It is absolutely formally necessary (absolute-$\Box_F$) that $p =_{def}$ were it not the case that *p*, nothing would be formally really possible ($\Diamond_F$).

We cannot say that the proposition *p* will be the proposition that the relevant first ground of real possibility *exists*, for not all kinds of real possibility have *existing* grounds (e.g. formal possibility – see earlier discussion); but we can say that *p* will be the proposition that the relevant first ground of real possibility (whatever it is) is actual or obtains (or something of that sort). And in the case of "immanent" real modalities – modalities with immanent grounds – we should expect that we can *know* what this absolutely necessary first ground of real possibility is, for these are precisely the kinds of real possibility for which we can identify the grounds.

And this expectation is satisfied, at least in the case of formal possibility. Kant identifies space as the absolutely necessary first ground of formal possibility for outer objects and explicitly compares space in this respect to the rational idea of God as the absolutely necessary first ground of real (noumenal) possibility:

> All reality must be completely given, and so some actuality is prior to possibility, just as space is not merely something possible, but the ground of all possible figures. (Refl. 4119, Ak. 17:424)
>
> From this it follows only that the *ens realissimum* must be given prior to real concepts of possibility, just as space cannot be thought antecedently as possible, but as given. But not as an actual [*wirklich*] object in itself, rather as a merely sensible form, in which alone all objects can be intuited. (Refl. 6290, Ak. 18:558)[40]

These texts suggest that some proposition about space will make the right-hand side of (1_F) true. However, it is not clear which proposition about space this is. In *OPG*, the absolutely necessary proposition is that God *exists*. This will not quite work for space, because space does not exist in the same way that causally efficacious substances do; space is a form of empirical object, not a substance in its own right (see Refl. 6324). The absolute formal necessity of space consists, then, not in the fact that space exists absolutely formally necessarily, but in the fact that it is absolutely formally necessary that we experience outer objects in space. If this is correct, then, *per impossible*, were it the case that we did not experience outer objects in space, then no outer object would be formally possible. It is hard to know how to evaluate counterpossible conditionals like this. I have argued elsewhere that we should evaluate claims like this by applying the following test: without appeal to the relevant *p* can we explain facts about the given kind of real possibility?[41] Applying this to space and formal possibility, we should ask, without appeal to space, can we explain facts about the formal possibility of outer empirical objects? It is a pillar of Kant's Critical theory of experience that we cannot. Therefore, for Kant, the representation of outer objects in space is absolutely formally necessary. Space is the first ground of the formal possibility of outer

[40] In the second passage, I take Kant's point to be that space does not exist as a substance, but as a form in which objects can be intuited. So he here uses "*wirklich*" in exactly the opposite sense of Refl. 6324, discussed previously. See also Refl. 4515, 4570, 5723, and 6285, as well as Ak. 28:1259.

[41] Stang (2010).

objects.[42] For further confirmation of my view I would need to search for evidence that something plays the role of an absolutely necessary first ground of real possibility for the other kind of "immanent" real possibility I distinguished earlier: empirical possibility (likewise, for any other kinds of immanent real possibility). I argue elsewhere that the absolutely necessary first ground of empirical possibility is the actual laws of nature and the past series of events, but I do not have the space here to go into that.[43]

I began by asking what we are doing when we use (unschematized) modal categories to think about noumena. I think we can now answer that question, as follows: we are applying the most general concept of real possibility, the unschematized category of possibility, of which the following are conceptual truths: (i) it is conceptually distinct from logical possibility, (ii) real possibilities are grounded in actuality, and (iii) it is interdefinable with real necessity. There are several different kinds of real possibility, lower-level species of the genius, (real) <*possibility*>$_{UC}$: formal real possibility, empirical real possibility, noumenal real possibility, etc. These conceptual truths (i)–(iii) also apply to each of these lower-level kinds of possibility. For each lower-level kind we also have the idea of a first ground of real possibility *of that kind* that exists with absolute necessity *with respect to that kind* (as defined previously). With respect to some of these kinds of real possibility, we know what their grounds are, and we can identify an absolutely necessary first real ground. For instance, we know that space, time, and the categories are the grounds of formal real possibility, and that space is absolutely formally necessary. But we do not know what grounds the real possibility of objects in general; in particular, we do not know any of the noumenal grounds of real possibility. For reasons I have not gone into in this chapter, we are rationally required to hypothesize that among the noumena there is such an absolutely necessary first ground of all noumenal real possibility. But we do not know whether such a being exists, and we can only have a quite indeterminate concept of it, and how it grounds real noumenal possibility.

[42] One would expect Kant to say similar things about time. However, on the basis of my overview of the relevant texts, I conclude that he more frequently draws the comparison between space and God as first grounds of (formal and noumenal, respectively) possibility.

[43] Cf. Stang (2016).

5. Abbreviations for Works of Kant

A/B	*Kritik der reinen Verunft*, Ak. 3 & 4. Cited by page number in 1st edition of 1781 (A) and 2nd edition of 1787 (B). Translations from Kant (1997).
Ak.	Kant (1900–). Cited by volume and page number (e.g. Ak. 29:1034).
CPR	*Kritik der reinen Verunft*, Ak. 3 & 4. Cited by page number in 1st edition of 1781 (A) and 2nd edition of 1787 (B). Translations from Kant (1997).
OPG	*Der einzig mögliche Beweisgrund zu einer Demonstration des Dasein Gottes*, Ak. 2:62–163. Translation: *The only possible ground of proof in support of a demonstration of the existence of God* in Kant (1992).
Pölitz RT	*Philosophische Religionslehre nach Pölitz*, Ak. 28:989–1126. Translation: *Lectures on the Philosophical Doctrine of Religion* in Kant (1996).
Volckmann RT	*Natürliche Theologie Volckmann nach Bambach*, Ak. 28: 1127–1226.
Danziger RT	*Danziger Rationaltheologie nach Bambach*, Ak. 28: 1227–1319.
Refl.	*Kants handschriftlicher Nachlass* ('Reflexionen') in Ak. 14–18. Cited by four-digit number, and volume and page number in Ak.

Where no translation is listed, translations are my own.

13 *Charles Sanders Peirce on Necessity*

CATHERINE LEGG AND CHERYL MISAK

1. Introduction

Necessity is a touchstone issue in the thought of Charles Peirce, not least because his pragmatist account of meaning relies upon modal terms. In contrast to the highly influential positivist currents in twentieth century philosophy in which "fact-stating" is paramount and put in the indicative mode, Peirce famously advised us to clarify the meaning of our terms using "would-bes," not "will-bes" (CP 5.453; 5.457, 1905). Peirce's pragmatic maxim is that we must look to the practical effects of our concepts, and the practical effects he is concerned with are those that would occur under certain circumstances, not those that will actually occur. This effectively translates every meaningful term into a set of hypothetical conditionals. Thus, for instance, the fact that a table is *hard* means (among other things) that if I were to rest a dinner plate on it, the plate would not fall through to the floor (Misak 2013, 29ff; Hookway 1985).

Peirce thought deeply about necessity in all its aspects, and this is one of the areas of his work with as-yet-untapped insights. Discussions of necessity show up from his earliest philosophical writings based on Kant's logic and its table of 12 categories (one row of which is, of course, explicitly modal) to the end of his life, when he was working out a forest of distinctions in semiotics. His thoughts on modality also span a multitude of philosophical areas – informing his work on mathematics, logic, metaphysics, philosophy of science, and ethics.

His views on the topic have some intriguing features. His modal epistemology understands logical form as essentially structural, thus requiring representation by iconic or diagrammatic signs. Peirce made good on this commitment, developing the Existential Graphs, a diagrammatic logical notation, which is still understudied by mainstream logicians. His metaphysics differs from that of many nineteenth century peers in flatly denying "necessitarianism" (determinism) and arguing for real chance. He also distinguished between a kind of necessity that corresponds to the compulsion given by one billiard ball to another (which we might refer to in Peircean terms as a "Secondness Necessity") and a kind of necessity that corresponds to a real universal ("Thirdness Necessity"), whereas in contemporary philosophy the two are routinely conflated.

Peirce's views on necessity also evolved through his life. It has been argued (Fisch 1967; Lane 2007) that although Peirce was always firmly opposed to nominalism, his scholastic realism became more "extreme," leading him to commit to at least some nonactual reality. In what follows, we shall give a snapshot of Peirce's important and complex thoughts about necessity.

2. Pragmatism, Modality, and Truth

Peirce is best known for his pragmatism and the view of truth he derives from it. Truth, for Peirce, is linked to what human inquirers would believe – truth is what is "indefeasible" or would best stand up to all experience and reasoning, were we to investigate the matter as far as we fruitfully could. Peirce says, "If Truth consists in satisfaction, it cannot be any actual satisfaction, but must be the satisfaction which would ultimately be found if the inquiry were pushed to its ultimate and indefeasible issue" (CP 5.569, 1901; 6.485, 1908). Peirce was a scientist as well as a philosopher and logician, and this experimental spirit led him wholeheartedly to embrace *fallibilism*: the idea that a thinker must be "at all times ready to dump his whole cart-load of beliefs, the moment experience is against them" (CP 1.55, 1896). He held that all of our beliefs form a corrigible whole, with none of them earmarked as certainly true, and all of them subject in principle to overthrow by experience, if we are willing to make sufficiently far-reaching revisions through our system of belief. C. I. Lewis is an explicit follower of Peirce on this point, and Quine is an implicit follower.

Yet how does this model fit with the idea of *necessary* truth, which is generally thought of as constituting certain and unrevisable knowledge? How can one "experiment" in logic and mathematics? Surely these sciences must be a priori? The concern deepens when it is noted that Peirce argued that there are real possibility and real necessity in the world (see §5), and that he had sophisticated modal notions at work in his exact logic. One might wonder how Peirce can hold together such metaphysical and logical researches and a pragmatist epistemology.

This challenge can be answered on a number of levels. The first is (somewhat ironically) *pragmatic*. For Peirce, as for Lewis and Quine, the necessity found in mathematical or logical truths is just the extent to which we find no reason to revise them. He explicitly addressed one supposed necessary truth of logic: the principle of bivalence (for any *p*, *p* is either true or false). Peirce argues that this is something we have to assume if we are to go on with a practice that is of great importance to us. To say that bivalence is a regulative assumption of inquiry is not a claim about special logical status (as a "logical truth"). Rather, the principle is taken to be a law of logic by a "saltus" – an unjustified leap (NE 4: xiii). Our reason for making such assumptions is that if we do not make them, we will "be quite unable to know anything of positive fact" (CP 5.603, 1903). It appears that here Peirce is arguing for bivalence as a kind of necessary presupposition for other beliefs we wish to hold. But this is an unusually weak kind of necessary presupposition, for at the same time he distinguishes his approach from that of the *transcendentalist*:

> A transcendentalist would claim that it is an indispensible "presupposition" that there is an ascertainable true answer to every intelligible question. I used to talk like that, myself; for when I was a babe in philosophy my bottle was filled from the udders of Kant. But by this time I have come to want something more substantial. [CP 2.13, 1902]

A second level on which we might address the challenge of consistency within Peirce's thinking is a *semiotic* one. Viewed from the perspective of inquiry, our beliefs are all fallible, but the *expression* of any given belief can be decomposed and analyzed in terms of Peirce's theory of signs. According to this theory, different kinds of signs with different functional roles – and thereby different modal purport – work together to compose any given proposition. Peirce believed that the necessarily true dimension of our thinking is conveyed by a particular kind of sign: the *icon*, one example of which is the *diagram*. Peirce addresses the question about fallibilism and necessary inference as follows:

> When we talk of deductive reasoning being necessary, we do not mean, of course, that it is infallible. But precisely what we do mean is that the

conclusion follows from the form of the relations set forth in the premiss. Now since a diagram ... is ... in the main an Icon of the forms of relations in the constitution of its Object, the appropriateness of it for the representation of necessary inference is easily seen. [CP 4.531, 1906]

But in a radically original move, Peirce claims that iconic signs can be experimented on, just as the physical world can be. The objects of these experiments should not be understood as objects in Plato's heaven, which would be rather mysterious, but merely the most general structural features of ordinary real world objects:

Although mathematics deals with ideas and not the world of sensory experience, its discoveries are not arbitrary dreams but something to which our minds are forced. [Peirce 2010, 41]

All deductive reasoning, even simple syllogism, involves an element of observation; namely, deduction consists in constructing an icon or diagram the relations of whose parts shall present a complete analogy with those of the parts of the object of reasoning, of experimenting upon this image in the imagination, and of observing the result so as to discover unnoticed and hidden relations among the parts. [CP 3.363]

In the next section we say more about Peirce's modally inflected theory of signs, first presenting its grounding in his philosophical categories.

3. Peirce's Modalized Semantics

3.1 Three Categories

A full understanding of any area of Peirce's thought requires familiarity with his three categories, which (in line with their derivation and ordering) he dubbed Firstness, Secondness, and Thirdness. Each of these categories, Peirce argued, is present in everything that appears before the mind or is experienced. In his early career he derived them – in the spirit of Kantian logic – from the structure of the proposition, whereby a predicate (First) and its subject (Second) are united by a copula (Third). The key paper here is "On a New List of Categories," in which Peirce was proud that he had reduced Kant's table of 12 by a factor of 4. Later in life he kept the same categories but pushed the derivation of them deeper, into phenomenology and mathematics, the latter being for Peirce the most fundamental science of all.

It is important to note that Peirce's categories are not properties but *modes of being*. What this means is that we cannot actually pull them apart or imagine a situation in which one or two stand alone. They are distinguished from one another by the Aristotelian and then Scholastic method of abstraction or "prescission," in which we may distinguish different elements of a concept by attending to those elements and neglecting others. Postulating such high-level structural features of being is a project somewhat unfamiliar in post-Quinean analytic philosophy, where one is encouraged to believe that reality is exhausted by a list of existent entities and their intrinsic properties.

A First is something simple, monadic: a quality of feeling, image, or mere possibility. It is indescribable: "It cannot be articulately thought: assert it, and it has already lost its characteristic innocence. … Stop to think of it, and it has flown!" (CP 1.357, 1890). The difficulty of pinning down a category that is "a special suchness" is not lost on Peirce (CP 1.303, 1894). It is perhaps helpful to focus on the idea that Firstness "comes first": it is "predominant" in being, in feeling, and in the ideas of life and freedom (CP 1.302, 1894). Even more importantly, given the continuity of Peirce's metaphysics with his logic of relations, a First is singular: "a pure nature … in itself without parts or features, and without embodiment" (CP 1.303, 1894). The First is a relatum prior to any relation, that which makes relations possible.

A Second is a dyadic element: the duality of action and reaction. It is the category that "the rough and tumble of life renders most familiarly prominent. We are continually bumping up against hard fact" (CP 1.324, 1903) and hence have revealed to us "something within and another something without" (CP 2.84, 1902). But as with the first category, this is all we can say of our encounters with hard fact. Any articulation or thought of what we experience takes us into the third category – the triadic realm of experience proper, which involves interpretation, signification, intention, and purpose realized across time. As soon as we try to describe our encounters with the world, thought and signs are involved. Any perception that we can think of ourselves as having is a perceptual *judgment*, something that requires, in Peirce's terms, "mediation" by some general concept. For instance, when I see a yellow chair, I see it *as* yellow, *as* a chair (CP 7.619–630, 1903).

These categories infuse and inform Peirce's thinking about modality in a host of ways. Indeed, the categories are full of modal notions – the description of Firstness invokes possibility, the description of Secondness mentions mechanical necessity as a key example, and the description of Thirdness corresponds to a kind of *semantic* or *intentional* necessity that unites all instances "with the same meaning" (so that, for example, if we understand the concept "white," we know that certain other things must also receive the label).

3.2 Triadic Sign Theory

Peirce puts his triadic categories to work in structuring his theory of signs, along a number of dimensions. First, the structure of representation itself is triadic, always involving a sign, an object, and an interpreter. By including both an object and an interpretation as a vital component of every sign, Peirce unites insights from traditions currently opposed under the labels "realism" and "antirealism."

Second, the relation between a sign and its object can take three forms, corresponding to *icons*, *indices*, and *symbols*. Icons are signs that represent their objects by virtue of a similarity or resemblance – a portrait is an icon of the person it portrays, a map is an icon of a certain geographic area, and, in exact logic, a diagram is an icon of validity's key forms. Peirce notes of the icon that it "represent[s] relations in the fact by analogous relations in the representation" (CP 3.641, 1902); thus, representing *structure* is one of the icon's greatest strengths (Stjernfelt 2007, 49–116, Legg 2008, 207). Indices are signs that point to their object by "being really connected with it" (W 2, 56, 1867) – a pointing finger, a demonstrative pronoun such as "this" draws the interpreter's attention to the object. Smoke, for instance, is an index of fire. A symbol is a word, proposition, or argument that depends on conventional or habitual rules; a symbol is a sign "because it is used and understood as such" (CP 2.307, 1901). It has the same effect from interpreter to interpreter.

This set of sign-types plays an important role in structuring living thought. Symbols in their role as habits established by convention pick out *general concepts*. Indices in their role as unmediated pointers pick out *specific individuals* (and locate us in the real world at a particular place and time). Icons in their role of conveying structure might be said to represent *logical form*. In "What Is a Sign" (EP 2, 4–10; 1894), Peirce explained these unique and mutually enabling roles as follows:

Suppose a man to reason as follows: The Bible says that Enoch and Elijah were caught up into heaven; then, either the Bible errs, or else it is not strictly true that all men are mortal. What the Bible is, and what the historic world of men is, to which this reasoning relates, must be shown by indices. The reasoner makes some sort of mental diagram by which he sees that his alternative conclusion must be true, if the premise is so; and this diagram is an icon or likeness. The rest is symbols; and the whole may be considered as a modified symbol.... The art of reasoning is the art of marshalling such signs, and of finding out the truth.

4. Peirce's Modal Epistemology

So far we have explored a little of how for Peirce all representation has modal purport. We now discuss Peirce's ideas about how philosophers should gather and evaluate modal knowledge in a deliberately focused manner.

4.1 Information-Relative Account

From his first published papers in 1867, Peirce analyzed modality in terms of states of information, or what is known (Morgan 1979; Lane 2007). A proposition is necessarily true if contained in – and possibly true if consistent with – some subject's state of information. This epistemic approach contrasts with contemporary modal semantics according to which a proposition is necessarily true if true in all possible worlds and possibly true if true in at least one possible world. It is worth noting the way in which the latter account decouples entailment from provability in principle. This allows a greater externalism (and thus, arguably, realism) into modal epistemology (Legg 2012, 6); but at the same time it challenges our ability to know modal truths (since if we do not know them by proving them, how might we know them?).

4.2 Kinds of Modality: Designated States of Information

Peirce used his information-relative approach to define different kinds of modality in terms of the kind of information the subject is postulated to have (their "designated state of information"). For instance, in 1893 he distinguished:

- *essential* or *logical* modality, where what is known is "*nothing*, except the meanings of words, and their consequences"
- *substantial* modality, where what is known is "everything now existing, whether particular fact or law, together with all their consequences"
- *physical* modality, where the subject is "thoroughly acquainted with all the laws of nature and their consequences, but ... ignorant of all particular facts"
- *practical* modality, which involves "imagining ourselves to know what the resources of men are, but not what their dispositions and desires are"
- *mathematical* and *metaphysical* modality, which involves imagining perfect knowledge in those areas

The first two, he notes, "are of special interest to the logician" (CP 4.66–67, 1893). The first appears to correspond to what we would regard today as logical necessity (i.e. tautology) plus analyticity, while the second corresponds to the actual, at a given time.

4.3 The "Modal Shift"

There are indications that from about 1896 onward, Peirce was working towards a greater realism in his modal epistemology. Lane 2007 has argued that his thinking evolved from a pure information-relative account toward an account in which we do not discern modal facts merely via inference from a given state of information. However, Peirce does not replace the information-relative account with the kind of externalism (frequently accompanied by metaphysical realism) seen in contemporary modal semantics. Rather, he thinks that the certain necessary truths are accessed through a kind of *direct perception*:

> It is not that certain things are possible because they are not known not to be true, but that they are not known not to be true because they are, more or less clearly, seen to be possible. [CP 6.367, 1902]

According to this account, we discern modal facts via some kind of unmediated intellectual vision of a "world of ideas" that is "but a fragment of the ideal world" (Lane 2007, 563). Peirce writes that with respect to this ideal world we are "virtually omniscient," and that by inspecting it one might be able to determine that a given proposition is impossible although the laws of logic do not show that it contains any contradiction. Peirce was here pursuing a particular example: whether two collections might both be sufficiently large that neither could be mapped into the other (CP 3.526, 1896; cited in Lane 2007, 561). He ended up deciding that the answer to this question could not be proven, but ultimately only "seen."

This new development in Peirce's modal epistemology separates out a "positive" sense of possibility from his previous "negative" sense of "not provable that not *p* in information state *S*." Lane argues that this shift was accompanied by an embrace of a new ("Aristotelian") conception of real possibility. Peirce calls this "the great step that was needed to render pragmaticism an intelligible doctrine" (MS 288, 129, 1897; cited in Lane 2007, 551–552). How does real possibility render pragmaticism intelligible? We noted in our Introduction that the pragmatic maxim may be understood to translate every meaningful term into a set of hypothetical conditionals. But in order to obtain a coherent account of meaning, these conditionals must be *subjunctive*.

In the early (1878) "How to Make Our Ideas Clear," Peirce had suggested that the meaning of "This diamond is hard" is in part "If scratched, it will resist." This entails that "there is absolutely no difference between a hard thing and a soft thing so long as they are not brought to the test." If we imagine, then, "that a diamond could be crystallized in the midst of a cushion of soft cotton, and should remain there until it was finally burned up" (W 3: 266, 1878), there is no truth of the matter about that diamond's hardness. Peirce was later anxious to correct his mistake:

> But I desire to express ... that in those original papers I fell into a grievous error. Looking into the future and thinking of what would take place, I overlooked the great truth that the future, as living, always is more or less general, and I thought that what *would be* could be resolved into what *will be*. [MS 289, 17–18]

Peirce's modal shift provides the epistemology for us to know "would-bes."

5. Peirce's Logic

Peirce was not only the most logically proficient of the founders of pragmatism, but one of the great logicians of the nineteenth century.

Among the enormous number and variety of his contributions, he invented a quantified, sound, and complete first-order logic in algebraic form independently of and at the same time as Frege[1] (a development that grew from more than a decade's preparatory work in the logic of relations),[2] discovered the Sheffer stroke decades before Sheffer, experimented with many-valued logics, and made significant advances in the logic of statistical reasoning. Possibly his most original contribution, and the one of which he was most proud, was his Existential Graphs (EG).

5.1 Overview of the Existential Graphs

In the EG Peirce sought a logical notation that would

> afford a method (1) as *simple* as possible (that is to say, with as small a number of arbitrary conventions as possible), for representing propositions (2) as *iconically*, or diagrammatically and (3) as *analytically* as possible. [CP 4.561n]

They were presented to the world in 1896, when Peirce was 57, although he had been working on them for the previous decade and a half. The Alpha graphs cover propositional logic, the Beta extension covers first-order predicate logic with identity, and the Gamma extension covers modal notions, abstractions, and a metalanguage for talking about the graphs themselves.

The EG have yet to receive significant attention in mainstream logic, for a range of reasons. Peirce was almost entirely locked out of academic philosophy, underread, and underappreciated in his lifetime. The EG have been seen as one of the least accessible areas of his work:[3] indeed, his penchant for expressing his thoughts in technical language was the source of considerable tension between Peirce and his fellow pragmatists, James and Dewey (Misak 2013). And finally, once an algebraic, stepwise[4] approach became dominant

1 "On the Algebra of Logic: A Contribution to the Philosophy of Notation" (1885).

2 See, for instance, "Description of a Notation for the Logic of Relatives, Resulting from an Amplification of the Conceptions of Boole's Calculus" (1870).

3 Although John Sowa writes, "Peirce's existential graphs ... are the simplest, most elegant, and easiest-to-learn system of logic ever invented" (Sowa 2011).

4 See Legg 2012 and 2014 for a discussion of how Peirce's diagrammatic proofs differ from "stepwise" proofs.

in the new logic, Peirce's diagrammatic system appeared unfamiliar and cumbersome.[5] It thus remained buried, except to a few select logicians, during and after Peirce's lifetime. It is only since Don Roberts published his book on the EG in 1973 that they have been available to a wider (if still specialized) audience.[6] John Sowa also deserves mention for introducing artificial intelligence researchers to "Conceptual Graphs," which derive from the EG.

Peirce asks us to think of the graphs as an inquiry between two individuals, or between one's current and future selves (MS 450, 3; MS 650). The inquiry starts with much that is taken for granted – a Sheet of Assertion that represents the universe of discourse. The Graphist then inscribes the graph, and the Interpreter draws inferences from what is inscribed, according to the rules of the system. They both see the results. Sometimes Peirce says that the Graphist is an ordinary logician or inquirer, and Grapheus is the creator of the universe of discourse, who also makes continuous additions to it from time to time (CP 4.431). Such ideas are now being worked out in Dialogical Logic, which understands a proof as equivalent to a "winning strategy" in a two-person, zero-sum game.[7]

Thus, Peirce's way of thinking about his Existential Graphs and their status in inquiry is of a piece with his pragmatist view of inquiry as conducted against a body of background belief, with an aim to settle belief by learning from the surprise of experience, leading to more successful action and thought.

5.2 Alpha

All Alpha Graphs consist merely of the *Sheet of Assertion* (a blank page), *proposition letters*, and two operators: *conjunction*, represented implicitly by writing proposition letters side by side, and *negation*, represented by a simple closed curve called a *cut*. Any well-formed part

[5] See Quine 1934a for the expression of this opinion.

[6] The material still awaits full publication. The first volume of Peirce's chronological papers appeared in 1982. The most recent appeared in 2009, with a projected twenty-three more to follow, including much of the Existential Graphs.

[7] For a lucid introduction to this area of logic, see Rückert 2001. For a rich treatment of its philosophical implications, see Marion 2009. One of the first to develop the idea that Peirce had covered this ground was Hilpinen 1982.

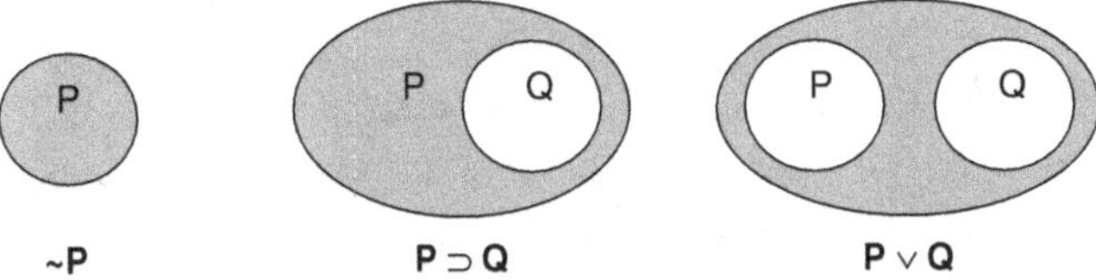

Figure 13.1 Alpha graphs: negation, implication, and conjunction.

of a graph is a subgraph. The conditional is represented by two nested cuts in a "*scroll.*" (Figure 13.1).

The proof rules depend crucially on whether the area in question is surrounded by an odd or even number of cuts (linking with "turn taking" in game-theoretic semantics and dialogical logic):

1. **Insertion.** In an odd-numbered depth, any subgraph may be inserted.
2. **Erasure.** In an even-numbered depth, any subgraph may be erased.
3. **Double Cut.** A pair of cuts with nothing in between may be: **i**) inserted in, or **ii**) erased from, any subgraph.
4. **Iteration/Deiteration.** Any subgraph may be: **i**) "iterated inward," into a cut within the scope of the area in which it appears, or **ii**) if such a subgraph might have been obtained by such iteration, it may be erased.

John Sowa argues that Peirce's system is especially elegant, by proving Frege's first axiom, a ⊃ (b ⊃ a), in five steps (Figure 13.2):

5.3 Beta

The beta graphs add only *lines of identity* that attach to the alphabetic portion of the graphs (at as many argument slots as are required), rendering them predicates rather than propositional variables:

In this way the Alpha graphs are revealed as a special case of the Beta Graphs (and propositional as a special case of predicate logic) where the "relations" have zero adicity (Peirce called these *medads*) (Sowa 2011).

Unattached ends of lines of identity serve as quantifiers, with their interactions with the cuts allowing scope distinctions (Figure 13.3). The lines of identity may themselves be negated, producing a more elegant treatment of exact number statements than standard approaches using "=" (Figure 13.4):

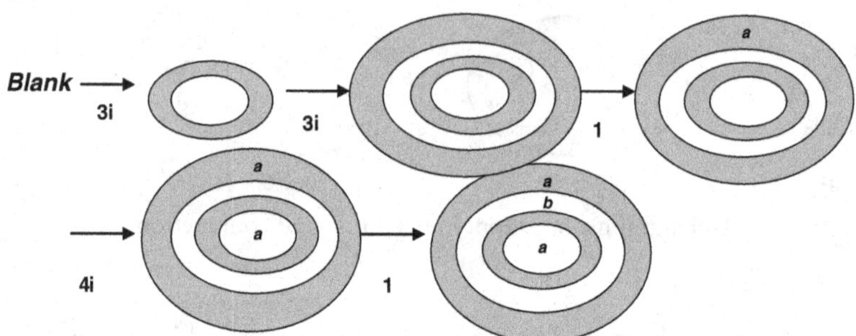

Figure 13.2 Alpha graphs: proof of (a ⊃ (b ⊃ a)).
[Sowa 2011, 368–369]

Figure 13.3 Beta graphs: existential quantification.

Figure 13.4 Beta graphs: numeric quantification.

The proof rules are those of Alpha with some extra permissions concerning the line of identity.

5.4 Gamma

What are today referred to as "the Gamma Graphs" consist of a wealth of fragments of a number of systems. Peirce never finished these, although his letters to Lady Welby[8] show he was pushing hard to do so toward the end of his life. With regard to what we have now, as Roberts says, "It is occasionally difficult to be sure of just what Peirce was up to" (1973, 64). The rather cumbersome machinery he worked with includes sheets of assertion, tints, cutouts, and bits of

[8] See the letters compiled in Hardwick (1977).

metal and fur. We therefore confine our discussion to a few simple steps Peirce took.

5.4.1 Change in Sheet of Assertion

The greatest shift in Gamma concerns reconceptualizing the Sheet of Assertion from "simply a universe of actual existent individuals, and ... facts or true assertions concerning that universe" (CP 4.512, 1903). Now Peirce planned to use two sides of the Sheet (*recto* and *verso*), also tacking sheets together. A top sheet might represent "a universe of existent individuals" and true propositions about them. The sheets underneath, which he cut out parts of the top sheet to expose, "represent altogether different universes with which our discourse has to do" (CP 4.514, 1903). These shifts may themselves give rise to further shifts:

> At the cuts we pass into other areas, areas of conceived propositions which are not realized. In these areas there may be cuts where we pass into worlds which, in the imaginary worlds of the outer cuts, are themselves represented to be imaginary and false, but which may, for all that, be true. [CP 4.512, 1903]

It seems that Peirce began to struggle with the two-dimensional limitations of the standard sheet of paper:

> The entire universe of logical possibilities might be conceived to be mapped upon a surface. Nevertheless, if we are going to represent to our minds the relation between the universe of possibilities and the universe of actual existent facts, if we are going to think of the latter as a surface, we must think of the former as three-dimensional space in which any surface would represent all the facts that might exist in one existential universe. [CP 4.514, 1903]

5.4.2 Modal Context

One key new symbol Peirce introduced on the Gamma graphs was the modal context, signified by a broken cut, which corresponds to "possibly not *P*" (Figure 13.5):[9]

Peirce developed rules for this new operator that allow derivation of "*g* is true" from "*g* is necessarily true"; "*g* is possible" from "*g* is necessary"; and "Every Catholic must adore some woman" from "There

[9] The exposition that follows is drawn from Pietarinen 2006a, 349.

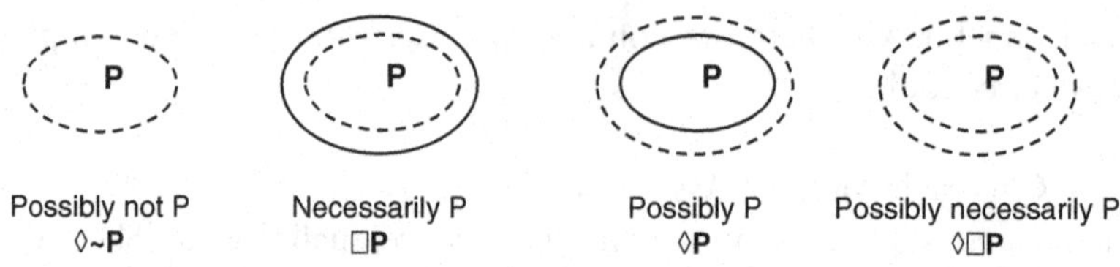

Figure 13.5 Gamma graphs: the broken cut.

is a woman, whom every catholic must adore" (CP 4.566).[10] However, the manipulations Peirce envisioned for modal logic are problematic in the state in which he left them.[11]

5.5 Varieties of Modality

5.5.1 Systems

A question currently of some interest among logicians is whether Peirce would have taken a stand on the many modal logic systems developed in the twentieth century, starting with the pioneering work of C. I. Lewis. Would he have had a preference? For instance, does the "epistemic" nature of his understanding of modality lead him to develop a version of S4?

It seems likely that Peirce meant the EG to be more flexible than that. Just as we would not expect a proof system such as semantic tableaux to take a position on this question, neither should we expect the EG. Thus Van Den Berg 1993 has worked out Gamma Graph systems for K, T, S4, and S5. (See also Braüner 1998 for a simpler version of S5, which draws graph-based analogs of standard Gentzen rules.) On the other hand, Pietarinen 2006a, 350 notes that Peirce's own description of the rules for the broken cut seems to rule out S5. Meanwhile, Zeman has devised a way to model accessibility relations between worlds in terms of Peirce's tinctures, using an R-G-B (red-green-blue) model of colors to place them in a partial ordering that enables S4 as its basic system (Zeman 1997b, 23; see also Zeman 1997c). Finally, it

[10] These transformations are well explicated in Øhrstrøm 1997.

[11] For one thing, they allow one to prove that for any q, it is necessary that q. He worried about the rule that allowed one to infer "There is a man x, such that if x is bankrupt, then x must commit suicide" from "There is a man x and a man y, such that if x is bankrupt, then y must commit suicide." Peirce in 1906 called this permission "quite out of place and inacceptable" (CP 4.580).

has been noted that at other times Peirce experimented with a specific sign (or "selective") in arrow form:

to represent a relation between two states of information that looks to be some form of accessibility (Pietarinen 2006a, 352; see also Roberts 1973).

5.5.2 Multimodality

As well as its rich structure of interrelated modal systems, modal logic divides into "kinds," such as temporal, epistemic, and deontic logic, the relationship among which is still far from well developed. It seems that Peirce was also planning to explore this issue in Gamma using his different colors and textures, but did not get far. Although Øhrstrøm has written, "There is ... no indication in his works of a study of the ways in which various kinds of modality may interact with one another" (Øhrstrøm 1997), a beginning has been made in Pietarinen 2006a, 350–351.

5.6 Peirce and Possible Worlds Semantics?

It has become popular in some circles to suggest that Peirce anticipates contemporary possible worlds semantics, even claiming that the latter "must be considered a rediscovery of something Peirce did a century ago" (Zeman 1997, 19; see also Pietarinen 2006a). Weak reasons to object to this include the fact that the connection of influence does not exist (logicians such as Saul Kripke and Ruth Barcan Marcus did not find inspiration in Peirce) and that Peirce rarely uses the expression "possible world," although it had been in use for a long time and he certainly knew his history of philosophy. On one of the very few occasions he does use it, he says, "We only wildly gabble about such things" (CP 6.509).

A thorough treatment of this question might usefully separate a *logical* from a *metaphysical* issue. The former asks whether in any of his logical analyses Peirce saw fit to quantify over sets of propositions. This he surely did (as we will see in the next section). The latter asks whether he might have considered such analyses to have existential import, for instance, in the sense of D. Lewis 1986, who argues that a plenitude of individual possible worlds exists concretely. This is much less likely, because the individualized nature of Lewis's possibilia, and their in-principle spatiotemporal disconnection from the actual render the theory a rather extreme form of nominalism by Peirce's lights. Toward the end of his life Peirce made the intriguing remark that

> modality is not, properly speaking, conceivable at all, but the difference, for example, between possibility and actuality is only recognizable much in the same way as we recognize the difference between a dream and waking experience. [CP 4.553ff, c. 1906]

5.7 *Conditionals*

How best to analyze the truth-conditions of the rich tapestry of "if … then …" usage in ordinary language is a deep matter that continues to engage logicians and philosophers. Peirce's writings on this topic reflect its difficulty, raising a number of tangled issues that he only gradually clarified through his career. Two important and related issues[12] are whether to understand conditionals as *de inesse* or "ordinary" and as Philonian or Diodoran.

5.7.1 Consequentia de Inesse

Peirce referred to material implication as a consequence *de inesse*. He notes that "such a proposition is altogether true or altogether false" (CP 4.376), essentially concerning only what happens here and now. So for instance, if I assert, "If I eat ice cream then grass is red," and I am not eating ice cream, then my assertion is true. As noted by bewildered first-year logic students worldwide, this seems to miss an important problem. Peirce here can be seen as entering the fray that later would play out in the work of Russell, C. I. Lewis, Quine, and Barcan Marcus.

Peirce contrasted *de inesse* conditionals with what he sometimes called "ordinary" conditionals, for their greater approximation to common usage, and sometimes called "hypotheticals," in homage to Kant. The latter pertain to what happens not just here and now, but across some "range of possibilities." This allows the ordinary conditional to be "sometimes" or "possibly" true across the set. Such quantitative judgments require quantifiers, and as others have noted (Zeman

[12] Some commentators run these issues together, e.g. Zeman 1997a. However, a good prima facie case for separating them is made in Lane 2007, 556n.

1997a), it was when Peirce developed his own quantification theory in 1885 that he made progress defining the ordinary conditional.[13]

5.7.2 Philonian or Diodoran?

At the same time Peirce was considering an issue that traces back to a dispute in Stoic logic between Philo and his teacher Diodorus over whether the antecedent of a conditional has existential import:

> Philo held that the proposition "if it is lightening it will thunder" was true if it is not lightening or if it will thunder and was only false if it is lightening but will not thunder. Diodorus objected to this.

Peirce notes his own preferences:

> Most of the strong logicians have been Philonians, and most of the weak ones have been Diodorans. For my part, I am a Philonian; but I do not think that justice has ever been done to the Diodoran side of the question. [NE 4,169, 1898]

In order to "fit him [the Diodoran] out with a better defense than he has ever been able to construct for himself," once again, Peirce draws on a quantified analysis of the conditional: "namely, that in our ordinary use of language we always understand the range of possibility in such a sense that in some possible case the antecedent shall be true" (NE 4, 169).

Peirce's mature theory of the conditional thereby resolves both issues at once. It goes beyond the material conditional in considering not just the here and now but a range of possible scenarios. It answers the Diodoran worry insofar as in order to decide the truth of a conditional it evaluates only possible scenarios where the antecedent is *true*. This would seem to have anticipated to some degree the truth-functions for the counterfactual conditional in, for example, D. Lewis 1973, although what Peirce would have had to say about the problem faced by Lewis of defining "nearness" across the evaluation set is an interesting question.

[13] Peirce himself admits this at CP 2.349. See CP 3.374 for an early discussion drawing on algebraic notation.

6. Peirce's Modal Metaphysics

Peirce was a naturalist in the broad sense that he held that the only way to find out what there is in the world is scientific inquiry (in a broad sense of science that means any inquiry that takes experience seriously). But he rejects any naturalism that gives ontological priority to matter or physicality – like all the pragmatists, he wants to consider whether value, generality, necessity, chance, etc., might be part of the natural world. His answer is yes to all of the above.

In phenomenology and logic Peirce showed the *irreducibility* of the three categories (as concepts, to one another). In metaphysics he considers their *reality*. Since the three have been shown to be mutually irreducible, it is possible to affirm or deny the reality of all of them independently. Peirce dubbed the doctrine of real Firstness Tychism, and real Thirdness Synechism. He devoted much less time explicitly to defending real Secondness – presumably because it is generally taken for granted by philosophers – but at one point refers to it as Anancism (CP 6.202, 1893).

6.1 *Tychism*

Recall that Firstness represents "what comes first." Peirce's Tychism holds that there "is an element of real chance in the universe." A key paper here is "The Doctrine of Necessity Examined," published in 1892. Here Peirce examines and rejects "the common belief that every single fact in the universe is precisely determined by law" (CP 6.36, 1892), which he calls *necessitarianism* (today's "determinism"). He does not deny that there are natural laws, but holds that they are violated in infinitesimal degree on sporadic occasions.

He considers a series of arguments for necessitarianism and rejects them all. First is the claim that it is a "postulate" of scientific reasoning. This is incorrect because in science, to "postulate" a proposition "is no more than to hope it is true." He then considers whether necessitarianism may be established empirically ("by observation of nature"). This is dubious because of the claim's apparent exactitude (deviations from natural law are *zero*), whereas scientists know that the strictest measurements "yet fall behind the accuracy of bank accounts" (CP 6.44, 1892). He also considers the argument that chance is *inconceivable*, to which he answers, essentially, "Try harder." Or perhaps chance is *unintelligible*, since the hypothesis of indeterminism is not explanatory. Here Peirce argues that on the contrary, real chance explains the growth of variety in the universe in a way mechanical law never could:

> The ordinary view has to admit the inexhaustible multitudinous variety of the world, has to admit that its mechanical law cannot account for this in the least, that variety can spring only from spontaneity, and yet denies without any evidence or reason the existence of this spontaneity, or else shoves it back to the beginning of time and supposes it dead ever since. [CP 6.59, 1892]

It is worth noting a certain subtlety of this view within Peirce's information-state analysis of modality. Recall that an important sense of modality for Peirce is *substantial modality*: what is necessary or possible according to "everything now existing … together with all their consequences." Peirce's Tychism holds that there is no *present* substantial contingency, but there *is future* substantial contingency, as laws of nature "swerve" across time (Lane 2007, 559). This differs interestingly from today's 4D, "tenseless" metaphysics of contingency.

6.2 *Anancism*

In §3.1 Secondness was defined in terms of "action and reaction." This category gives rise to a kind of necessity that is manifested in brute forcefulness (for instance, between billiard balls, pace Hume). Peirce tosses brickbats at other philosophers – most notably Hegel – for not recognizing real Secondness – which he sometimes refers to as "the outward clash."[14]

6.3 *Synechism*

We saw that Thirdness pertains to our seeing a yellow chair *as yellow*: a general concept that in order to honor its meaning *must* be extended to all relevantly similar instances. Peirce's argument for real Thirdness attempts to show that we live in a universe that is *genuinely intelligible* – that the ways in which we extend our predicates give us at least some predictive power. In his 1903 Harvard lectures, Peirce

[14] See Misak (forthcoming).

made a particularly vivid and dramatic case for this. He held a stone in the air in front of his audience and challenged them to admit that they knew that when he dropped it, the stone would fall, rather than flying upwards

> Suppose we attack the question experimentally. Here is a stone. Now I place that stone where there will be no obstacle between it and the floor, and I will predict with confidence that as soon as I let go my hold upon the stone it will fall to the floor. I will prove that I can make a correct prediction by actual trial if you like. But I can see by your faces that you all think it will be a very silly experiment. [CP 5.93, 1903]

He pointed out that "it would be quite absurd to say that ... I can so peer into the future merely on the strength of any acquaintance with any pure fiction" (CP 5.94, 1903). He concluded that his audience possessed knowledge of an "active general principle," exerting influence on physical objects. Peirce thought that such a commitment to general principles should be understood as a form of Scholastic realism or realism about universals, and that the kind of influence exerted by such universals was different from the "efficient causal biff" manifested by one individual billiard ball striking another (although of course the two coexist: both real Secondness and real Thirdness are required to build a full picture of reality).

7. Conclusion: Issues for Further Work

It should be clear from this snapshot of the many things Peirce said about the many aspects of necessity that there is grist for a number of mills. Epistemologists might want to explore Peirce's iconic approach to knowledge of necessity, since it offers an interesting alternative to the current landscape. It might allow us to move beyond a certain tradition (deeply influenced by Hume, and then the logical positivists) of throwing one's hands up at modality as a major skeptical problem; and either taking a eliminativist or projectivist stance toward it, or embracing mysterious metaphysical realisms such as D. Lewis 1986.

Formal logicians might to want to continue the work of excavating – using modern tools and conceptions – both Gamma graphs and Peirce's analysis of the conditional. The task is not easy because Peirce's logic is so finely sifted with the rest of his philosophy, but could be hugely rewarding. Metaphysicians might want to explore Peirce's denial of necessitarianism, especially in the light of certain recent developments in physics. Synechism, too, might be worth a new look, especially by those who are frustrated with the twentieth century's founding of metaphysics on set theory, since Peirce claimed that true "generals" or universals are not reducible to the extension of any set.

It is unsurprising that the two early twentieth century philosophers most heavily influenced by Peirce were also philosopher–logicians – C. I. Lewis and Frank Ramsey. Ramsey's view of laws as rules with which we meet the future is explicitly drawn from Peirce. Lewis attributes to Peirce many ideas he used in founding modal logic for the twentieth century (e.g. C. I. Lewis and Langford 1959). And like Peirce, Lewis is first and foremost a pragmatist. We have a certain amount of freedom in our choice of a set of logical rules – freedom within the constraints of hard facts about deducibility that need to be captured by a logic. Edwin Mares in Chapter 14, this volume, reminds us that Lewis took a priori truths to be true partly because of stipulations we make and that logical truths are both a priori and analytic. Necessity is thus to be interpreted as an epistemological operator.

But at the same time we have seen that Peirce is also a realist about modality. In this chapter we offer a brief overview of how he sought to combine the two into one harmonious position. The attempt is complex, and we have merely gestured at it here. But it is important to see the magnitude of Peirce's ambition. He wanted to put forward an account of necessity on which our actual practices are taken seriously, which can be fully formalized and clarified, and which allows for real chance, direct forceful interactions between individuals, and intelligible patterns across reality as a whole. He wanted to combine ideas from current antirealist and realist theories with the original thought that "generals" are real, but not existent. Some may think that Peirce was reaching for the impossible, but here we have tried to go some distance toward showing how he might have succeeded.

8. A Note on Citations

In-text references to the works of Charles Sanders Peirce are in the following form:

The Collected Papers of Charles Sanders Peirce: CP n.m, date; where n is the volume number and m is the paragraph number.

The Writings of Charles S. Peirce: A Chronological Edition: W n, m, date; where n is the volume number and m is the page number.
The Essential Peirce: EP n, m, date; where n is the volume number and m is the page number
The New Elements of Mathematics: NE n, m, date; where n is the volume number and m is the page number.

References to the Charles S. Peirce Papers at Houghton Library, Harvard University, are cited as MS n, m, date; where n is the manuscript number and m is the page number.

14 *The Development of C. I. Lewis's Philosophy of Modal Logic*

EDWIN MARES

1. Introduction

C. I. Lewis is sometimes accused of taking a purely syntactic approach to logic. His favourite logics, S2 and S3, are sometimes claimed to be mere collections of axioms without representing any coherent notion of logical necessity, possibility, or deducibility (Anderson and Belnap, 1975, p 117). Lewis did, however, say a great deal about the meaning of the modal operators. In this chapter I examine what Lewis actually says about the modal operators and his philosophy of logic more generally and argue that there is much more to these systems, S2 and S3 in particular, than mere collections of axioms. I claim that certain of the features of these logics that are often criticized are in fact well motivated by his epistemological and semantical views.

I do not claim, however, that Lewis had a well-developed semantics of modality when he first put forward his logic of strict implication in 1911. Most of his remarks on semantical issues occur much later than that. Which, if any, of his later views he had in mind when he originally constructed his modal logic is impossible to tell. He does not make any remarks about this either in his published writings or (as far as I know) in the manuscripts in his *Nachlass*.

Moreover, his philosophical views changed, in some ways rather radically, over the course of his career. In his PhD thesis and his early philosophical writings, he is a Kantian idealist, but he adopts a form of pragmatism in the 1920s that has important consequences for his views about modality. In his more mature writings – especially in the 1940s and later – the influence of Kant is much less pronounced and his metaphysical views seem more realist. These more realist metaphysical views are reflected in his theory of meaning and his philosophy of logic.

We see Lewis's philosophy of logic develop along with his other philosophical views. In the 1910s, Lewis presented his modal logic largely as a response to classical logic. In the 1920s he provided an epistemological interpretation of modal logic. In the mid-1930s, he attempted to link his logic of strict implication with the notion of a tautology from Wittgenstein's *Tractatus*. In the 1940s, this connection took a more metaphysical and algebraic cast.

The structure of this chapter is as follows. After reminding the reader of certain central features of Lewis's logical systems, I examine the evolution of his philosophy of logic in three phases:

1. The Deducibility Phase (1911–1920). This phase was marked by the original formal development of Lewis's philosophical logic, his initial criticisms of classical logic, and his proposal for a logic of strict implication as a means of formalizing the notion of deducibility.
2. The Epistemological Phase (1920–1935). In this phase, the role and interpretation of logic were largely epistemological. Necessity was understood in terms of what can be known a priori. The central role of strict implication was to give closure conditions for theories or sets of beliefs.
3. The Metaphysical Phase (1935–1964). In this final phase, Lewis adopted a more complicated ontology of meaning. The logical relationships between statements and between other phrases were understood in terms of relations between their intensions.

I look at the development of Lewis's logical and, in particular his semantical ideas, through these three phases to see how he interprets his theories of strict implication. The reason this is interesting is that Lewis's preferred logical systems have largely fallen out of favour with logicians, especially after the introduction of modern possible world semantics, for reasons that I discuss in Section 2. I argue that Lewis has good reasons for accepting certain of the controversial features of his logics. In the epistemological phase of his philosophical development, Lewis constructs a conventionalist theory of logic and of necessity. This conventionalism motivates his rejection of necessitation. In his metaphysical phase, Lewis tempers his conventionalism with a more absolutist and realist metaphysics of meaning. But there are still conventionalist elements in his philosophy of logic and these can be used to explain his retention of the logic S2, and his continued rejection of S4 and S5.

2. The Formal Evolution of Lewis's Logics

This is a brief section on the formal development of Lewis's logics. A more complete version is given by William Parry (Parry, 1968).

Lewis first presents a logic for strict implication in his "Implication and the Algebra of Logic" (1912). There he introduces the idea of defining implication in terms of an intensional disjunction. An intensional disjunction, 'p or q', is true if and only if, if p were false, then q would be true (p 355). Lewis did not analyze the subjunctive mood of intensional disjunctions; nor does he formalize the logic of intensional disjunction. Lewis gives slightly different formalizations of the logic of strict implication in (1913b) and (1914), but he claims that his first "complete" presentation is in the *Survey of Symbolic Logic* (1918). That system, however, collapses into classical logic, as Emil Post showed. Lewis corrects the system in (1920), where it becomes what we know today as S3. Until at least 1927, Lewis maintains that this logic is the logic of strict implication. Circa 1930, Lewis uses Oskar Becker's (1930) rules and postulates, as well as some of his own, to formulate five different logical systems, which he calls S1, S2, S3, S4, and S5. These are presented in appendix II of *Symbolic Logic* (1959, pp 500–502).

From their initial presentation, Lewis adopted very different attitudes towards his different logics. He was always disparaging towards S4 and S5. In the appendix in which he formulates the logics, Lewis says that whereas S5 (and classical logic) may appeal more to mathematicians, the weaker systems are of more interest to those who wish to study reasoning (1959, p 502).

In appendix II of *Symbolic Logic*, Lewis gives the following axiomatization of his weakest logic, S1. He begins by defining the connectives in terms of $\Diamond$, ~, and $\wedge$. In particular $p \prec q$ is defined as $\sim\Diamond(p \wedge \sim q)$. The axioms for S1 are:

- $(p \wedge q) \prec (q \wedge p)$;
- $(p \wedge q) \prec p$;
- $p \prec (p \wedge p)$;
- $((p \wedge q) \wedge r) \prec (p \wedge (q \wedge r))$;
- $((p \prec q) \wedge (q \prec r)) \prec (p \prec r)$;
- $(p \wedge (p \prec q)) \prec q$.

The rules are modus ponens for strict implication, adjunction, uniform substitution, and the replacement of strict equivalents.

Lewis developed S1 as an emergency measure. He would adopt S1 if S2 turned out to contain a version of the transitivity of implication that is too strong. After it was shown that S2 did not contain that version of transitivity (1959, pp 503–514), Lewis never discussed S1 in print again.

S2 results from the addition of the axiom $\Diamond(p \wedge q) \prec \Diamond p$ to S1. By the time that he and Langford wrote *Symbolic Logic*, S2 had become Lewis's choice for the logic of strict implication. Lewis's change in preference from S3 to S2 seems to have taken place some time between 1927 and 1932. In his notes for a lecture to students at Columbia University in 1927, the logic of strict implication is still presented as S3.[1] The reason for his rejection of S3 is related to the transitivity of strict implication. Lewis thinks that the strong form of transitivity is problematic. The strong form is $(p \prec q) \prec ((q \prec r) \prec (p \prec r))$, and this is derivable in S3. Suppose that $p \prec q$, then according to S3, $(q \prec r) \prec (p \prec r)$. On Lewis's view, it is not really the case that $p \prec r$ follows from $q \prec r$, but rather that $p \prec q$ and $q \prec r$ together entail $p \prec r$ (1959, p 496). S3 results from the addition of $(p \prec q) \prec (\sim\Diamond q \prec \sim\Diamond p)$ to S2. Adding the axiom $\Box p \prec \Box\Box p$ to S3 yields S4 and adding Becker's Brouwersche axiom – $p \prec \Box\Diamond p$ – to S4 gives us S5.

One feature of S1, S2, and S3 that causes problems when we attempt to give them semantics is their rejection of the rule of necessitation, $\vdash A \Rightarrow \vdash \Box A$. One reason for this rejection is Lewis's interpretation of the turnstile, $\vdash$. He understands the turnstile to be an assertion sign. '$\vdash A$' merely says that A is asserted. It does not say that A is necessarily true. To say '$\vdash\Box A$' is to assert that A is necessarily true. As I shall argue, Lewis's mature view is that the axioms of S2 are true, but not necessarily true.

3. The Deducibility Phase

My main interest is in the development of Lewis's philosophy of logic in his epistemological and metaphysical phases. In this section I set the stage for the epistemological and metaphysical phases with a brief exposition of Lewis's philosophical position prior to 1920 and the role of logic in it. In this period, Lewis identifies himself as an idealist

[1] C. I. Lewis, Columbia Lectures on Symbolic Logic, Stanford University Library, Department of Special Collections, M174, Box 20, folder 20–21, 1923–1927.

(1912b).[2] This idealism seems to be a form of Kantianism in which the knowing subject constructs the world from sensations. Lewis never explicitly ties his logic to his idealism, and it is not clear that he had any serious connection between the two in mind.

In his earliest papers on logic, Lewis does not present his theory of implication as an element in a grand philosophical project. Rather, Lewis says that his logic of strict implication is a way of "bringing the meaning of implication into accord with that of ordinary inference and proof".[3] It is the ordinary notion of a valid inference, as it is used in particular in science and mathematics that Lewis wishes to capture in his system.

Lewis motivates his system of strict implication by contrasting it with classical logic. He claims that classical logic does not formalize a real notion of deducibility at all (1913, p 242). Consider a typical theorem of classical propositional logic, $(p \wedge q) \supset p$. This statement only tells us that either the antecedent is true or the conclusion is false. It does not tell us that the consequent follows from the antecedent. Of course, one might point out that classical logicians do not merely assert theorems; they assert that they are theorems, in particular, by prefixing them with a turnstile: $\vdash (p \wedge q) \supset p$. But for Lewis, the turnstile is just an assertion sign. Prefixing a formula with a turnstile merely says that one is asserting it. What we need is a way of saying that, even if the antecedent does not happen to be true, if it were true than the consequent would be true. This is exactly what the strict implication $(p \wedge q) \prec p$ is meant to express (1917, p 355).

In Lewis's published writings and from the very few items in his *Nachlass* from this period, Lewis said very little about the meaning of the modal operators or about how he chose the axioms that he did for his system of strict implication. I move on now to the epistemological and metaphysical phases of his career to examine the positions that he held then.

4. The Epistemological Phase

Upon his return to Harvard from Berkeley in 1920, Lewis shared an office with the unpublished papers of C. S. Peirce. Lewis claims that

[2] For a detailed discussion of Lewis's idealism, beginning with his PhD thesis, see Murphy 2005.

[3] Letter to Josiah Royce, 15 October 1911, Papers of Josiah Royce, Harvard University Archives, box 122, folder 44.

reading these influenced him in part to abandon Kantian idealism and adopt a version of pragmatism (1968, p 16). Likewise, in the 1920s he began to develop a pragmatic view of logic.

In his epistemological period, Lewis adopts the aim of trying to find the structures that we do in fact use to understand experience instead of those that necessarily underlie experience. For this reason, Lewis says that his conceptual pragmatism is not a "transcendental" philosophy but rather a "reflective" one (1929, ch 1). Unlike Kant, Lewis thinks that the concepts that we use to interpret the world are not fixed. For example, our concept of cause has changed over time and differs from culture to culture. Some cultures think that the stars or planets can affect our lives or that talismans can affect events that occur at some distance in space or time. Lewis says that we use the word 'cause' to express both their concept and our own, but that these concepts really have little in common (1929, pp 234–235). Lewis applies his conceptual relativism to logic. He holds that we have a certain amount of freedom in our choice of a set of logical rules. In holding this position, Lewis is clearly a pluralist about logic in the contemporary sense. But, as we shall see in Section 4.3, he thinks that this pluralism has limits.

Lewis's voluntarist pluralism about logic dovetails with two other positions that Lewis holds. He claims that all a priori truths are true because of stipulations that we make. This is a form of conventionalism about the a priori that is familiar to many of us from the logical positivists. Moreover, all a priori truths, according to Lewis, are analytic. In fact, he holds that all logical truths are both a priori and analytic. He interprets '□A' as saying that A is a priori. In his epistemological phase, then, Lewis holds that necessity is to be interpreted as an epistemological operator. It allows us to sort true sentences into those that are empirical and those that are a priori. As we shall see, this treatment of necessity commits Lewis to the view that certain logical theorems are contingent and empirical.

The epistemological phase is epistemological in other respects. In this period, Lewis thinks of logic as a whole and strict implication in particular as having an epistemological role. As I set out in Section 4.1, Lewis thinks that a central role of formal logic is to set out closure conditions for theories or belief sets. As we shall see, examining this function of logic helps us to understand the nature of the a priori in Lewis.

4.1 *Systems and Worlds*

A central function of a logic, at this point in Lewis's thought, is to provide closure principles for systems. In order to understand what a system is, we need to delve a bit more into the epistemological views that Lewis holds at this time.

In "Facts, Systems, and the Unity of the World" (1923) Lewis puts forward a Kantian constructivist theory of systems and possible worlds. Although Lewis rejects the synthetic a priori by the early 1920s, his philosophy is still very Kantian. In this paper, he considers the universe to be a construct from experience and our beliefs about experience. The universe itself is an idea in the Kantian sense of an ideal goal; it is what we want to construct in forming beliefs.

A system is what is represented by a theory or ideal belief set. A system is a particular sort of set of facts. A fact is an extra-linguistic entity. Lewisian facts need not be true. A possible world is a negation complete and consistent system. For each world w and fact F, either F is in w or the negation of F is in w, but not both.[4]

A system is a set of facts that is closed under the true strict implications and under conjunction. Thus, if $\vdash A \prec B$ and A is in a system s, then B is also in s. Moreover, if A and B are in s, then $A \wedge B$ is in s. Since $(A \wedge B) \prec A$ and $(A \wedge B) \prec B$ are both theorems of all of Lewis's logics, we can strengthen the conjunction condition to say that $A \wedge B$ is in s if and only if A and B are both in s, for all systems s.

We can derive from any of Lewis's modal logics that any statement A that is necessarily true, $B \prec$ A for any statement B. Thus, all necessary truths are true in all systems. In his notes for a series of lectures to students at Columbia University in 1923, Lewis explicitly links necessity and possibility to his notion of a system:

> A proposition is possible if it is true in some system or state of affairs.
>
> A proposition is necessary if it is true in every system.
>
> Two propositions are consistent if they are jointly true is some system.[5]

[4] In Mares 2013, I construct a model theory from Lewis's theory of systems and worlds for his logic S2.

[5] C. I. Lewis, Columbia Lectures on Symbolic Logic, Stanford University Library, Department of Special Collections, M174, box 20, folder 20–21, 1923–1927.

We can derive from these clauses that $p \prec q$ if q is in every system that contains p.

At first glance, these clauses do not seem to fit with Lewis's other logical views. They seem to characterize the logic S5.[6] At this stage Lewis had not yet formulated S5, and he still accepted S3. But Lewis does not seem to be constructing an indexical semantics in the sense of Kripke semantics. He is telling us what it is for a proposition to be actually necessary or possibly true. He is not telling us what it is for a proposition to be necessarily or possibly true in an arbitrary system, but only in the actual world.

In his 1927 lectures at Columbia, the semantic role of systems is largely replaced by that of possible worlds. In his lecture notes he writes:

> To be genuinely possible or self-consistent, a proposition or situation or system must be compatible with one or other of every contradictory pair of propositions.
>
> This is to say, it must be true of a 'conceivable world' W, and that:
>
> 1. If p is in W, and p $\prec$ q,¬q is not in W. 2. If p is not in W, ¬p is in W. 3. If p and q are in W, p∧q is in W.
>
> The actual is not the only such conceivable world. There are as many as there are mutually incompatible but self-consistent systems.
>
> These are infinite in number. Every two conceivable worlds have something in common. Meaning of 'necessary', 'possible', 'impossible' and 'inconsistent', in terms of the set of all conceivable worlds. Meaning of 'implies' in terms of the set of conceivable worlds. The nature of inference. Contrast with material implication.[7]

In these notes, Lewis uses possible worlds instead of systems as the central elements in his theory of meaning for the modal connectives. But there appears in both these views about meaning to be a serious problem about circularity. Lewis defined systems and worlds in terms of the true strict implications and promised a theory of meaning of the modal connectives including implication in terms of systems or worlds.

[6] See Sedlar 2009.

[7] C. I. Lewis. Columbia Lectures on Symbolic Logic, Stanford University Library, Department of Special Collections, M174, box 20, folder 20–21, 1923–1927.

This circularity, however, may not be vicious. The true implications need to be set in order to determine the class of worlds. Lewis claims that the meaning of these implications is determined by the set of worlds. He is not saying directly that the truth conditions of statements that include those connectives are determined by the set of worlds. If Lewis has a truth-conditional theory of meaning, then the circle is complete, and vicious.

A vicious circle can be avoided here if we view Lewis as having a pragmatic, rather than a truth-conditional, theory of meaning. According to the pragmatic theory of meaning, a concept or proposition gets its meaning from the way in which it is verified or assimilated into the life of one's mind. For Lewis, one of the important modes of assimilation is the way in which a concept or proposition is used by an agent to justify, impel, or constrain him or her to do particular actions. In this case, the use of the modal propositions is to constrain the agent to construct worlds of certain sorts. For example, if one holds that *A* is true, then she is constrained to construct only worlds in which A is true.

Although Lewis might have avoided a vicious circle in his theory of meaning by rejecting a truth-conditional theory of meaning, he still needed a theory of truth conditions for modal statements. In the next section, I examine what Lewis said about truth and modality in his epistemological period.

4.2 Analyticity, the A Priori, and Necessity

According to Lewis, what is necessarily true is what is a priori and what is a priori is what is analytically true. In the 1920s, Lewis is a conventionalist about analyticity. Thus, what is necessarily true is determined to be so by convention. In various works in the 1920s, such as "A Pragmatic Conception of the A Priori" (1923) and *Mind and the World Order* (1929), Lewis sets out a conventionalist theory of analyticity.

A priori truths are chosen by us in the sense that we can stipulate definitions of terms. Like his contemporaries, the logical positivists, Lewis rejects the synthetic a priori. He holds that the only truths that we can know a priori are also analytic. We even choose the axioms of our logic. In doing so we define the logical connectives. As we shall see, Lewis does not mean that we produce an implicit definition of the connectives by devising a logical theory. Rather, we associate the connectives with ways of applying them. I will explain this at some length in Section 4.3.

There are restrictions on what can count as a definition. Definitions cannot determine the nature of possible experience. By adopting a definition, we cannot determine what can or cannot be perceived (1923b, p 233). This view of definitions has difficulties. Clearly, because of adoption of a particular definition of 'bachelor' we cannot perceive any married bachelors. For Lewis, this is just a matter of how we talk or think about what we perceive. We can distinguish between what we actually perceive and how we think about it by means of a form of philosophical analysis (1926, p 240). For Lewis, thinking about experience requires that we have categories and concepts, but these concepts and the analytic links between them are for us to choose (Misak 2013, pp 192–196).

Among the analytic truths are the truths of pure mathematics. Here we see Lewis abandon his attachment to Royce's philosophy of mathematics and adopt Russellian logicism (1929, pp 244–251]. Lewis thinks that the truths of pure mathematics are theorems of logic and that the logical truths are analytic. What Lewis means by the "logical truths" here, however, is not straightforward. Let us consider the assertion of one axiom, rewritten slightly:

$$\vdash \Box\,((p \wedge q) \supset p)$$

This is an axiom of S1, S2, and S3. For Lewis, $(p \wedge q) \supset p$ is a theorem of logic, and is analytic. But the axiom itself, $\Box\,((p \wedge q) \supset p)$, is not analytic. It is an empirical truth. It tells us that $(p \wedge q) \supset p$ is stipulated as an analytic truth. We stipulate $(p \wedge q) \supset p$ to determine, in part, the meanings of $\wedge$ and $\supset$.

Moreover, the necessitation of this axiom, viz. $\Box\Box\,((p \wedge q) \supset p)$, is not a theorem of the logic. In other words, the logic does not assert that this axiom is analytic.

So, in effect, Lewis claims that all the truths of logic are analytic but denies that the axioms of logic are analytic. Surely the axioms of his logics are logical truths. In order to make sense of his holding both these views, we have to distinguish between two senses of 'logical truth' in Lewis. The axioms of a logic together with the rules determine a set of analytic logical truths in a strict sense, but need not themselves be logical truths in this sense. That is, if $\Box A$ is a theorem of S2, then A is a logical truth in the strict sense but $\Box A$ itself may not be a logical truth in the strict sense.

As we saw in Section 3.1, the turnstile for Lewis is only an assertion sign. It does not tell us that what follows the assertion sign is a logical truth in the strict sense. The assertion sign does not stand only before statements that are necessary, analytic, or a priori. Some of the theorems of logic are, for Lewis, contingent, synthetic, and a posteriori.

4.3 Lewis's Pluralism

As I said, Lewis in the 1920s and early 1930s was a conventionalist about logic. His pluralism in the 1920s was quite extreme. Although Lewis realized that it is a precondition of rational discussion that the discussants apply and appeal to the same rules of inference, he claimed that there were no facts that determine which rules of inference. He said:

> It is easy to define "rationality" in one's own terms. But that can only lead to the familiar conclusion, "All the world is strange save thee and me – and thee's a little strange." With respect to our ideals, we all of us stand in the egocentric predicament; we can only assert our own and hope for agreement. (1921, p 379)

Here Lewis said that it was legitimate for different people to adopt different rules of inference.

Lewis retained a form of logical pluralism in the early 1930s, but it is less radical. He kept the view that different logical systems can be equally correct. He also maintained that the choice of a logic is a pragmatic matter. What was new in the writings of the 1930s, especially *Symbolic Logic* and "Alternative Systems of Logic" (1932), was that each logical system he considered is given a different interpretation. In (1932), Lewis considered three logics. Classical logic is understood as the logic of truth and falsity. Łukasiewicz three-valued logic is understood in epistemic terms: a statement is given the value 1 if it is known to be true, 0 if it is known to be false and 0.5 if it is neither known to be true or false. In different logical systems, the logical connectives were given different meanings. Parry's four-valued logic had the values 1, 2, 3, and 4. Lewis read 1 as 'necessarily true',

2 as 'is true but not necessary', 3 as 'false but not impossible', and 4 as 'impossible'.

In Lewis's position of the early 1930s, it was no longer clear that the dispute between logical systems is a real dispute at all. Lewis says that 'a "system of logic" is nothing more than a convenient collection of such concepts [negation, disjunction, conjunction, etc.], together with the principles to which they give rise by the analysis of their meaning' (1932, p 401). We can view Lewis's position in this period as a precursor to Quine's later view that when people adopt alternative logical systems, they are not in direct dispute with classical logic; rather, they "only change the subject" (Quine 1970, p 80). But in choosing a logic, one is choosing a basic structure for all of her systems (her theories and her set of beliefs). And this can only be done on the basis of the pragmatic virtues of the different logics.

Once we give an interpretation of this sort to the connectives of a logic, we understand how to apply the logic to determine the truth or falsity of its sentences, determine the validity of its inferences, and so on. That we are able to do this is essential for Lewis. As a pragmatist, Lewis thinks that in order for anything to be a proper concept, we must know how to apply it. As we shall see when we discuss his later theory of meaning in Section 5.2, Lewis does not think that the application of a concept exhausts its meaning, but its ability to be applied is essential to its having meaning.

Despite Lewis's pluralism, he thinks that there are fixed facts about deducibility. He says that "there must be a truth about deducibility which is not relative to this variety and difference of systems and their relations" (1959, p 236). In chapter 8 of *Symbolic Logic*, Lewis defends both pluralism and the view that there is a single correct system of deducibility.

Within a logical system certain formulas entail other formulas. For example, it is a fact that in classical logic any proposition entails every instance of the law of excluded middle. In Ł$_3$, on the other hand, it is not the case that, for propositional variables p and q, p entails $AqNq$ ('q or not-q' in Polish notation). These facts about what entails what in a given system are themselves in some sense true absolutely. On Lewis's view, we can represent these two facts using strict implication. That is, $p \prec (q \vee {\sim}q)$ obtains but $p \prec (AqNq)$ is fails. Note that the principle of excluded middle is represented using different notation in classical logic and in Ł$_3$. This is needed to prevent contradiction (see

Lewis and Langford 1959, p 242n). In this way, we can represent the implications of all logics within one logical framework (Lewis and Langford 1959, p 244).

Lewis, however, never attempted to formulate a theory of strict implication that captures the implications of all logical systems. To do so might have been rather tricky or even impossible if he allowed formulas that mix notation from different systems. It is known that adding new notation to a logic does not always lead to a conservative extension of the original logic. If we restrict our formation rules, however, to bar the mixing of notations from system to system, what we get is, in effect, many different theories of strict implication – one to capture the notion of deducibility for each of the original logics.

Even in the logic of strict implication, however, there is some room for convention. Consider one of Lewis's favourite principles, $((p \prec q) \wedge (q \prec r)) \prec (p \prec r)$. Theorems like this, with nested strict implications, are not forced on the logic of deducibility by the base logic. The fishhook does not belong to the vocabulary of the base logic and so its consequence relation cannot tell us which implications are deducible from which. Where do these theorems originate? The only answer that Lewis can give is that they result from the notion of deducibility that we choose to express with the fishhook. In other words, they arise from convention.

5 The Metaphysical Phase

In the 1940s, Lewis adopted a much more elaborate ontology. As we shall see, he adopted an ontology of intensions and attributes. How much distance there was between this more elaborate metaphysical view and his earlier Kantian constructivism is somewhat controversial. Whether Lewis thinks that abstract objects, such as attributes of individuals, are mental constructs or mind-independent, I do not know. Moreover, nothing that I say in what follows depends on taking a side in this issue. I only claim that Lewis's theory of meaning in his metaphysical phase has many more commitments. The status of the elements of meaning is not important to what I say.

Lewis needed a metaphysics of meaning, at least in part, to explain his views about strict implication. The idea that there are facts of the matter about what is deducible from what raises a question: on what features of reality are these facts based? In the period from 1935 until almost the end of his life, Lewis addressed this question in a series of works, some of them still unpublished. This turn in Lewis's thinking is marked by his 1935 manuscript, "A Note on Strict Implication."

5.1 "A Note on Strict Implication"

The manuscript "A Note on Strict Implication" dates from 1935. It has both Lewis and Langford as authors, but in a note Lewis claims that Langford's contribution was "nominal." It was withdrawn by Lewis from *Mind* after the proofs had been sent to him. His reasons for withdrawing the paper are not important to my present purposes, since none of them is related to his commitment to the central position of that paper.

In constructing a metaphysics of meaning, "A Note" belongs squarely to Lewis's metaphysical phase. He used a structure of intensions as the basis for his theory of meaning for the modal connectives. A statement A is necessarily true, according to this view, if and only if the intension of A is the same as the intension of some tautology. By 'tautology' here, Lewis meant a truth functional tautology, which he interpreted following the *Tractatus* as a statement that is true in all possible situations.

An intension (which is also called a "connotation") is here taken to be a universal, a property of some sort. Lewis said:

> "Socrates is a man" implies "Socrates is an animal" because the connotation of "man" includes the connotation of "animal". Connotations are (in the language of the older logic) universals; and any relation of connotations as such is a universal relation. If q is deducible from p, this is because an analysis of the meanings of the propositions p and q will reveal a universal relation of connotation or intension.

As we shall see in Section 5.3, in his later work he continues to think of a certain sort of meaning as a property. In his later works, he claims that a sentence signifies a property of worlds. But he does not say in "A Note" what sorts of properties are expressed by sentences.

As I have said, a statement is necessarily true if its connotation is the same as that of some tautology. Formally, this is written:

$$\Box A = \exists\theta\ \exists x(A = (\sim\theta x \vee \theta x))$$

On this theory, a strict equivalence, $A = B$, holds if and only if A and B have the same intension. Except in certain special cases, such as when A and B are the same sentence, it is a contingent matter as to whether A and B have the same intension. It depends on which intensions speakers attribute to them. Let us say that we have chosen to attribute to p the same intension as a tautology, $\sim q \vee q$. The choice that we have made determines that $\Box p$ is true, but it is a contingent matter that we have made this choice. Thus, it would seem that whereas $\Box p$ is true, $\Box\Box p$ is false. This gives us a counterexample to the S4 axiom, $\Box p \prec \Box\Box p$. We can use similar reasoning to show that the rule of necessitation fails.

5.2 The Theory of Meaning in the 1940s and 1950s

In the 1940s, Lewis develops the ideas of "A Note" into a more elaborate metaphysics of meaning. The central distinction that Lewis draws with regard to meaning is between linguistic meaning and sense meaning. The linguistic meaning of a phrase is given by the relationships that phrase has to other phrases. The sense meaning of a phrase is a rule that allows us to apply that phrase to things.

Linguistic meaning is the more important notion for logic, but we need to understand sense meaning too, since Lewis thinks that it is more basic (1946, p 141). The sense meaning of a phrase is rather like a Kantian schema. It is a rule that allows the mind to apply a concept to things and events in experience. Lewis gives the example of the concept of a chiliagon, a 1,000-sided figure. We cannot picture a chiliagon accurately, but we can imagine counting the sides of a figure and coming up with 1,000. We know a rule for determining whether a thing is a chiliagon. This is the sense meaning of the word 'chiliagon' (1946, p 134).

Lewis's solution to the empiricists' problem of abstract objects involves sense meaning. Berkeley was right that we cannot have an image of a triangle in general, but we can understand a rule that allows us to determine whether or not a shape is a triangle (ibid.).

The sense meaning of a word or phrase determines a relationship between that word or phrase and its extension. As in Lewis and Langford 2014, the extension of a word or phrase in manner that it is required to understand sense meaning does not only contain actual things, but also contains merely possible things as well. In order to understand a meaning rule fully, we need to take into account not only how the phrase applies to actual things, but to any conceivable thing. For example, 'centaur' has a sense meaning: "We should recognise a centaur if we saw one" (1946, p 137).

Turning to linguistic meaning, Lewis's notion of an intension or connotation from "A Note" became bifurcated in *Analysis* into two concepts: the idea of signification and the notion of an intension (1946, p 39). An intension here is a set of linguistic expressions. In the intension of a sentence *A* are all the sentences that follow from *A*. In the intension of a predicate $\varphi\hat{x}$ are all the predicates that are analytically contained in $\varphi\hat{x}$. For example, in the intension of 'is a dog' are 'is a mammal' and 'is an animal'. The signification of an expression is an attribute. The signification of a predicate is a property of things. The signification of a statement is a property of worlds (1946, p 52).

The addition of the notion of signification in particular to Lewis's theory of meaning gave it a strong realist cast and is in sharp contrast with his earlier conventionalism. Logical systems, if they are adequate, express pre-existing relationships between attributes. Lewis said:

> Until symbols have fixed and specific meanings, a relation of them conveys no relation of meanings; and as soon as the meanings attaching to symbols are fixed, the question whether a relation of symbols expresses truth depends upon the relation of the meanings expressed. Also, there would be these relationships of meanings which hold, whether they be verbally symbolized or not, and irrespective of the manner of their symbolic representation. Thus the conventions of language determine no analytic truth but only how it may be expressed. (1946, p 156)

Analyticity, a priority, and necessity are not determined merely by choices. They are determined at least in part by relationships in the universe of meanings. This sounds very Fregean, and not very much like the Lewis of the 1920s.

Despite Lewis's deemphasizing of the role of convention, he still maintained that convention played a role in semantics. He still thought that statements of necessities express what were in part conventional truths. '$\Box A$' said that '*A*' signified an entity of a particular kind. The fact that the entity was of the kind that determined an analytic truth was not a fact of convention. But the fact that '*A*' signified this entity was conventional. Thus, '□A' still represented a truth that was (in part) conventional.

In an appendix to the second edition of *Symbolic Logic* (1959), Lewis set out an algebraic semantics that captured much of the theory of meaning of "A Note" (2014) and *Analysis* (1946).The base of this semantics was a Boolean algebra. The technical details of Boolean algebras are not important for us here. What is important is that they have a top element, 1, and a bottom element, 0. They may have other elements as well. All tautologies were given the value 1 and all contradictions were given the value 0. $\Box A$ is true if and only the value of A is 1. $\Diamond A$ is true if and only if the value of A is not 0. And so on. Unfortunately, this semantics did not provide a way of calculating truth-values for nested or iterated modal formulas, and its publication so close to the appearance of Kripke's semantics caused it to be overlooked by the logic community.

5.3 Use and Mention

The fact that Lewis treated the truth conditions for modal statements to include both conventional and non-conventional elements raises the question about whether he was guilty, as Quine claimed, of confusing use and mention.

Lewis did not view modal statements as telling us merely about the metaphysical status of some extra-linguistic entity such as a property. On the other hand, modal statements were not, in his metaphysical period, merely about linguistic convention. We can represent the modal operators as operators on properties. We can say, for instance, that $\Box P$ if and only if any predicate that signifies P will be necessary because P is contained in every other property. But thinking of necessity as a property of properties in this way is rather contrived. Similarly, thinking of necessity as a predicate of predicates is fully possible but overemphasizes its linguistic aspects. In attempting to capture both linguistic and non-linguistic facts by means of the modal operators, facts about both the expressions that appear behind the operators and what these expressions mean are indicated by modal statements. This makes the semantics of modal statements complicated, but it does not seem to constitute a confusion of any kind.

6 Conclusion: Evaluating Lewis's Semantic Views

It is clear that Lewis's semantical programmes have significant positive and negative aspects. I will treat the negative aspects first.

I see the central failings of Lewis's semantics as technical. First, although in his epistemological period Lewis has a metaphysics of systems and possible worlds he does not postulate semantically salient relationships between them. In particular, Lewis does not discover the sort of accessibility relations that are of central importance to modern possible world semantics. Without accessibility relations or some similar device, Lewis cannot give a semantics that accounts for nested modal operators. Second, a closely connected point, but perhaps a more important one, is that Lewis makes no attempt in his epistemological period to give any sort of truth or containment condition for modal formulas in non-actual worlds or systems. I think it is fair to say that Lewis does not really grasp the full notion of an indexical semantics in its modern sense.

Similarly, in his metaphysical period, although Lewis adopts a rather elaborate metaphysics of meaning, he does not provide a treatment of the meaning of sentences with nested modal operators. Although Lewis makes some suggestions concerning an algebraic semantics, he realises that these remarks are inadequate and he does not even point to a way to solve the problem of nested modalities.

These failings, however, are tempered by the fact that Lewis's suggestions concerning the meanings of the modal operators can be used as the basis for a coherent formal semantics. For example, I have taken Lewis's views from his epistemological period and constructed a class of models that characterises S2 (Mares 2013).

In Lewis's epistemological phase, the purpose of the system of strict implication is to give us rules for the closure of our theories. These logical norms are chosen by us. Even in Lewis's metaphysical phase, we see that there is room for convention and conventions still define the meanings of the modal operators. This conventionalism may not at the moment seem acceptable to most philosophers. But Lewis's work does raise the possibility that much of our current debate about the choice of alternative logics is at least in part an argument about a decision we need to make rather than about a fact of the matter. I think this is likely true.

Moreover, in modern terms, logicians since Lewis have tended to treat linguistic conventions as constraints on models rather than as elements that appear within models. Carnap, for example, treats his meaning postulates as helping to define linguistic frameworks, which in turn define possible worlds models. For Carnap and more modern semanticists, linguistic conventions are true at every world within those structures that model those conventions. Lewis's view encourages us to think in about conventions very differently.

Lewis's view, however, does more than force us to confront convention. Rather, his early work attempts to combine an interesting form of a Kantian constructive approach to epistemology with formal logic, and the later work attempts to place a slightly more realist cast on the epistemology while retaining a similar place for formal logic. In my opinion, both of these projects seem promising and should be updated and developed using more contemporary logical and epistemological tools.

The negative aspects of Lewis's semantical thinking, then, concern his place in the history of formal semantics. Lewis is not a plausible candidate for the title of "inventor of possible world semantics"; his use of possible worlds is solidly prefigured by Leibniz, Peirce, Russell, and Moore. The positive aspects of Lewis's semantical theory relate to the philosophical interpretation of the modal logic and its role in epistemology and the theory of meaning. In my opinion, it is the philosophical aspects that are more important right now. Logicians have discovered possible worlds semantics. But we can always use more philosophies of logic.

15 *Carnap's Modal Predicate Logic*

MAX CRESSWELL

1. Introduction

Why did Carnap want to develop a modal logic? Here I think we have to note the influence of Wittgenstein's *Tractatus*. In his philosophical autobiography Carnap writes:

> For me personally, Wittgenstein was perhaps the philosopher who, besides Russell and Frege, had the greatest influence on my thinking. The most important insight I gained from his work was the conception that the truth of logical statements is based only on their logical structure and on the meaning of the terms. Logical statements are true under all conceivable circumstances; thus their truth is independent of the contingent facts of the world. On the other hand, it follows that these statements do not say anything about the world and thus have no factual content. (Carnap 1963, p. 25)

Wittgenstein's account of logical truth depended on the view that every (cognitively meaningful) sentence has truth conditions (Wittgenstein 1922, 4.024).[1] Carnap certainly appears to have taken Wittgenstein's remark as endorsing the truth-conditional theory of meaning. (See, for instance, Carnap 1947, p. 9.) If all logical truths are tautologies, and all tautologies are contentless, then you do not need metaphysics to explain (logical) necessity.

[1] "To understand a proposition means to know what is the case if it is true. (One can understand it, therefore, without knowing whether it is true.) It is understood by anyone who understands its constituents."

The research involved in this paper was begun between September and December of 2010, when I held a residential Fellowship at the Flemish Institute for Advanced Studies (VLAC) of the Royal Flemish Academy of Belgium for Science and the Arts, on a project with Dr A. A. Rini, called 'Flight from Intension', investigating early attempts to produce a semantics for modal logic without using any intensional entities. The research is also supported by a Marsden grant with E. D. Mares and A. A. Rini on a natural history of necessity.

The truth-conditional theory of meaning was crucial to the Vienna Circle. In particular, you need to be able to provide truth-conditions *in advance* of whether you judge a statement to be true or false. This is the point that Carnap seems to have felt Tarski and Quine never appreciated:

I discovered that in these questions, even though my thinking on semantics had originally started from Tarski's ideas, a clear discrepancy existed between my position and that of Tarski and Quine, who rejected the sharp distinction I wished to make between logical and factual truth. (Carnap, 1963, p. 36)

In his autobiography (Carnap 1963, p. 62) Carnap makes the link with modality explicit:

After defining semantical concepts like logical truth and related ones, I proposed to interpret the modalities as those properties of propositions which correspond to certain semantical properties of the sentences expressing the propositions. For example, a proposition is logically necessary if and only if a sentence expressing it is logically true.

On that basis, Carnap proposed a semantics for modal logic.[2] Not all philosophers were sympathetic to this project. In Quine 1953b, W. V. Quine distinguishes three grades of what he calls 'modal involvement'. The first grade he regards as innocuous. It is no more than the meta-linguistic attribution of validity to a formula of non-modal logic. In the second grade we say that where α is any sentence then $\Box\alpha$ is true iff α itself is valid – or logically true. On pp. 166–169 Quine argues that while such a procedure is possible it is unilluminating and misleading. The third grade allows free variables to occur in the scope of modal operators, and it is this grade that Quine finds objectionable. The aim of this chapter is to look at Carnap's attempts to produce a modal logic which can survive Quine's criticisms.

Wittgenstein's theory does seem to presuppose metaphysics. But one point which becomes clear very quickly to anyone working on the *Tractatus* is that the way to see what that metaphysics is to look at

[2] See Carnap 1946, Carnap 1947 pp, 173–177 and Carnap 1963 pp. 889–905. Modal logic is also discussed on pp. 250–260 of Carnap 1937. It does seem that Carnap felt that you could only make sense of the logical modalities by understanding them as 'quasi-syntactical' (1937, pp. 246 and 250).

the structure of what Russell called a 'logically perfect language', which reflects the structure of the world (Russell 1918, p. 197). So suppose that we *begin* with this language, and instead of asking whether it reflects the structure of the world, we ask whether it is a useful language for describing the world. From Carnap's perspective (Carnap 1950), one might describe the matter in such a way as this: Given a language L we may ask whether L is adequate, or perhaps merely useful, for describing the world as we experience it. It is incoherent to speak about what the world in itself is like without presupposing that one is describing it. What makes L a Carnapian equivalent of a logically perfect language would be that every possible world can be described by what he called a 'state description'. But if we are describing the world, we are already using a language, and then one might say that what passes for 'metaphysics' is in reality no more than the description of a language.

2. Carnap's Non-Modal Predicate Logic

Although Carnap's modal propositional logic is not without interest (see Cresswell 2013) the purpose of this chapter is to look at his modal predicate logic. In preparation for this I shall first set out the fundamentals of Carnap's non-modal predicate logic, based on the 1946 article. One of the points at issue between Quine and Carnap arises when we introduce definite descriptions into the language. In my view the question of how to interpret descriptions can only be addressed if we are already clear about the semantics of a predicate logic without them, and that it only muddies the waters if they are introduced too soon. I shall therefore initially concentrate on the 1946 paper, where these are not involved. In setting out the semantics found in Carnap's 1946 paper for first-order (non-modal) predicate logic, I shall avoid his use of German letters, which make his work difficult for current readers, and will use a more standard labelling of his axioms and rules. I shall base my presentation of Carnap's logic on $\bot$ and $\supset$ rather than on Carnap's ~, ∨ and '.'.[3] For quantifiers I use $\forall x$ in place of (x), and $\exists x$ in place of $(\exists x)$. Carnap's language of FL (first order logic) contains a denumerable infinity of individual variables and constants, which can be referred

[3] The other truth functions are all definable in terms of $\supset$ and $\bot$. in particular $\alpha \vee \beta$ may be defined as $(\alpha \supset \beta) \supset \beta$, $\sim\alpha$ as $\alpha \supset \bot$ and $\alpha \wedge \beta$ as $\sim(\sim\alpha \vee \sim\beta)$. Carnap takes ~, ∨ and '.' (∧) as primitive, but for the points made here these differences do not matter.

to simply as 'variables' or 'constants'. Carnap uses the term 'matrix' for a (well-formed) formula (wff), and the term 'sentence' for closed wff, i.e., wff with no free variables, When it comes to interpreting FL Carnap assumes denumerable domains. For this reason he is able to represent a domain by a denumerable set of individual constants – a distinct constant for each individual. The presence of constants enables atomic sentences such as ϕab. Truth and falsity in FL are relative to a *state-description* (Carnap 1946, p. 50, and 1947, p. 9). Formally, a state-description s for a language L is a set of atomic sentences or their negations, such that for every n-place predicate ϕ of L, and every n-tuple of constants $a_1,\ldots,a_n$ either $\phi a_1\ldots a_n \in s$ or $\sim\phi a_1\ldots a_n \in s$. This way of talking highlights the fact that state-descriptions represent possible worlds.[4] If s is any state description, I write $s \models \alpha$ to mean that α is true in s and $s \mathrel{=\!\!|} \alpha$ to mean that α is false in s. The truth of a sentence α in a state-description s may be defined formally by the following valuation rules:

[Vϕ] $s \models \phi a_1\ldots a_n$ iff $\phi a_1\ldots a_n \in s$

[V=] $s \models a = b$ iff a and b are the same constant

[V$\bot$] $s \mathrel{=\!\!|} \bot$

[V$\supset$] $s \models \alpha \supset \beta$ iff $s \mathrel{=\!\!|} \alpha$ or $s \models \beta$

[V$\forall$] $s \models \forall x\alpha$ iff $s \models \alpha[a/x]$ for every constant a, where $\alpha[a/x]$ is α with a replacing every free x.

$s \models \alpha$ means that α is true in s. Carnap does not speak of truth in a state-description, but speaks of the *range* of a sentence – which is a set of state-descriptions. Thus, we may either say that $s \models \alpha$, or say that s is in the range of α. Where we say that $s \models \alpha$, Carnap will say that s is in the range of α. Carnap's 'rules of ranges' (p. 50) then become our truth-valuation rules [Vϕ]–[V$\forall$], allowing for our adoption of $\bot$ and $\supset$ as the primitive truth functional operators.

3. Carnap's 1946 Modal Predicate Logic

We may extend L to a modal language L_M. L_M is simply the modal predicate language which extends L by the addition of the (monadic)

[4] Carnap's phrase in 1946, p. 50, is that a state description 'represents a possible specific state of affairs'. There is of course a serious problem in this way of looking at matters. If a state-description represents a possible world, how can we tell whether the state-description for one language represents the same possible world as the state-description for another language?

necessity operator $\Box$, and the formation rule that if α is a wff then so is $\Box\alpha$. The idea behind Carnap's account of modality is that $\Box$ (which he writes as *N*) is to be interpreted in such a way that $\Box\alpha$ is to be true iff α is *valid* – or, as Carnap puts it, iff α is L-true. We shall make this precise as follows. Where *s* is a state-description [Vϕ]–[V$\forall$] shew how to extend $\models$ to all LPC wff. To extend $\models$ to all wff of L_M we add the condition

[V$\Box_{MFL}$] $s \models \Box\alpha$ iff $s' \models \alpha$ for every state-description s'.

Say that a wff is *C-valid* iff it is true in every state-description, when $\models$ satisfies [Vϕ]–[V$\forall$] and [V$\Box_{MFL}$]. One result is that if α is S5-valid then α is C-valid. However, (quantified) S5 is not complete for C-validity. This is because, where α is a wff of FL – i.e., a non-modal wff – $\sim\Box\alpha$ is C-valid iff α is not valid. This feature of Carnap's modal predicate logic makes it unaxiomatisable.[5]

4. Analyticity

Part at least of Quine's objections to modal logic stems from his rejection of the analytic/synthetic distinction.[6] As we have seen, this rejection was, in Carnap's mind, a crucial point of difference between him and Quine and Tarski. For Carnap's views on analyticity we can look at a number of sources. All these postdate Carnap 1946 and 1947, and, except for Carnap 1963, do not deal with modal logic. For that reason the application of the later ideas to modal logic must be somewhat conjectural. Carnap 1963 replies to Quine's attack on analyticity. This reply occurs just after the section (pp. 889–899) where he has been expounding his latest conception of the logic of modalities. While Quine can accept that Carnap may have defined a formal concept which he calls 'analyticity in *L*', where *L* is a formal language, Quine cannot accept that there is any intuitive natural language concept which can be the explicandum (Carnap 1963, p. 919). Carnap actually agrees that it is necessary to provide such if his account of analyticity is to have any genuine content, and that it is legitimate for Quine to demand that

[5] For more on the details of the completeness of Carnap's predicate logic see Cresswell 2014.

[6] For this rejection see especially Quine 1953c.

there be an empirical criterion for intensional notions in natural language. But he claims that it is not difficult to shew that one can be given. He alludes to the fact that empirical linguists have little trouble in making claims about the meanings of words and phrases:

It seemed rather plausible to me from the beginning that there should be an empirical criterion for the concept of the meaning of a word or phrase in view of the fact that linguists traditionally determine empirically the meanings, meaning differences, and shifts of meanings of words, and that with respect to these determinations they reach a measure of agreement among themselves which is often considerably higher than that reached for results in most of the other fields of the social sciences. Quine's arguments to the effect that the lexicographers actually have no criterion for their determinations did not seem at all convincing to me. (Carnap 1963, p. 919f)

He follows this up on p. 920 of Carnap 1963 by addressing the blackness of ravens:

Let us suppose that two linguists study the natural language *L* as used by the person X. Let us suppose that *L* consists of some English words and English sentences, among them the following sentence:

(S_1) 'All ravens are black'.

We assume that the two linguists agree on the basis of previous experiments that X uses the words 'all' and 'are' in the ordinary sense, and that X has repeatedly affirmed the sentence S_1; and hence presumably regards it as true. Now the first linguist states the following hypothesis:

(5) 'The sentence S_1 is analytic in language *L* for person X.'.

The other linguist denies this hypothesis. In order to obtain evidence relevant for (5), the linguists say to X: 'Mr. Smith told us that he had found a raven which is not black but white, and that he will show it to you tomorrow. Will you then revoke your assertion of S_1?' Let us consider the following two of many possible responses by X:

(6) 'I would never have believed that there are white ravens; and I still do not believe it until I see one myself. In that case I shall, of course, have to revoke my assertion'.

(7) 'There cannot be white ravens. If a bird is not black, then I just would not call it a raven. If Mr. Smith says that his raven is not black, then (assuming that he is not lying or joking) his use either of the word 'raven' or of the word 'black' must be different from my use'.

It seems obvious to me that a response like (6) would be disconfirming evidence for hypothesis (5), while a response like (7) would be confirming evidence for it. Thus it is clear that (5) is an empirical hypothesis which can be tested by observations of the speaking behavior of X. (Carnap 1963, p. 920, example numbers are Carnap's)

For Carnap the most important feature of analytic sentences is that they are non-contingent – they have no empirical content. So what these conditions shew is that we are talking about a class of sentences whose lack of empirical content depends on the meaning of particular words. It is the way 'raven' and 'black' are being used which determines whether S_1 is contingent.

One point which seems clear from Carnap's discussion of his examples is that he takes it to be obvious that it makes sense to speak of how a person *would* apply a word in circumstances different from the actual circumstances. That is to say, Carnap is accepting that modal locutions can have definite truth values. This acceptance does not commit him to any *metaphysical* account of modality, but it does distinguish his position from that of Quine. It is for this reason that I began the chapter by stressing the importance Carnap attached to the difference between him and Quine in this respect. Carnap may not have an answer to where necessity comes from, but he certainly accepts that it is a phenomenon which can be formally studied and formally structured.

5. Meaning Postulates

We had better not attribute to Carnap the acceptance of what he would undoubtedly regard as dubious entities such as possible worlds – or at least not without some way of shewing how to reconcile such entities with the views on ontology found in Carnap 1950.[7] The place to look would seem to be in Carnap 1952, where the notion of a meaning

[7] On pp. 64–68 of Carnap 1947, Carnap discuses a view of C. I. Lewis which uses non-actual individuals. (Unfortunately it is a view which links non-actual possibilia with impossibilia as if they were susceptible of the same treatment.) Characteristically Carnap concedes that such a view is probably capable of a consistent formalisation, so the question is solely about its utility. He argues that his own view handles the required cases, and so a view such as Lewis's is not necessary, and because of its strangeness is therefore probably not useful. The closest he comes to the argument that 'there *are* no such things' is when he refers to the intuition that an *X* and an actual *X* are no different.

postulate is introduced. Meaning postulates do not appear in Carnap 1946 or 1947 so we must do some conjectural reconstruction.

How can meaning postulates lead to a modal semantics? In Carnap 1952 (p. 222) Carnap is concerned to allow a logic in which sentences such as

(1) If Jack is a bachelor, then Jack is unmarried

or

(2) All ravens are black

can be displayed as non-contingent. Since neither (1) nor (2) will come out as L-true according to Carnap's usual definition – since they will not be true in all state-descriptions – a theorist who wants either of them to be analytic needs to add it as a *meaning postulate*. With R for 'is a raven' and B for 'is black' we get for (2),

(3) $\forall x(Rx \supset Bx)$

Carnap is concerned to argue that it is up to the theorist to stipulate which sentences are to be regarded as analytic. So it is up to the theorist whether to include (3) or not. The theorist who thinks (2) to be a matter of meaning will, while the theorist who thinks it is not will not. Imagine a theorist who has a language with predicates referring to various ornithological features and is concerned to establish which of the interpretations for this language actually match up with the facts. The theorist knows in advance that all ravens are black, so there is no point in considering interpretations which do not satisfy (3). Adopting (3) could then help cut down the search time. But this in no way makes (3) *necessary* in any sense other than that its truth is not at issue. That is why Carnap had to take up Quine's challenge to give an empirical criterion for analyticity, because he has to be able to distinguish a theorist who stipulates (3) to be *analytic* from one who requires it merely to be true. Whether Carnap has adequately met Quine's challenge is not my question. It is enough that we can see how he thinks that he has. In a sense this is simply the theorist stipulating how the language is to be understood. In a formal language a meaning postulate will be a wff of that language. In his paper Carnap is not explicit about just which formal languages he has in mind, though none of his examples involves modal languages, and indeed his aim of explaining modality in non-modal terms would make this plausible.[8] Given a set of meaning postulates Carnap (p. 225) describes a wff as L-true with respect to that set of meaning postulates iff it is 'L-implied' by the conjunction of the meaning postulates, which is to say that it is true in all state-descriptions which satisfy the meaning postulates.[9] Let Π be a set of (non-modal) LPC sentences, and call a state-description a Π-*description* if every wff in Π is true in it. Where s is any state-description [$V\Box_{MFL}$] is replaced by

(4) $s \models \Box\alpha$ iff $s' \models \alpha$ for every Π-description

Say that α is CΠ-valid iff $s \models \alpha$ for every Π-description. Obviously if α is a member of Π then $\Box\alpha$ is Π-valid. Say that α is CM-valid iff α is CΠ-valid for every set Π of meaning postulates. It is an open question (as far as I am aware) whether the standard S5 modal predicate logic is complete for CM-validity.

6. *De Re* Modality

Although Quine objects to analyticity per se, his harshest criticism of modal logic is reserved for *de re* formulae. What can be said on the basis of Carnap 1946 to answer Quine's complaints about 9 and the number of planets?[10] The example occurs in Quine 1943, pp. 119–121, but Quine repeats versions of this argument many times, most famously perhaps in Quine 1947, 1953a, 1953b and 1960. It goes like this:

(5) 9 is necessarily greater than 7
(6) The number of planets = 9

[8] For problems which arise if modal operators are allowed in meaning postulates see Cresswell 2013, p. 62f.

[9] This does have the consequence that not all state-descriptions are permitted, so that the state-descriptions no longer have the metaphysical power of the elementary propositions of the *Tractatus*. This links with the problem mentioned in footnote 4. It also means that, unlike Wittgenstein, Carnap cannot help himself to an atomistic account of logical possibility. In turn that means that he has to take modality as primitive. As I suggest in this chapter, maybe this is something he would be comfortable with.

[10] Recall that when Quine was writing Pluto counted as a planet and there were indeed held to be nine of them. We will assume that in this discussion.

therefore

(7) The number of planets is necessarily greater than 7

Quine's original argument was couched in terms of definite descriptions, and Smullyan 1948 shewed that paying careful attention to scope provides an answer to arguments put this way.[11] Carnap 1946 does not introduce definite descriptions into his language, and I shall present the argument in a formalisation which only uses the resources of that paper. It will also simplify the discussion to use a predicate O, where Ox means 'x is odd', rather than the complex predicate 'is greater than 7'. This will enable us to avoid reference to '7', which is of no relevance to Quine's argument. P means 'is the number of the planets', so that Px means 'there are x-many planets'. With this in mind we may take '9' to be an individual constant, and use O and P to express (5) and (6) by

(8) $\Box O9$

(9) $\exists x(Px \wedge x = 9)$

One could account for (8) by adding $O9$ as a meaning postulate, though from some remarks on p. 201 of Carnap 1947 it seems that Carnap might have regarded both O and 9 as complex expressions defined by the resources of the Frege/Russell account of the natural numbers and their arithmetical properties. It also seems that he might have treated the numbers as higher-order entities referred to by higher-order expressions. If so then the necessity of arithmetical truths such as (8) would derive from their analysis into logical truths. In our exposition we shall take the numerals as individual constants, and assume somehow that $O9$ is a logical truth, true in every state-description, and that therefore (8) is true.

[11] Note that Smullyan's invocation of scope assumes that we can make sense of quantifying in. Quine 1969, p. 338, is correct that if Smullyan thought that this *justifies* quantifying in, then Smullyan's argument is question-begging. But of course Smullyan's aim was simply to answer Quine's arguments *against* quantifying in. Quine himself in fact sums this up by saying:

> My objection to quantifying in to non-substitutive positions dates from 1942. In response Arthur Smullyan invoked Russell's distinction of scopes of descriptions to show that the failure of substitutivity on the part of descriptions is no valid objection to quantification. (*loc cit.*)

In this formalisation we are ignoring the claim that the description 'the number of the planets' is intended to claim that there is only one such number. So much for the premises. But what about the conclusion? The problem is where to put the □. There are at least three possibilities:

(10) $\Box\exists x(Px \wedge Ox)$

(11) $\exists x\Box(Px \wedge Ox)$

(12) $\exists x(Px \wedge \Box Ox)$

We first show that (10) and (11) do not follow from (8) and (9). Since (8) is true in every state-description it is sufficient to find a state-description s which satisfies (9) but satisfies neither (10) nor (11). Let s and s' be two state-descriptions in which in s, P is true only of 9, and in s' P is true only of 6. s clearly satisfies (9), since both $P9$ and $9 = 9$ will be in s. But if c is odd, then Pc is not in s' and if c is not odd, then Oc is not in s', and so $\exists x(Px \wedge Ox)$ fails in s', and so (10) fails in s. (A similar argument works for (11).) In contrast to (10) and (11), (12) does follow from (8) and (9). But there is no problem here, since (12) says that there is a necessarily odd number which is such that there happen to be that many planets. And this is true, because 9 is necessarily odd, and there are 9 planets. All of this should make clear how the phenomenon which upset Quine can be presented in the formal language of the 1946 article. Quine thought that the only answer to his objections was to espouse 'essentialism'. He also characterised essentialism as 'the adoption of an asymmetrical attitude to different ways of specifying the same object' (Quine 1969, p. 339). One can see a difference between being odd and being the number of the planets. Let Ox mean that x is odd, and let Px mean that there are x many planets. The intuition at work here is that the predicate O satisfies the condition

(13) $\forall x(Ox \supset \Box Ox)$

but the predicate P does *not* satisfy the condition

(14) $\forall x(Px \supset \Box Px)$.

In the argument (8)(9)(12) we need the premise $\Box O9$. This is a consequence of the principle (13). There is no analogous principle for P. That is to say, the formula

(15) $\exists x(Ox \wedge \Box Px)$

is not true according to the 1946 semantics, since in fact $\Box Pa$ is not true for any constant a, since it is not true for 9, and even Pa is not true for any *a other* than 9. So (15) is false in the state-description which represents the actual world. There is certainly an asymmetry here, but it is an asymmetry between two kinds of predicate, and says nothing about how the subject is identified. Furthermore, it is the *very same number* which satisfies $\Box Ox$ and Px, but does not satisfy $\Box Px$, That a number x is odd or even is not a contingent matter, but whether or not there are x planets is. Quine may disagree, by claiming that there is no coherent distinction between predicates such as O which hold or fail to hold by necessity, and predicates such as P which hold or fail to hold contingently; but that cannot be used as an *argument* against essentialism. If, *pace* Quine, we assume that there is such a difference between a number's being odd or even, and there being that many planets, then we have a phenomenon to account for, and our present task is to consider how Carnap's later views might account for it.

7. Meaning Postulates and *De Re* Modality

I have suggested that (8) could be guaranteed by postulating $O9$ as a meaning postulate. So let us see how the method of meaning postulates can help in addressing Quine's problem about *de re* formulae. Reflect carefully on what a meaning postulate does. In adopting (3) as a meaning postulate we are stating that it is *analytic* – that is to say, it is in part constitutive of the meanings of the predicates R and B that they are so related that (3) is true. This is indicated by the response attributed to X in the quotation provided above:

> There cannot be white ravens. If a bird is not black, then I just would not call it a raven. If Mr. Smith says that his raven is not black, then (assuming that he is not lying or joking) his use either of the word 'raven' or of the word 'black' must be different from my use.

The claim is that a sentence like

(16) All ravens are black –

whether regarded as a sentence in English, as in (16), or an interpreted sentence of LPC, as in (3) – is analytic, is a legitimate tool in partially specifying the meanings of the terms in it. That is not to say that (16) or (3) is *itself a* statement about language. It is the metalinguistic claim that they are analytic which makes a statement about language.

Carnap's examples of bachelors and ravens concern facts which constrain meanings in terms of relations between two predicates. But one could argue that the crucial fact about meanings is that they relate words and *things*. And indeed, if we look at p. 238 of Carnap 1955 we see that one of the options for eliciting a judgement about whether anything is a 'pferd' is for the theorist to

> point toward a thing and then describe the intended modification in words, e.g.: 'a thing like this one but having one horn in the middle of the forehead'. Or, finally, he might just point to a picture representing a unicorn. Then he asks Karl whether he is willing to apply the word 'pferd' to a thing of this kind.

This seems to imply that one can constrain what is meant by a predicate by specifying not only what things it does apply to, but also what things it would apply to.

As we have seen, in Carnap's 1946 predicate logic members of the domain are uniquely denoted by individual constants. While this may appear to make it a linguistic matter what individual is pointed to, we can, I think, easily construe a constant as merely standing in for an individual. Quine characterises essentialism as privileging certain ways of referring to an individual, and he might be supposing that the use of an individual constant is one of these privileged ways; but there seems no problem in construing a language-user as referring to a particular *thing*, and as claiming that a predicate ϕ is being used in such a way that it has to apply to that *thing*.

Let us apply this to (8). There are two ways of requiring (8) to be true. One is to assume, as I suggested Carnap may have wanted to do, that (8) is guaranteed in some way by the Frege/Russell account of number. The other is to add O9 as a meaning postulate. For if O9 is one of the meaning postulates, then $\Box$O9, and therefore a *de re* wff $\exists x \Box Ox$, will be true in every state-description which respects those postulates. What this second way of handling the matter amounts to could be held to reflect some such position as this:

By 'odd' I *mean* a predicate, which, among other things, applies to 9. If Mr. Smith says that 9 is not odd, then (assuming that he is not lying or joking) his use of the word 'odd' must be different from my use.

The treatment of the oddness of 9 by meaning postulates, unlike the procedure of accounting for it in terms of the Frege/Russell analysis of number, can apply to any predicate in which its application to an individual follows from the meaning of the predicate alone. In this treatment, while the theorist may adopt O9, the theorist will, presumably, not wish to adopt *P*9, since the theorist will, presumably, not take it to be part of the meaning of 'is the number of planets', that there be any particular number of planets. That is to say, in the case of the predicate *P*, even if there should be n planets, so that n in fact satisfies *P*, the theorist is not going to say, 'If someone says that there are not n planets that person means something different from what I do by the predicate *P*.'

In treating

(17) 9 is odd

as a meaning postulate we are taking it to be analytic, and in doing so we are perhaps stretching the meaning of 'analytic', but we are still claiming that its truth is in some sense a consequence of the meaning of the word 'odd'. In this particular case the constraint involves just that one predicate, but of course there are many predicates such as *P* of which there may be nothing to which they apply solely in virtue of their meaning. In fact, cases such as (17) are not Carnap's standard examples of meaning postulates. In examples such as (1) and (2) the meaning postulate imposes a constraint on a connection between several predicates. One can see this as adopting a holistic approach to meaning. (2) would say that you so use the words 'raven' and 'black' that when you apply 'raven' to something you will also apply 'black' to it. Carnap's description of what it is to claim that they are analytic ensures that it is a consequence of the way these predicates *R* and *B* are being used that any time there is an individual *a* to which *R* is applied, then *B* can also be applied to *a*, and this is so *whatever the world may happen to be like*. Note that although this has been stated in terms of possible worlds, there is no claim here that such entities as 'worlds' appear in Carnap's ontology. All that can be claimed is what Carnap claimed – that there are non-contingent sentences – sentences whose meaning alone guarantees their truth. So, in place of talking about the acceptance of a formula such as (3), which raises issues about the status of 'logical' symbols such as ∀ or ⊃, we can simply say that, as far as speaker X is concerned, every state-description of L compatible with X's use of 'raven' and 'black' would be such that if *Ra* is in it, so is *Ba*. In other words, X's linguistic usage is such as to disallow any state-description which contains *Ra* but not *Ba*. More generally one could regard a speaker as saying something like this:

The way I use the words of my language is such that thus and such a state description is not ruled out.

Of course the usage of any particular speaker, or even a total linguistic population, will not be sufficient to determine a precise set of allowable state-descriptions; but that is simply the general fact that language use underdetermines the model theory. What *can* be said is that a class of state-descriptions determines an S5-model, in which the class of logically necessary truths is the same throughout the model, and so provides a standard semantics for quantified S5 modal logic, including *de re* formulae.

8. Individual Concepts

Although I suggest that something like what has been said in the last section might enable Carnap's 1952 paper to answer Quine's complaints about *de re* modality, it seems clear that Carnap had not availed himself of it in the 1947 book, and I shall now look at the modal logic presented in Carnap 1947. Carnap 1947, p. 193f cites the argument (5)(6)(7) from Quine 1943 discussed earlier. Carnap does not appear to recognise any potential ambiguity in the conclusion and characterises (7) as false. Carnap does not consider (12), and on p. 194 simply says:

"we obtain the false statement [(7)]"

In Carnap's view the problem with Quine's argument is that it assumes an unrestricted version of what is sometimes called 'Leibniz' Law':

I2 $\forall x \forall y(x = y \supset (\alpha \supset \beta))$, where α and β differ only in that α has free x in 0 or more places where β has free y[12]

In the 1946 paper this law holds in full generality, as does a consequence of it which asserts the necessity of true identities.

LI $\forall x \forall y(x = y \supset \Box x = y)$

Carnap 1947 holds that I2 must be restricted so that neither x nor y occurs free in the scope of a modal operator. In particular the following would be ruled out as an allowable instance of I2:

(18) $x = y \supset (\Box Ox \supset \Box Oy)$

In order to explain how this failure comes about and solve the problems posed by co-referring singular terms, Carnap modifies the semantics of the 1946 paper. The principal difference from the modal logic of the 1946 paper, as Carnap tells us on p. 183, is that the domain of quantification for individual variables now consists of individual concepts, where an individual concept i is a function from state-descriptions to individual constants. Where s is a state-description we shall let i_s denote the constant which is the value of the function i for the state-description s. Carnap is clear that the quantifiers range over *all* individual concepts, not just those expressible in the language.

Using this semantics it is easy to see how (18) can fail. For let x have as its value the individual concept i, which is the function such that i_s is 9 for every state description s, while the value of y is the function j such that, in any state description s, j_s is the individual which is the number of the planets in s, i.e., j_s is the (unique) constant a such that Pa is in s. (Assume that in each state-description there is a unique number, possibly 0, which satisfies P.) We then assume that $x = y$ is true in any

[12] In this schema and in what follows we are allowing formulae with free variables as theorem schemas. This is in line with the revised semantics of Carnap 1947, p. 183f, in which Carnap relativises truth to an assignment of values to the individual variables, and so we may call a formula valid iff it is true in every state description with respect to every assignment (of individual constants) to its variables. State-descriptions, however, still consist of a collection of atomic *sentences*, with no free variables, which includes either α or $\sim\alpha$ for every atomic α.

state-description s iff, where i is the individual concept which is the value of x, and j is the individual concept which is the value of y, then i_s is the same individual constant as j_s. In the present example it happens that, when s is the state-description which represents the actual world, i_s and j_s are indeed the same, for in s there are nine planets, making $x = y$ true at s. Now $\Box Ox$ will be true if Ox is true in every state-description s', which is to say if $i_{s'}$ satisfies O in every s'. Since $i_{s'}$ is 9 in every state description, then $i_{s'}$ does satisfy O in every s', and so $\Box Ox$ is true at s. But suppose s' represents a situation in which there are six planets. Then $j_{s'}$ will be 6 and so Oy will be false in s', and for that reason $\Box Oy$ will be false in s, thus falsifying (18). (It is also easy to see that LI is not valid, since it is easy to have $i_s = j_s$ even though $i \neq j$.)

The difference between the modal semantics of Carnap 1946 and Carnap 1947 is that in the former the only individuals are the genuine individuals, represented by the constants of the language L. In the proof of the invalidity of (18) it is essential that the semantics of identity require that when x is assigned an individual concept i and y is assigned an individual concept j that $x = y$ be true at a state-description s iff i_s and j_s are the same individual. And now we arrive at Quine's complaint (Quine 1953a, p. 152f). It is that Carnap replaces the domain of things as the range of the quantifiers with a domain of individual concepts. Quine then points out that the very same paradoxes arise again at the level of individual concepts. Thus for instance it might be that the individual concept which represents the number of planets in each state-description is identical with the first individual concept introduced on p. 193 of *Meaning and Necessity*. Carnap is alive to Quine's criticism that ordinary individuals have been replaced in his ontology by individual concepts. In essence Carnap's reply to Quine on pp. 198–200 of Carnap 1947 is that if we restrict ourselves to purely extensional contexts then the entities which enter into the semantics are *precisely the same entities* as are the extensions of the intensions involved. What this amounts to is that although the domain of quantification consists of individual concepts, the arguments of the *predicates* are only the genuine individuals. For suppose, as Quine appears to have in mind, we permit predicates which apply to individual concepts. Then suppose that i and j are distinct individual concepts. Let ϕ be a predicate which can apply to individual concepts, and let s be a state-description in which ϕ applies to i but not to j but in which i_s

and j_s are the same individual. We now have two options depending on how = is to be understood. If we take $x = y$ to be true in s when i_s and j_s are the same individual, then if x is assigned i and y is assigned j we would have that $x = y$ and ϕx are both true in s, but ϕy is not. So that even the simplest instance of I2

I2$_\phi$ $x = y \supset (\phi x \supset \phi y)$

fails, and here there are no modal operators involved. The second option is to treat = as expressing genuine identity. That is to say $x = y$ is true only when the individual concept assigned to x is the same individual concept as the one assigned to y. In the example we have been discussing, since i and j are distinct individual concepts, if i is assigned to x and j to y, then $x = y$ will be false. But on this option the full version of I2 becomes valid, even when α and β contain modal operators. This is just another version of Quine's complaint that if an operator expresses *identity* then the terms of a true identity formula must be interchangeable in all contexts.

9. Conclusion

In this chapter I have attempted to shew that Carnap's modal logic has a way of responding to Quine's criticisms if you use features of his later work in ways in which he himself did not. This applies particularly to the use of meaning postulates in accounting for *de re* modality, rather than using Carnap's own device of individual concepts. Why then did Carnap not take the former way out? My conjecture is this. The response I have argued available to Carnap is one which ignores definite descriptions. In dealing with the problems in primitive notation one is in effect opting for a Russellian solution rather than a Fregean solution. A large amount of Carnap 1947 is concerned to defend Carnap's own take – the method of intension and extension – on a broadly Fregean account of the various paradoxes of reference. In fact on p. 35 of 1947 he actually *criticises* Russell's theory of descriptions on the ground that it allows the very scope distinctions which many of us see as its glory. Carnap's distinction between intensions and extensions makes good sense when applied, for example, to sentences and predicates, whose intensions are functions from indices to truth values or sets of n-tuples of individuals. But in the case of individual concepts the values are individuals themselves, and if you are to answer Quine by restricting α and β in I2 to non-modal wff, then you have to accept a domain of quantification which admits individual concepts, yet at the same time restricts the arguments of the predicates to the genuine individuals which are the arguments of those concepts.[13]

[13] A recent critical account of Carnap's use of individual concepts occurs in Williamson 2013, pp. 46–60. See also the discussion in Heylen 2010.

Bibliography

Authors have been referred to by surname and initials except in the case of some historical figures up to the early modern period.

Abaci, U., 2014, "Kant's *Only Possible Argument* and Chignell's Real Harmony," *Kantian Review* 19(1), 1–25.

Abaelard 1970, *Petrus Abaelardus, Dialectica*, 2nd ed., ed. L. M. De Rijk, Assen: Van Gorcum, Tract. II.2, 191–210.

2010, "Petrus Abaelardus, *Glossae Super Peri Hermenias*," in *Corpus Christianorum Continuatio Mediaevalis 206*, ed. K. Jacobi and C. Strub Brepols: Turnhout, cap. 12, 391–433.

Ackrill, J. L., 1963, *Aristotle's Categories and De Interpretatione*, Oxford: Clarendon Press.

Adam, Ch and P. Tannery, 1964–1976, (AT), *Oeuvres de Descartes*, 11 vols., Paris: J. Vrin.

Adams, M. 1973, "Did Ockham Know of Material and Strict Implication," *Franciscan Studies* 33, 5–37.

Adams, R. M., 1994. *Leibniz: Determinist, Theist, Idealist*, Oxford: Oxford University Press.

2000, "God, Possibility and Kant," *Faith and Philosophy* 17, 425–440.

Adams R. M., and N. Kretzmann, 1983, *Ockham, William, Predestination, God's Foreknowledge and Future Contingents*, tr. with intro by M. M. Adams and N. Kretzmann 2nd edition, Hackett Classics.

Adickes, E., 1924, *Kant und das Ding an Sich*, Berlin: Pan Verlag.

Alexander of Aphrodisias, 1883, *On Aristotle's Prior Analytics*, ed. M. Wallies, in Diels, 1882–1909, Vol. 2.1.

1892, *On Fate*. ed. I. Bruns, *Supplementum Aristotelicum*, Vol. 2.2, Berlin: Reimer.

1991, *On Aristotle's Topics*. ed. M. Wallies, in Diels 1882–1909, Vol. 2.2.

Allais, L., 2004, "Kant's 'One World': Interpreting Transcendental Idealism," *British Journal for the History of Philosophy* 12, 655–684.

Allison, H., 1983, *Kant's Transcendental Idealism*, New Haven, CT: Yale University Press. (2004). Revised and Enlarged Edition.

1994, "Causality and Causal Laws in Kant: A Critique of Michael Friedman" in Parrini, P. (1994), 291–307.

Allwein, G. and Barwise, J., ed., 1996, *Logical Reasoning with Diagrams*, Oxford University Press.

Ameriks, K., 1982, "Recent Work on Kant's Theoretical Philosophy," *The Philosophical Quarterly* 19, 1–24.

2003, *Interpreting Kant's Three Critiques*. Oxford: Oxford University Press.

Ammonius, 1897, *Ammonius in Aristotelis De Interpretatione Commentarius*, ed. A. Busse, in Diels, 1882–1909, Vol. 4.5.

Anderson, A. R., and N. D. Belnap, 1975, *Entailment: Logic of Relevance and Necessity*, Vol. I. Princeton University Press.

Anderson, R. L., 2004, "It Adds Up After All: Kant's Philosophy of Arithmetic in Light of the Traditional Logic," *Philosophy and Phenomenological Research* 69, 501–540.

Angelelli, I., 1979, "The Aristotelian Modal Syllogistic in Modern Modal Logic," in *Konstruktionen versus Positionen II (P. Lorenzen zum 60. Geburtstag)*, ed. K. Lorenz, Berlin: de Gruyter, 176–215.

Anselm of Canterbury, 1936, "Ein neues unvollendetes Werk des hl. Anselm von Canterbury," *Beiträge zur Geschichte der Philosophie und Theologie des Mittelalters* 33(3), 23–48.

1968, *Anselmi Cantuariensis Archiepiscopi Opera Omnia*, ed. F. S. Schmitt, Stuttgart: Friedrich Fromann Verlag.

Anstey, P. R., 2011, *John Locke and Natural Philosophy*, Oxford: Oxford University Press.

(forthcoming) "Locke and non-propositional knowledge."

Aquila, R., 1979, ""Things in Themselves: Intentionality and Reality in Kant," *Archiv Fur Geschichte der Philosophie* 61, 293–307.

Aristotle, 1985, *The Complete Works of Aristotle: The Revised Oxford Translation*, ed. Jonathan Barnes, 2 vols., Princeton, NJ: Princeton University Press.

1994, *Posterior Analytics*, Translated with an Introduction and Commentary by Jonathan Barnes, Oxford, Clarendon Press, 2nd ed.

1998, *Aristotle, Metaphysics, trans*. and with an introduction and commentary by Stephen Makin, Clarendon Aristotle Series Oxford: Clarendon Press.

2009, *Aristote, Topiques tome I, livres I-IV* ed. Jacques Brunschwig, Paris: Les belles lettres.

Arnauld, A. and P. Nicole, 1996, *Logic or the Art of Thinking*, trans. J. V. Buroker, Cambridge University Press.

Arnim, J. von, *Stoicorum Veterum Fragmenta*, Vols. I-IV. Leipzig: Teubner. 1902.

Ashworth, E. J., 1974, *Language and Logic in the Post-Medieval Period*, Dordrecht: D. Reidel.

Astroh, M., 2001, "Petrus Abaelardus on Modalities de re and de dicto," *Potentialität und Possibilität: Modalaussagen in der Geschichte der Metaphysik*, ed. T. Buchheim, Stuttgart: Frommann-Holzboog, 79–95.

Averroes (Ibn Rușd), 1562. Quaesita octo in librum Priorum, in Aristotelis omnia quae extant opera – Averrois Cordubensis in ea opera omnes, qui ad haec usque tempora peruenere, commentarii (Venetiis apud Iunctas).

1954, *Tah al-tah: The Incoherence of the "Incoherence,"* trans. Simon Van Den Bergh, 2 vols., Oxford University Press.

Avicenna (Ibn Sīnā), 1977, *Liber de philosophia prima sive scientia divina* ed. S. van Riet, Louvain: E. Peters.

1984, *Remarks and Admonitions* [*Pointers and Reminders*] translated from the original Arabic with an introduction and notes by Shams Constantine Inati, Toronto: Pontifical Institute of Mediaeval Studies.

2004, *The Metaphysics of the Healing*, trans. Michael Marmura, Provo, Utah: Brigham Young University Press.

2011, *Deliverance Logic*, trans. Asad Q. Ahmed (Oxford University Press).

Bacon, F., 2004, *The Instauratio magna Part II: Novum organum and Associated Texts, Oxford Francis Bacon*, trans. and ed. Graham Rees, Vol. XI, Oxford: Clarendon Press.

Barreau, H., 1978, "Cléanthe et Chrysippe face au maitre-argument de Diodore," in Brunschwig 1978, 20–41.

Barrow, I., 1734, *The Usefulness of Mathematical Learning*, trans. John Kirby, London: Stephen Austen.

Becker, A., 1933, *Die Aristotelische Theorie der Möglichkeitsschlüsse*, Berlin: Junker und Dünnhaupt.

Becker, O., 1930, "Zur Logic der Modalitäten," *Jahrbuch fur Philosophie und phänomenologische Forschung* 11, 497–548.

Beeley, P., 2002. "Leibniz on Wachter's *Elucidarius cabalisticus*: A Critical Edition of the So-Called 'Refutation of Spinoza.'" *The Leibniz Review* 12, 1–14.

Beere, J., 2009, *Doing and Being: An Interpretation of Aristotle's Metaphysics Theta*, Oxford University Press [Oxford Aristotle Series].

Beiser, F., 2002, *German Idealism: The Struggle against Subjectivism 1781–1801*, Cambridge, MA: Harvard University Press.

Bennett, J. F., 2000, "Infallibility and Modal Knowledge in Some Early Modern Philosophers," in *Mathematics and Necessity: Essays in the History of Philosophy*, ed. Timothy Smiley, Oxford: Oxford University Press, 139–166.

Bird, G., 1962, *Kant's Theory of Knowledge*, London: Routledge & Kegan Paul.

Blackburn P., and K. Jørgensen, 2012, "Indexical Hybrid Tense Logic," in *Advances in Modal Logic*, London: King's College, 144–160.

Bluck, R. S., 1963, "On the Interpretation of Aristotle, *De Interpretatione* 12–13," *Classical Quarterly* 13, 214–222.
Blumenfeld, D., 1988. "Freedom, Contingency, and Things Possible in Themselves." *Philosophy and Phenomenological Research* 49(1): 81–101. doi:10.2307/2107993.
Blundeville, T., 1617, *The Arte of Logick*, London [1st edition 1599].
Bobzien, S., 1986, *Die stoische Modallogik*. Königshausen & Neumann.
1993, "Chrysippus' modal logic and its relation to Philo and Diodorus," in K. Döring and Th. Ebert (eds) *Dialektiker und Stoiker*, Stuttgart, 63–84.
1999, "Logic: The Megarics, the Stoics," in *The Cambridge History of Hellenistic Philosophy*, ed. K. ALgra, J. Barnes. J. Mansfeld, M. Schofield, Cambridge: Cambridge University Press, 83–157.
2003, "Stoic Logic," in *The Cambridge Companion to the Stoics*, ed. B. Inwood, Cambridge University Press, 85–123.
Bochenski, J. M., 1951, *Ancient Formal Logic*, Amsterdam: North-Holland.
Boethius, 1877, *Commentarii in Librum Aristotelis Peri Hermenias*, ed. C. Meiser, Leipzig: Teubner.
1969, *De Hypotheticis Syllogismis*, ed. L. Obertello, Brescia: Paidea.
2008, *De Syllogismo Categorico*, ed. C. Thomsen Thörnqvist, Gothenburg: University of Gothenburg.
Bonitz, H., 1870, *Index Aristotelicus*, Berlin: Reimer.
Boyle, R., 1999–2000, *Works: The Works of Robert Boyle*, ed. Michael Hunter and Edward B. Davis, 14 vols., London: Pickering and Chatto.
2006, *Boyle on Atheism*, transcribed by and ed. J. J. MacIntosh, Toronto: University of Toronto Press.
Brading, K., 2011, "On Composite Systems: Descartes, Newton, and the Law-Constitutive Approach," in *Vanishing Matter and the Laws of Motion: Descartes and Beyond*, ed. Dana Jalobeanu and P. R. Anstey, New York: Routledge, 130–152.
2012, "Newton's Law-Constitutive Approach to Bodies: A Response to Descartes," in *Interpreting Newton: Critical Essays*, ed. Andrew Janiak and Eric Schliesser, Cambridge: Cambridge University Press, 13–32.
Brandon, E. P., 1978, "Hintikka on ἀκολουθεῖν," *Phronesis* 23, 173–178.
Braüner, T., 1998, "Peircean Graphs for the Modal Logic S5," in *Conceptual Structures: Theory, Tools and Applications*, ed. Marie-Laure Mugnier and Michel Chein, Berlin and Heidelberg: Springer, 255–269.
2006, "Hybrid Logic," The *Stanford Encyclopedia of Philosophy*, http://plato.stanford.edu/entries/logic-hybrid/.
Brennan, T., 1994, "Two Modal Theses in the Second Half of *Metaphysics* Theta 4." *Phronesis* 39, 160–173.
Brown, G., 1987. "Compossibility, Harmony, and Perfection in Leibniz." *The Philosophical Review* 96(2), 173–203.

Brunschwig, J., 1969, "La proposition particulière et les preuves de non-concluance chez Aristote," *Cahiers pour l'Analyse* 10, 3–26.

1978, *Les stoïciens et leur logique*. Vrin. Second edition 2006.

Buddensiek, F., 1994, *Die Modallogik des Aristoteles in den Analytica Priora A*, Hildesheim: Olms.

Burch, R. W., 1994, "Game-Theoretical Semantics for Peirce's Existential Graphs," *Synthese* 99, 361–375.

Burnyeat, M. F., 1981, "Aristotle on Understanding Knowledge," in *Aristotle on Science: The Posterior Analytics*, Proceedings of the Eighth Symposium Aristotelicum held in Pisa from September 2 to 15, 1978, E. Berti, ed., Padua: Antenore, 97-139.

Carnap, R., 1937, *The Logical Syntax of Language*, London: Kegan Paul, Trench Truber.

1946, "Modalities and Quantification," *The Journal of Symbolic Logic* 11, 33–64.

1947, *Meaning and Necessity*, Chicago: University of Chicago Press. Second edition 1956, references are to the second edition.

1950, "Empiricism, Semantics and Ontology," *Revue Intern de Phil* 4, 20–40. Reprinted in the second edition of Carnap 1947, 205–221. Page references are to this reprint.

1952, "Meaning Postulates," *Philosophical Studies* 3, 65–73. Reprinted in the second edition of Carnap 1947, 222–229. Page references are to this reprint.

1955, "Meaning and Synonymy in Natural Languages," *Philosophical Studies* 7. Reprinted in the second edition of Carnap 1947, 233–250. Page references are to this reprint.

1963, "Carnap's Intellectual Biography," *The Philosophy of Rudolf Carnap*, ed. P. A. Schilp, La Salle, IL: Open Court, 3–84.

Chignell, A., 2007a, "Belief in Kant," *The Philosophical Review* 116(3), 323–360.

2007b, "Kant's Concepts of Justification," *Noûs* 41(2), 33–63.

2009, "Kant, Modality, and the Most Real Being," *Archiv für Geschichte der Philosophie* 91(2), 157–192.

2010, "Real Repugnance and Belief about Things in Themselves: A Problem and Kant's Three Solutions," in *Kant's Moral Metaphysics*, ed. J. Krueger and B. Lipscomb, Berlin: Walter DeGruyter.

2012, "Kant, Real Possibility, and the Threat of Spinoza," *The Monist* 121(483), 635–675.

2014, "Kant and the 'Monstrous' Ground of Possibility: A Reply to Abaci and Yong." *Kantian Review*, 19(1), 53–69.

Cicero, 1975, *De divinatione; De fato; Timaeus*, ed. R. Giomini, Leipzig: Teubner.

1933, *On the Nature of the Gods, Academica*, trans. H. Rackham, London: Heinemann.

Conway, A., 1692, *The Principles of the Most Ancient and Modern Philosophy*, trans. J. C. (Jodocus Crull?), London.

Cottingham, J., R. Stoothoff and D. Murdoch, trans., 1984–1991, *The Philosophical Writings of Descartes*, 3 vols., Vols. I and II, Cambridge: Cambridge University Press, 1985, and Vol. III, The Correspondence, by the same translators plus A. Kenny, Cambridge: Cambridge University Press, 1991.

Cresswell, M. J., 2013, "Carnap and Mckinsey: Topics in the Pre-History of Possible Worlds Semantics," *Proceedings of the 12th Asian Logic Conference*, ed. J. Brendle, R. Downey, R. Goldblatt and B. Kim Singapore: World Scientific, 53–75.

2014, "The Completeness of Carnap's Predicate Logic," *Australasian Journal of Logic* 11, 47–63.

Crivelli, P., 2004, *Aristotle on Truth*, Cambridge: Cambridge Universit Press.

Cudworth, R., 1678, *The True Intellectual System of the Universe*, London: Richard Royston.

1996, *A Treatise Concerning Eternal and Immutable Morality*, ed. Sarah Hutton, Cambridge: Cambridge University Press. Circa 1664, 1st ed. 1731.

Cunning, D., 2014, "Descartes' modal metaphysics," The Stanford encyclopedia of philosophy, Spring 2014 ed., ed. Edward N. Zalta, URL = <http://plato.stanford.edu/archives/spr2014/entries/descartes-modal/>

Curley, E. M., 1984, "Descartes on the Creation of the Eternal Truths," *The Philosophical Review* 93, 569–597.

D'Alembert, J., 1743, *Traité de dynamique*, Paris.

de Harven, V., 2015, "How Nothing Can Be Something: The Stoic Theory of Void," *Ancient Philosophy* 35, 405-29.

Denyer, N., 1981, "Time and Modality in Diodorus Cronus," *Theoria* 47, 31–53.

Descartes, R., 1985. *The Philosophical Writings of Descartes*, ed. John Cottingham, Robert Stoothoff, Dugald Murdoch, and Anthony Kenny, 3 vols., Cambridge: Cambridge University Press.

1996. *Oeuvres de Descartes*, ed. Charles Adam and Paul Tannery, 11 vols., Paris: J. Vrin.

Diels, H., ed., 1882–1909, *Commentaria in Aristotelem Graeca*, Berlin: Reimer.

Diogenes Laërtius, 1964, *Diogenis Laertii Vitae Philosphorum*, ed. H. S. Long, Oxford Classical Texts, 2 vols., Oxford: Clarendon Press.

Döring, K., 1989, "Gab es eine Dialektische Schule?" *Phronesis* 34, 293–310.

Downing, L., 1992, "Are Corpuscles Unobservable in Principle for Locke?," *Journal of the History of Philosophy* 30, 33–52.

Du Moulin, P., 1624, *The Elements of Logick*, trans. Nathanael De-Lawne, London: Printed by I. D. for Nicolas Bourne.

Ebert, T. and U. Nortmann, 2007, *Aristoteles: Analytica Priora, Buch I*, Berlin: Akademie Verlag.

El-Rouayheb, K., 2009. "Impossible Antecedents and Their Consequences: Some Thirteenth-Century Arabic Discussions," *History and Philosophy of Logic* 30, 209–225.

Epictetus, 1916, *Epicteti Dissertationes ab Arriano digestae*, ed. H. Shenkl, Leipzig: Teubner.

Erdmann, B., 1878, *Kants Kriticismus in der ersten und in der zweiten Auflage der Kritik der reinen Vernunft*, Leipzig: Voss.

Fisch, M., 1967, "Peirce's Progress from Nominalism toward Realism," *The Monist* 51(2), 159–178.

1986, *Peirce, Semeiotic, and Pragmatism*, Bloomington, IN: Hilpinen, Risto (1982, "On C. S. Peirce's theory of the proposition: Peirce as a Precursor of Game-Theoretical Semantics), *The Monist* 62, 182–189.

Fisher, A. R. J., 2011, "Causal and Logical Necessity in Malebranche's Occasionalism," *Canadian Journal of Philosophy* 41, 523–548.

Fitting, M. and R. L. Mendelsohn, 1998, *First-Order Modal Logic*, Dordrecht: Kluwer.

Frankfurt, H., 1977, "Descartes on the Creation of the Eternal Truths," *The Philosophical Review* 86, 36–57.

Frede, D., 1976, "Review of Hintikka 1973," *Philosophische Rundschau* 22, 237–242.

1990, "Fatalism and Future Truth," *Proceedings of the Boston Area Colloquium in Ancient Philosophy* 6, 195–227.

2003, "Stoic Determinism" in ed. B. Inwood *The Cambridge Companion to the Stoics*. Cambridge University Press. 179–205.

Frede, M., 1974, *Die stoische Logik*. Vandenhoeck und Ruprecht.

Galileo Galilei, 1989, *Two New Sciences, Including Centers of Gravity and Force of Gravity and Force of Percussion*, Toronto, Wall and Thompson.

1992, "Treatise on Demonstration," trans. William A. Wallace in *Galileo's Logical Treatises*, Dordrecht: Kluwer.

Garland, (Garlandus Compotista), 1959, *Dialectica*, ed. L. M. De Rijk, Assen: Van Gorcum.

Garsia, Petrus, 1489, *Determinationes Magistrales*, Rome.

Gassendi, P., 1981, *Institutio Logica*, trans. Howard Jones [1658], Assen: van Gorcum.

Gellius, 1968, *Aulus Gellius Noctes Atticae*, ed. P. K. Marshall, Oxford Classical Texts, Vols. 1–2, Oxford: Clarendon Press.

Gendler, T., and Hawthorne, J., eds. *Conceivability and Possibility*, Oxford: Clarendon Press.

Gödel, K., and S. Feferman, 1995. *Kurt Gödel: Collected Works.* Vol. III. *Unpublished Essays and Lectures*, Oxford: Oxford University Press.

Goheen, J. D., and J. L. Mothershead, eds., 1970, *Collected Papers of Clarence Irving Lewis*, Stanford University Press, (cited as CP).

Gohlke, P., 1936, *Die Entstehung der Aristotelischen Logik*, Berlin: Junker und Dünnhaupt.

Gourinat, J-B., 2000, *La Dialectique des Stoïciens*, Vrin.

Gregory of Rimini, *Gregorii Ariminensis OESA Lectura super Primum et Secundum Sententiarum* ed. A. Trapp et al., Berlin and New York: Walter de Gruyter.

1981, Tomus I: Super Primum (Dist. 1–6).

1982, Tomus II: Super Primum (Dist. 17–17).

1984, Tomus III: Super Primum (Dist. 19–48).

1979a, Tomus IV: Super Secundum (Dist. 1–5).

1979b, Tomus V: Super Secundum (Dist. 6–18).

1980, Tomus VI: Super Secundum (Dist. 24–44).

1987, Tomus VII: Indices.

Griffin, M. V., 2013. *Leibniz, God and Necessity*, Cambridge: Cambridge University Press.

Grosseteste, Robert, 1912, *Die Philosophishen Werke des Robert Grossetest: Bischofs von Lincoln, Beitraäge zur Geschichte de Philosophie des Mittelalters*, ed L. Baur, 9, Münster. W.

Hacking, I., 1967, "Possibility," *The Philosophical Review* 76, 143–168.

1980, "Proof and Eternal Truths: Descartes and Leibniz," in Gaukroger, Stephen, ed., *Descartes: Philosophy, Mathematics and Physics*, Sussex: The Harvester Press.

Hankinson, R. J., 1999, "Determinism and Indeterminism" in eds. K. ALgra, J. Barnes. J. Mansfeld, M. Schofield. *The Cambridge History of Hellenistic Philosophy*. Cambridge University Press. 1999. 513–541.

Hanna, R., 2006, *Kant, Science, and Human Nature*. Oxford University Press.

Hardwick, Charles (ed.), 1977, Semiotics and Signifies: The Correspondence between Charles S. Peirce and Victoria Lady Welby. Bloomington: Indiana University Press.

Hawthorne, J. and J. A. Cover, 2000, "Infinite Analysis and the Problem of the Lucky Proof," *Studia Leibnitiana* XXXII(2), 151–165.

Heylen, J., 2010, "Carnap's Theory of Descriptions and Its Problems," *Studia Logica*, 94, 355–380.

Hilpinen, R. 1982, "On C. S. Peirce's Theory of the Proposition: Peirce as a Precursor of Game-Theoretical Semantics," *The Monist* 62, 182–189.

Hintikka, K. J. J., 1969, *Models for Modalities: Selected Essays*, Dordrecht: Reidel.
1973, *Time and Necessity: Studies in Aristotle's Theory of Modality*, Oxford: Clarendon Press.
1977, *Aristotle on Modality and Determinism*, in collaboration with U. Remes and S. Knuuttila, *Acta Philosophica Fennica* 29.1, Amsterdam: North Holland.
1997, "The Place of C. S. Peirce in the History of Logical Theory," in *The Rule of Reason*, ed. J. Brunning and P. Forester, Toronto: University of Toronto Press, 140–161.
Hodges, W., 2010, "Ibn Sīnā on understood time and aspect (Qiyās 1.3)," http://wilfridhodges.co.uk/arabic08.pdf.
2012a, "Ibn Sīnā's modal logic," http://wilfridhodges.co.uk/arabic20a.pdf slide 31.
2012b, Translation of Ibn Ruṣd, *Mas'ala* i.3 p.122, in (Hodges, "Handout for 'Necessary and permanent in Ibn Sīnā'," http://wilfridhodges.co.uk/arabic20b.pdf.
2014, *Mathematical Background to the Logic of Ibn Sīnā* Draft, http://wilfridhodges.co.uk/arabic44.pdf.
Hookway, C., 1985, *Peirce*, London: Routledge.
2008, "Pragmatism and the Given: C. I. Lewis, Quine, and Peirce" in *The Oxford Handbook of American Philosophy*, ed. C. J. Misak, Oxford: Oxford University Press.
Horn, L. R., 1973, "Greek Grice: A Brief Survey of Proto-Conversational Rules in the History of Logic," *Proceedings of the Chicago Linguistic Society* 9, 205–214.
1989, *A Natural History of Negation*, Chicago: University of Chicago Press.
Houser, N. and D. Roberts, J. Van Evra, eds., 1997, *Studies in the Logic of Charles Sanders Peirce*, Bloomington: Indiana University Press.
Huby, P. M., 2002, "Did Aristotle Reply to Eudemus and Theophrastus on Some Logical Issues?" In *Eudemus of Rhodes*, Rutgers University Studies in Classical Humanities, Vol. 11, ed. I. Bodnár and W. W. Fortenbaugh, New Brunswick: Transaction, 85–106.
Hume, D., 1902, *An Enquiry Concerning the Human Understanding*, Ed. L. A. Selby-Bigge, Oxford, Clarendon Press.
Hunter, M., and E. B. Davis, 1999–2000, *The Works of Robert Boyle*, ed., 14 vols., London: Pickering and Chatto.
Inwood, B., and L. P. Gerson, 1997, *Hellenistic Philosophy, Introductory Readings*, Indianapolis: Hackett. 1988. 1997.
Ishiguro, H., 1990. *Leibniz's Philosophy of Logic and Language*, Cambridge: Cambridge University Press.

Jesseph, D. M., 1999, *Squaring the Circle: the War Between Hobbes and Wallis*, Chicago: University of Chicago Press.

2013, "Logic and Demonstrative Knowledge," in Peter Anstey, ed., *The Oxford Handbook of British Philosophy in the Seventeenth Century*, Oxford University Press,373–390.

Johnson, F., 1989, "Models for Modal Syllogisms," *Notre Dame Journal of Formal Logic* 30(2), 271–284.

1995, "Apodeictic Syllogisms: Deductions and Decision Procedures," *History and Philosophy of Logic* 16(1), 1–18.

Kant, I., 1900-, *Kants gesammelte Schriften*, ed. the Berlin- Brandenburg (formerly Royal Prussian) Academy of Sciences, Berlin: Walter de Gruyter.

1992, *Theoretical Philosophy 1775–1770*, trans. and ed. D. Walford and R. Meerbote, Cambridge: Cambridge University Press.

1996, *Religion and Rational Theology*, trans. and ed., A. Wood and G. di Giovanni, Cambridge: Cambridge University Press.

1997, *Critique of Pure Reason*, trans. and ed. P. Guyer and A. Wood, Cambridge: Cambridge University Press.

Karttunen, L., 1972, "Possible and Must," in *Syntax and Semantics*, ed. J. Kimball, vol. 1, New York: Academic Press, 1–20.

King, C. G., 2014, "Review of Striker 2009," *Gnomon* 86, 484–488.

King, P., 2015, "Pseudo-Joscelin: The Treatise on Genera and Species," *Oxford Studies in Medieval Philosophy* 2, 104–211.

Kneale, M., 1937, "Logical and Metaphysical Necessity," *Proceedings of the Aristotelian Society*. Harrison & Sons, 253–268.

Kneale, W., and M. Kneale, 1962, *The Development of Logic*. Oxford University Press. 1962.

Knuuttila, S., 1980, "Time and Modality in Scholasticism," in S. Knuuttila ed. *Reforging the Great Chain of Being*, Dordrecht: Reidel, 163–257.

Kripke, S. A., 1972, *Naming and Necessity*, Cambridge MA: Harvard University Press.

Kung, J., 1978, "*Metaphysics* 8.4: Can Be but Will Not Be," *Apeiron* 12, 32–36.

Lane, R., 2007, "Peirce's Modal Shift: From Set Theory to Pragmaticism," *Journal of the History of Philosophy* 45(4), 551–576.

Langton, R., 1998, *Kantian Humility*, Oxford: Oxford University Press.

Lear, J., 1979, "Aristotle's Compactness Proof," *Journal of Philosophy* 74(4), 198–215.

Legg, C., 2008, "The Problem of the Essential Icon," *American Philosophical Quarterly* 45(3), 207–232.

2012, "The Hardness of the Iconic Must: Can Peirce's Existential Graphs Assist Modal Epistemology?," *Philosophia Mathematica* 20(1), 1–24.

2014, "Things Unreasonably Compulsory": A Peircean Challenge to a Human Theory of Perception, Particularly with Respect to Perceiving Necessary Truths," *Cognitio* 15(1), 89–112.

Leibniz, G. W., 1875–1890, 1961. *Die philosophischen Schriften von Gottfried Wilhelm Leibniz*, ed. Carl Immanuel Gerhardt, Berlin: Weidman. [Hildesheim: Georg Olms, 1961, Authors who cite Leibniz 1961 are referring to this reprint]

1903, *Opuscules et Fragments Inédits de Leibniz*, ed. Louis Couturat, Paris: Félix Alcan.

1923-. *Sämtliche Schriften und Briefe*, ed. Deutsche Akademie der Wissenschaften zu Berlin, Darmstadt and Berlin: Akademie Verlag.

1948, *Textes inédits: d'après les manuscrits de la Bibliothèque provincale de Hanovre*. ed. Gaston Grua, 2 vols., Paris: Presses Universitaires de France.

1966, *Logical Papers: A Selection*, ed. G. H. R. Parkinson, Oxford: Clarendon Press.

1969, *Philosophical Papers and Letters*, ed. Leroy E. Loemker, 2nd ed., Dordrecht: Reidel.

1981, *New Essays on Human Understanding*, Peter Remnant and Jonathan Bennett, Cambridge: Cambridge University Press.

1981a, *Theodicy*, trans. E. M. Huggard, La Salle: Open Court, 1985.

New Essays on Human Understanding, trans. and ed. Peter Remnant and Jonathan Bennett, Cambridge University Press.

1985, *Theodicy*, trans. E. M. Huggard, La Salle, IL: Open Court.

1989, *Philosophical Essays*, ed. Roger Ariew and Daniel Garber, Indianapolis: Hackett.

1992, *De Summa Rerum: Metaphysical Papers, 1675–76*, ed. G. H. R. Parkinson, New Haven, CT: Yale Univ. Press.

1995, *Philosophical Writings*, ed. Mary Morris and G. H. R. Parkinson, London: Everyman.

1996, *New Essays on Human Understanding*, ed. Peter Remnant and Jonathan Bennett, Cambridge: Cambridge University Press.

2006, *Confessio Philosophi: Papers Concerning the Problem of Evil, 1671–1678*, ed. R. C. Sleigh, New Haven, CT: Yale University Press.

Levinson, S. C., 1983, *Pragmatics*, Cambridge: Cambridge University Press.

Lewis, C. I., 1912, "Implication and the Algebra of Logic," *Mind* 21, 522–531.

1912b, "Naturalism and Idealism," *University of California Chronicle*, 14, 192–211. Reprinted in Lewis 1970.

1913, "Interesting Theorems in Symbolic Logic," *Journal of Philosophy, Psychology and Scientific Methods* 10, 239–342.

1913b, "A New Algebra of Implications and Some Consequences," *Journal of Philosophy, Psychology and Scientific Methods* 10, 428–438.

1914, "The Matrix Algebra for Implications," *The Journal of Philosophy, Psychology and Scientific Methods* 11, 589–600.

1916, "Types of Order and the System Σ," *The Philosophical Review* 25, 407– 419. Reprinted in Lewis 1970, 360–370.

1917, "The Issues Concerning Material Implication," *The Journal of Philosophy, Psychology and Scientific Methods* 14, 350–356.

1918, *A Survey of Symbolic Logic*, Berkeley, University of California Press, first edition.

1920, "Strict Implication – an Emendation," *The Journal of Philosophy, Psychology and Scientific Methods* 17, 300–302.

1921, "The Structure of Logic and Its Relation to Other Systems," *The Journal of Philosophy* 18, 505–516. Reprinted in Lewis 1970, pp 371–382.

1923, "Facts, Systems, and the Unity of the World," *The Journal of Philosophy* 20, 141–151. Reprinted in Lewis 1970, pp 383–393.

1923b, "A Pragmatic Conception of the A Priori," *The Journal of Philosophy* 20, 169–177. Reprinted in Lewis 1970, pp 231–239.

1926, "The Pragmatic Element in Knowledge," *University of California Publications in Philosophy* 6, 205–227.

1929, *Mind and the World Order: Outline of a Theory of Knowledge*, New York: Charles Scribner and Sons.

1932, "Alternative Systems of Logic," *The Monist* 17, 481–507. Reprinted in Lewis 1970, pp 400–419.

1946, *Analysis of Knowledge and Valuation*, LaSalle, IL: Open Court.

1955, "Realism or Phenomenalism?," *The Philosophical Review* 64, 233–247. Reprinted in Lewis 1970, pp 335–347.

1968, "Autobiography," in *The Philosophy of C. I. Lewis*, ed. P. A. Schilpp, LaSalle, IL: Open Court, 1–21

1970. *Collected Papers of Clarence Irving Lewis*, ed. John D. Goheen and John L. Mothershead, Jr., Stanford, CA: Stanford University Press.

Lewis, C. I., and C. H. Langford, 1932, *Symbolic Logic*. New York: Dover, second edition, 1959 [Note: references to Lewis and Langford 1959, are to the second edition of Lewis and Langford 1932]

2014, "A Note on Strict Implication," *History and Philosophy of Logic* 35, 44–49. The manuscript of this article was written before or in 1935.

Lewis, D. K. 1973, "Causation," The Journal of Philosophy 70(17), 556–567.

1986, *On the Plurality of Worlds*, Oxford: Blackwell.

Lin, M., 2012. "Rationalism and Necessitarianism," *Noûs* 46(3): 418–448.

Locke, J., 1823, *The Works of John Locke*, 12th edition, 10 vols., London: Tegg.

1975, *An Essay concerning Human Understanding*, 4th ed., ed. P. H, Nidditch, Oxford: Clarendon Press, 1975 [1st edition 1690; 4th edition 1700].

1989, *Some Thoughts Concerning Education*, eds John W. Yolton and Jean S. Yolton, Oxford: Clarendon Press.

Long, A. A., 1970, "Stoic Determinism and Alexander of Aphrodisias *De Fato* (I-Xiv)," *Archiv für Geschichte der Philosophie* 52, 247–268.

1971, "Freedom and Determinism in the Stoic Theory of Human Action," *Problems in Stoicism*. The Athlone Press. 173–199.

Long, A. A., and D. Sedley, 1987, *The Hellenistic Philosophers*. Cambridge University Press.

Look, B. C., 2006. "Some Remarks on the Ontological Arguments of Gödel and Leibniz," in *Einheit in der Vielheit: Akten des VIII. Internationalen Leibniz-Kongresses*, ed. H. Breger et al., Hanover: Hartmann, 510–517.

(forthcoming) "Leibniz's Arguments for the Existence of God." In *The Oxford Handbook to Leibniz*, ed. Maria Rosa Antognazza, Oxford: Oxford University Press.

Lovejoy, A. O., "Kant's Antithesis of Dogmatism and Criticism," *Mind* 15, 1906, 191–214.

Lowde, J., 1694, *A Discourse Concerning the Nature of Man*, London: Walter Kettilby.

Łukasiewicz, J., 1957, *Aristotle's Syllogistic from the Standpoint of Modern Formal Logic*, 2nd ed., Oxford: Clarendon Press.

1972. *On a Controversial Demonstration of Aristotle's Modal Syllogistic: An Enquiry on Prior Analytics A.15*. Padua.

McCall, S., 1963. *Aristotle's Modal Syllogisms*, Amsterdam: North-Holland Publishing Company.

MacDonald, S., 1993, "Theory of Knowledge," in *The Cambridge Companion to Aquinas*, eds. N. Kretzmann and E. Stump, Cambridge University Press, 160–195.

MacIntosh, J. J., 1991, "Theological Question Begging," *Dialogue* 30, 531–547.

1997, "The Argument from the Need for Similar or 'Higher' Qualities: Cudworth, Locke, and Clarke on God's Existence," *Enlightenment and Dissent* 16, 29–59.

1998a, "Aquinas and Ockham on Time, Predestination and the Unexpected Examination," *Franciscan Studies* 55, 181–220.

1998b, "Aquinas on *Necessity*," *American Catholic Philosophical Quarterly* 72, 470–503.

2006, *Boyle on Atheism*, transcribed and edited, Toronto, University of Toronto Press.

Mackie, J. L., 1974, "*De* What *re* Is *de re* Modality?" *Journal of Philosophy* 71(16), 1974, 551–561.

McKirahan, R. D., 1992, *Principles and Proofs: Aristotle's Theory of Demonstrative Science*, Princeton, NJ: Princeton University Press.

Maier, H., 1900, *Die Syllogistik des Aristoteles II.1: Formenlehre und Technik des Syllogismus*, Tübingen: Laupp.

Malcolm, N., 1960. "Anselm's Ontological Arguments," *The Philosophical Review* 49, 41-62.

Makin, S., 1998, "Aristotle's Two Modal Theses Again," *Phronesis* 44, 114–126.

Malink, M., 2013, *Aristotle's Modal Syllogistic*, Cambridge, MA: Harvard University Press.

Malink, M., and Rosen, J., 2013, "Proof by Assumption of the Possible in *Prior Analytics* 1. 15," *Mind* 122, 953–986.

Mancosu, P., 1996, *Philosophy of Mathematics and Mathematical Practice in the Seventeenth Century*, New York: Oxford University Press.

Mares, E. D., 2013, "A Lewisian Semantics for S2," *History and Philosophy of Logic* 34, 53–67.

2014, "A Survey of Lewisian Logic: The Development of C. I. Lewis's Philosophy of Modal Logic," (This volume).

Marion, M., 2009, "Why Play Logical Games?," *Logic, Epistemology, and the Unity of Science* 15, 3–26.

Martin, C. J., 2001, "Abaelard on Modality, Some Possibilities and Some Puzzles," in *Potentialität und Possibilität: Modalaussagen in der Geschichte der Metaphysik, Stuttgart*, ed. T. Buchheim et al., Frommann-Holzboog, 97–125.

2004, "Logic," in *The Cambridge Companion to Abelard*, ed. J. Brower and K. Guilfoy Cambridge: Cambridge University Press, 158–199.

2009 "The Logical Textbooks and their Influence," in *The Cambridge Companion to Boethius*, Cambridge: Cambridge University Press, 2009, 56–84.

Mates, B., 1953, *Stoic Logic*. University of California Press.

1986. *The Philosophy of Leibniz: Metaphysics and Language*, Oxford: Oxford University Press.

Menn, S., 1999, "The Stoic Theory of Categories", *Oxford Studies in Ancient Philosophy*, XVII, 215-47.

Mignucci, M., 1969, *Aristotele: Gli Analitici Primi*, Napoli: Loffredo.

"Sur la logique modale des stoïciens," in *Les stoïciens et leur logique*, ed. J. Brunschwig, Vrin. 1978. 303–332.

Milton, J. R., 2012, "Locke and the *Elements of Natural Philosophy*: Some Problems of Attribution," *Intellectual History Review* 22, 199–219.

Misak, C. J., 1991, *Truth and the End of Inquiry: A Peircean Account of Truth*, Oxford: Oxford University Press.

2013, *The American Pragmatists*, Oxford: Oxford University Press.

(forthcoming). *Cambridge Pragmatism: The Influence of Peirce and James on Ramsey and Wittgenstein*, Oxford: Oxford University Press.

More, H., 1662, *An Antidote Against Atheism, in A Collection of Several Philosophical Writings of Dr Henry More*, London: William Borden, 2nd ed.

Morgan, C. G., 1979, "Modality, Analogy, and Ideal Experiments According to C. S. Peirce," *Synthese* 41, 65–83.

Murphy, M. G., 2005, *C. I. Lewis: The Last Great Pragmatist*, Albany, NY: SUNY Press.

Newlands, S., 2010. "The Harmony of Spinoza and Leibniz," *Philosophy and Phenomenological Research* 81(1): 64–104.

Newton, I., 1730, *Opticks or a Treatise of the Reflections, Refractions, Inflections and Colours of Light*, London: William Innys, 4th ed.

Normore C. G., 1984, "Future Contingents," in *The Cambridge History of Later Medieval Philosophy* ed. N. Kretzmann and J. Pinborg, Cambridge University Press.

1993, "The Necessity in Deduction: Cartesian Inference and Its Medieval Background," *Synthese* 96, 437–454.

Nortmann, U., 1990, "Über die Stärke der aristotelischen Modallogik," *Erkenntnis* 32, 61–82.

1996, *Modale Syllogismen, mögliche Welten, Essentialismus: eine Analyse der aristotelischen Modallogik*, Berlin and New York: de Gruyter.

2006, "Against Appearances True: On a Controversial Modal Theorem in *Metaphysics* Theta 4," *Zeitschrift für philosophische Forschung* 60: 380–393.

Ockham, William, 1970, *Scriptum in Librum Primum Sententiarum (Ordinatio)*, in *Opera Theologica II*, ed. Stephanus Brown, New York: St. Bonaventure.

1974a, *Ockham's Theory of Terms, Part I of the Summa Logicae*, translated and introduced by Michael J. Loux, Notre Dame, University of Notre Dame Press.

1974b, *Opera philosophica.* Vol 1. *Summa logicae*, St Bonaventure, NY: Editiones Instituti Franciscani Universitatis S. Bonaventurae.

1978, "*Tractatus de Praedestinatione*," in *Opera Philosophica II*, ed. Philotheus Boehner, New York: St. Bonaventure.

1980a, *Ockham's Theory of Propositions: Part II of the Summa Logicae*, translated by Alfred J. Freddoso and Henry Schuurman, introduction by Alfred J. Freddoso, Notre Dame: University of Notre Dame Press.

1980b, *Quodlibeta Septem*, ed J. C. Wey, *Opera Theologica*, Vol. 9, St. Bonaventure, NY: The Franciscan Institute.

Øhrstrøm, P., 1982, "'Temporalis' in Medieval Logic," *Franciscan Studies* 42, 166–179.

1997, "C. S. Peirce and the Quest for Gamma Graphs" in *Conceptual Structures: Fulfilling Peirce's Dream*, Berlin and Heidelberg: Springer 357–370.

Oppy, G., 1995. *Ontological Arguments and Belief in God*. Cambridge: Cambridge University Press.

Pacius, J., 1597a, *In Porphyrii Isagogen et Aristotelis Organum Commentarius Analyticus*, Frankfurt: Heredes Andreae Wecheli.

1597b, *In Porphyrii Isagogen et Aristotelis Organum Commentarius Analyticus*, Frankfurt: Heredes Andreae Wecheli.

Parkinson, G. H. R., 1965, *Logic and Reality in Leibniz's Metaphysics*, Oxford: Clarendon Press.

ed., 1973, *Leibniz, Philosophical Writings*, trans. M. Morris and G. H. R. Parkinson, London: Dent.

Parrini, P., ed., 1994, *Kant and Contemporary Epistemology*, Dordrecht: Kluwer.

Parry, W. T., 1968, "The Logic of C. I. Lewis," in *The Philosophy of C. I. Lewis*, ed. P. A. Schilpp, LaSalle, IL: Open Court, 15–54.

Partington, J. R., 1961, *A History of Chemistry*, 4 vols., London: Macmillan.

Pascal, B. 1954, "*L'Esprit Geometrique*," in *Blaise Pascal, Œuvres Complètes*, ed. Jacques Chevalier, Paris: Librairie Gallimard.

Patterson, R., 1995, *Aristotle's Modal Logic: Essence and Entailment in the Organon*, Cambridge: Cambridge University Press.

Patzig, G. 1968, *Aristotle's Theory of the Syllogism*, (Translated by J. Barnes), Dordrecht: D. Reidel.

Peirce, C. S., 1870, "Description of a Notation for the Logic of Relatives, Resulting from an Amplification of the Conceptions of Boole's Calculus," *Memoirs of the American Academy of Arts and Sciences* 9(2), 317–378.

1885, "On the Algebra of Logic: A Contribution to the Philosophy of Notation," *American Journal of Mathematics* 7(2), 180–196.

1958–1966, *The Collected Papers of Charles Sanders Peirce*, Vols. 1–6, ed. Charles Hartshorne and Paul Weiss, Vols. 7–8, ed. A. W. Burks, Cambridge: Belknap Press of Harvard University Press.

1976, *The New Elements of Mathematics*, Vols. 1–3, ed. Carolyn Eisele, Berlin: Mouton.

1976a, *The Writings of Charles S. Peirce: A Chronological Edition*, ed. Nathan Houser, 6 vols., et al. Indianapolis: Peirce Edition Project, Indiana University-Purdue University Indianapolis.

1997, *Pragmatism as a Principle and Method of Right Thinking: The 1903 Harvard lectures on pragmatism*, ed. P. A. Turrisi: SUNY Press.

1998, *The Essential Peirce: Selected Philosophical Writings*, Vol. 2 (1893–1913), ed. The Peirce Edition Project, Bloomington: Indiana University Press.

2010, *Philosophy of Mathematics: Selected Writings*, ed. M. E. Moore, Bloomington: Indiana University Press.

n.d. Charles Sanders Peirce Papers, Houghton Library, Harvard University.

Pereboom, D. (2007, "Kant on Transcendental Freedom," *Philosophy and Phenomenological Research* 73, 537–567.

Perelmuter, Z., 2010, "*Nous* and Two Kinds of *Epistêmê* in Aristotle's *Posterior Analytics*," *Phronesis* 55, 228–254.

Philodemus, 1968, *Philodemus: On the methods of inference*, ed. P., and E. De Lacy, Naples: Bibliopolis. 2nd ed., 1978.

Pietarinen, A., 2005, "Compositionality, Relevance, and Peirce's Logic of Existential Graphs," *Axiomathes* 15(4), 513–540.

2006a, "Peirce's Contributions to Possible-Worlds Semantics," *Studia Logica* 82(3), 345–369.

2006b, *Signs of Logic: Peircean Themes on the Philosophy of Language, Games, and Communication*, Dordrecht: Springer.

Pizzi, C., 2008, "Aristotle's Cubes and Consequential Implication," *Logica universalis* 2, 143–153.

Plantinga, A., 1978. *The Nature of Necessity*, Cambridge: Clarendon Press.

Plutarch 1976a, *De communibus notitiis contra Stoicos. Moralia XIII.2*. ed. H. Cherniss, Cambridge, MA, and London: Loeb.

1976b, *De Stoicroum repugnantiis, Moralia XIII.2*. ed. H. Cherniss, Cambridge, MA, and London: Loeb.

Prauss, G., 1974, *Kant und das Problem der Dinge an Sich*, Bonn: Grundman.

Prior, A. N., 1955, "Diodoran Modalities," *The Philosophical Quarterly* 5, 205–213.

1960, "Identifiable Individuals," *Review of Metaphysics*, 13, 684–696. Reprinted in A. N. Prior 1968, ch. 7, 66–77.

1967, *Past Present and Future*, Oxford: Clarendon Press.

1968, *Papers on Time and Tense*, Oxford: Clarendon Press.

Putnam, H., 1975,"Is Semantics Possible," *Mind, Language and Reality, Philosophical Papers,* Vol. 2, Cambridge University Press.

Quine, W. V. O., 1934a, "Review of the Collected Papers of Charles Sanders Peirce. Vol. 4. The Simplest Mathematics," *Isis* 22, 551–553.

1934b, "Ontological Remarks on the Propositional Calculus," *Mind* 433, 473–476.

1943, "Notes on Existence and Necessity," *The Journal of Philosophy* 40, 113–127.

1947, "The Problem of Interpreting Modal Logic," *The Journal of Symbolic Logic*, 12, 43–48.

1953a, "Reference and Modality," *From a Logical Point of View*, Cambridge, MA: Harvard University Press, second edition 1961, 139–159 (references are to the second edition.).

1953b, "Three Grades of Modal Involvement," *The Ways of Paradox*, Cambridge MA: Harvard University Press, 1976, 158–176.

1953c, "Two Dogmas of Empiricism," *From a Logical Point of View*, Cambridge, MA: Harvard University Press, second edition 1961, 20–46 (references are to the second edition.).

1960, *Word and Object*, Cambridge, MA: MIT Press.

1969, "Reply to Sellars," in *Words and Objections*, ed. D. Davidson and K. J. J. Hintikka, Dordrecht: Reidel, 1969, 337–340.

1970, *Philosophy of Logic*, Englewood Cliffs, NJ: Prentice-Hall.

Reesor, M., 1965, "Fate and Possibility in Early Stoic Philosophy," *Phoenix* 19, 285–297.

1978, "Necessity and Fate in Stoic Philosophy" in *The Stoics*, ed. John M. Rist, University of California Press. 187–202.

Reynolds, A., 2002, *Peirce's Scientific Metaphysics*, Nashville: Vanderbilt University Press.

Rijen, J. van, 1989, *Aspects of Aristotle's Logic of Modalities*, Dordrecht: Kluwer.

Rini, A. A. 2011, *Aristotle's Modal Proofs: Prior Analytics A8-22 in Predicate Logic*, The New Synthese Historical Library 68, Dordrecht: Springer.

2013, "The Birth of Proof: Modality and Deductive Reasoning," *Logic and Its Applications*, ed. Kamal Lodaya, Springer Lecture Notes in Computer Science LNCS 7750, Springer: Berlin, 34–49.

Rist, J. M., 1969, "Fate and Necessity," in *Stoic Philosophy*, ed. John M. Rist, Cambridge University Press. 112–132.

1978, "Necessity and Fate in Stoic Philosophy," in *The Stoics*, ed. John M. Rist, University of California Press.

Robers, D., 1973, *The Existential Graphs of Charles S. Peirce*, The Hague: Mouton.

Rosen, J., and Malink, M., 2012, "A Method of Modal Proof in Aristotle," *Oxford Studies in Ancient Philosophy* 42, 179–261.

Ross, W. D., 1949, *Aristotle's Prior and Posterior Analytics: A Revised Text with Introduction and Commentary*, Oxford: Clarendon Press. Reprinted with corrections 1957, 1965. Page references to the 1957 printing.

Rückert, H., 2001, "Why Dialogical Logic?," *Essays on Non-classical Logic* 1, 165–182.

Russell, B. A. W., 1918, "The Philosophy of Logical Atomism," in *Logic and Knowledge*, Ed R. C. Marsh, New York: Capricorn Books, 1955, 177–281.

1937. *A Critical Exposition of the Philosophy of Leibniz*. 2nd ed., London: Allen & Unwin.

1976, *The Problems of Philosophy*, Oxford University Press [1st edition 1912].

Salles, R., 2005, *The Stoics on Determinism and Compatibilism*, Ashgate.

Sambursky, S., 1959, *Physics of the Stoics*. Princeton University Press.

Scanlan, M., 1983, "On Finding Compactness in Aristotle," *History and Philosophy of Logic* 4(1 and 2), 1–8.
Schafer, K., (forthcoming). "*Kant's Conception of Cognition*," In Schafer, K., and Stang, N. *The Sensible and Intelligible Worlds: New Essays on Kant's Metaphysics and Epistemology*, Oxford: Oxford University Press.
Sedlar, I., 2009, "C. I. Lewis on Possible Worlds," *History and Philosophy of Logic* 30, 283–291.
Sedley, D., 1977, "Diodorus Cronus and Hellenistic Philosophy," *Proceedings of the Cambridge Philological Society* 203, 74–120.
1982, 'The Stoic Criterion of Identity', *Phronesis*, 27, 255–75.
Seel, G., 1982, *Die Aristotelische Modaltheorie*, Berlin and New York: de Gruyter.
Serjeantson, R., 2006, "Proof and persuasion," in *The Cambridge History of History of Science*. Vol. III. *Early Modern Science*, eds Katharine Park and Lorraine Daston, Cambridge: Cambridge University Press, 132–175.
Sextus Empiricus, 1939–1949, *Sextus Empiricus*, ed. G. Bury, Cambridge, MA: Loeb.
Sharples, R. W., 1983, *Alexander of Aphrodisias on Fate*, London: Duckworth.
Shehaby, N., 1973, *The propositional logic of Avicenna*, Dordrecht: Riedel.
Shin, S., 2002, *The Iconic Logic of Peirce's Graphs*, Cambridge MA: MIT Press.
Shin, S., and E. Hammer, 2013, "Peirce's Deductive Logic," *Stanford Encyclopedia of Philosophy* http://plato.stanford.edu/entries/peirce-logic/ (downloaded September 10, 2014).
Sider, T., 2001, *Four Dimensionalism: A Study in Persistence and Ontology*, Oxford: Oxford University Press.
Sleigh, R. C., 1990. *Leibniz and Arnauld: A Commentary on Their Correspondence*, New Haven, CT: Yale University Press.
Smith, R. A. 1989, *Aristotle: Prior Analytics*, Indianapolis and Cambridge: Hackett.
Smullyan, A. F., 1948, "Modality and Description," *The Journal of Symbolic Logic* 13, 31–37.
Sorabji, R., 1980a, "Causation, Laws, and Necessity" in *Doubt and Dogmatism*, ed. M. Schofield, M. Burnyeat, J. Barnes. Oxford: Clarendon Press.
1980b, *Necessity, Cause, and Blame: Perspectives on Aristotle's Theory*, Ithaca, NY: Cornell University Press.
1988, Matter, Space and Motion, Ithaca NY: Cornell University Press.
Sowa, J. F., 2008, "Conceptual Graphs," *Foundations of Artificial Intelligence* 3, 213–237.
2011, "Peirce's Tutorial on Existential Graphs," *Semiotica* 186, 347–394.
Stang, N. F., 2010, "Kant's Possibility Proof," *History of Philosophy Quarterly* 27, 275–299.

2011, "Did Kant Conflate the Necessary and the *A Priori*?" *Noûs* 45, 3, 443–471.

2012, "Kant on Complete Determination and Infinite Judgment," *British Journal of the History of Philosophy* 20(6), 1117–1139.

2016, *Kant's Modal Metaphysics*. Oxford: Oxford University Press.

(forthcoming) "Kant's Transcendental Idealism," *Stanford Encyclopedia of Philosophy*.

Stern, R., 2007, "Peirce, Hegel, and the Category of Secondness," *Inquiry* 50(2), 123–155.

Stewart, D., 1805, *A Short Statement…*, Edinburgh: William Creech, and Arch. Constable & Co.

Stjernfelt, F. 2007. *Diagrammatology: An Investigation on the Borderlines of Phenomenology, Ontology and Semiotics*, Dordrecht: Springer.

Stough, C., 1978, "Stoic Determinism and Moral Responsibility," *The Stoics*, ed. John M. Rist, University of California Press, 1978, 203–232.

Street, T., 2002, "An Outline of Avicenna's Syllogistic," *Archiv für Geschichte der Philosophie* 84, 129–160.

2004, "Arabic Logic," in *Handbook of the History of Logic. Vol.1. Greek, Indian and Arabic Logic,* ed. Dov M. Gabbay and John Woods, Elsevier, 523–596.

2010, "Appendix: Readings of the Subject Term," *Arabic Sciences and Philosophy* 20, 119–124.

2013a, "Arabic and Islamic Philosophy of Language and Logic," The Stanford Encyclopedia of Philosophy (Winter 2013 Edition), ed. Edward N. Zalta, http://plato.stanford.edu/archives/win2013/entries/arabic-islamic-language/

2013b, "Avicenna on the Syllogism," in *Interpreting Avicenna: Critical Essays*, ed. Peter Adamson, Cambridge: Cambridge University Press, 48–70.

2014, "Afḍal al-Dīn al-Khūnajī (d. 1248) on the Conversion of Modal Propositions," *Oriens* 42, 454–513.

Striker, G., 1987, "Origins of the Concept of Natural Law," *Proceedings of the Boston Area Colloquium in Ancient Philosophy* 2, 79–94.

1994, "Assertoric vs. Modal Syllogistic," *Ancient Philosophy* 14, special issue, 39–51.

2009, *Aristotle's Prior Analytics, Book 1*, Oxford: Clarendon Press.

Stuart, M., 1996, "Locke's Geometrical Analogy," *History of Philosophy Quarterly* 13, 451–467.

2013, *Locke's Metaphysics*, Oxford: Clarendon Press.

Thom, P., 1991, "The Two Barbaras," *History and Philosophy of Logic* 12, 135–149.

1996. *The Logic of Essentialism. An Interpretation of Aristotle's Modal Syllogistic*, Dordrecht: Boston and London: Kluwer.

2001, *Mediaeval Modal Systems*, Aldershot: Ashgate.

2008, "Logic and Metaphysics in Avicenna's Modal Syllogistic," in *The Unity of Science in the Arabic Tradition: science, Logic, Epistemology and Their Interactions*, ed. Shahid Rahman, Tony Street and Hassan Tahiri, Dordrecht: Kluwer-Springer, 283–295.

2010, "Abharī on the Logic of Conjunctive Terms," *Arabic Sciences and Philosophy* 20, 105–117.

2012, "Syllogisms about Possibility and Necessity in Avicenna and Tusi," in *Insolubles and Consequences*, ed. Catarina Dutilh Novaes and Ole Hjortland, London: College Publications, 239–248.

Thomas Aquinas, 1946, *Summa contra gentiles*, Romae: Apud sedem commissionis Leoninae.

1949, *Quaestiones Disputatae de Anima*, translated by, J. P. Rowan, St. Louis and London: B. Herder Book Co.

1952a, 1953, 1954, *Questiones Disputatae de Veritate*, Questions 1–9 translated by Robert W. Mulligan, S. J., Questions 10–20 translated by James V. McGlynn, S. J., Questions 21–29 translated by Robert W. Schmidt, S. J., Chicago, Henry Regnery.

1952b, *Quaestiones Disputatae de Potentia Dei*, translated by the English Dominican Fathers Westminster, MD: Newman Press, 1952, reprint of 1932.

1962, *Expositio libri Peryermeneias*, Commentary by Thomas Aquinas finished by Cardinal Cajetan, translated by Jean T. Oesterle, Milwaukee: Marquette University Press.

1964–1981, *Summa theologiae. Latin text and English translation, introductions, notes, appendices, and glossaries*, London: Blackfriars in conjunction with Eyre and Spottiswoode.

Van Den Berg, Harmen (1993, "Modal Logics for Conceptual Graphs, in *Proceedings of First International Conference on Conceptual Structures*, Vol. 699 of LNCS, Berlin: Springer-Verlag.

van der Auwera, J., 1996, "Modality: The Three-layered Scalar Square," *Journal of Semantics* 13, 181–195.

Waitz, T., 1844, *Aristotelis organon graece I*, Leipzig: Hahn.

Wallace, W. A., 1992, *Galileo's Logic of Discovery and Proof*, Dordrecht: Kluwer.

Wansing, H., 2014, "Connexive Logic," The Stanford Encyclopedia of Philosophy (Fall 2014 Edition), Edward N. Zalta (ed.), URL = http://plato.stanford.edu/archives/fall2014/entries/logic-connexive/

Watkins, E., and M. Fisher, 1998, "Kant on the Material Grounds of Possibility," *Review of Metaphysics* 52, 369–395.

Watts, I., 1726, *Logick*, London: John Clark.

Weidemann, H., 2002, *Aristoteles: Peri Hermeneias*, 2nd ed., Berlin: Akademie Verlag.

2012, "*De interpretatione*," in *The Oxford Handbook of Aristotle*, ed. C. Shields, Oxford: Oxford University Press, 81–112.

Whitaker, C.W.A., 1996, Aristotle's *De Interpretatione: Contradiction and Dialectic*, Oxford, Clarendon Press.

Williamson, T., 2013, *Modal Logic as Matphysica*, Oxford University Press.

Wittgenstein, L., 1922, *Tractatus Logic-Philosophicus*. (Translated by C. K. Ogden), London: Kegan Paul.

Wolter, A. B., 2003, *Selected Essays on Scotus and Ockham*, Franciscan Institute Publications.

Wood, A., 1984, "Kant's Compatibilism," in *Self and Nature in Kant's Philosophy*, ed. A. Wood, Ithaca, NY: Cornell University Press.

Yong, P., 2014, "God, Totality and Possibility in Kant's *Only Possible Argument*," *Kantian Review*, 19(1), 27–51.

Zeman, J. J., 1986, "Peirce's Philosophy of Logic," *Transactions of the Charles S. Peirce Society* 22, 1–22.

1997a, "Peirce and Philo," *Studies in the Logic of Charles Sanders Peirce*, ed. Nathan Houser, Roberts, D. D., and J. Van Evra, Bloomington: Indiana University Press, 402–417.

1997b, "Peirce's Graphs," in *Conceptual Structures: Fulfilling Peirce's Dream*, ed. D. Lukose, et al., Berlin: Springer.

1997c, "The Tinctures and Implicit Quantification over Worlds, in *The Rule of Reason*, ed. J. Brunning and P. Forster, Toronto: University of Toronto Press, 96–119.

Index

Printed in the United States
By Bookmasters